Antedependence Models for Longitudinal Data

MONOGRAPHS ON STATISTICS AND APPLIED PROBABILITY

General Editors

J. Fan, V. Isham, N. Keiding, T. Louis, R. L. Smith, and H. Tong

Monographs on Statistics and Applied Probability 112

Antedependence Models for Longitudinal Data

Dale L. Zimmerman
University of Iowa
Iowa City, Iowa, U.S.A.

Vicente A. Núñez-Antón
The University of the Basque Country (UPV/EHU)
Bilbao, Spain

CRC Press
Taylor & Francis Group
B   Rt   L d   N   Y k

Chapman & Hall/CRC
Taylor & Francis Group
6000 Broken Sound Parkway NW, Suite 300
Boca Raton, FL 33487-2742

First issued in paperback 2017

© 2010 by Taylor and Francis Group, LLC
Chapman & Hall/CRC is an imprint of Taylor & Francis Group, an Informa business

No claim to original U.S. Government works

ISBN 13: 978-1-138-11362-6 (pbk)
ISBN 13: 978-1-4200-6426-1 (hbk)

Library of Congress Cataloging-in-Publication Data

Zimmerman, Dale L.
    Antedependence models for longitudinal data / Dale L. Zimmerman, Vicente A. Nunez-Anton.
        p. cm. -- (Monographs on statistics and applied probability ; 112)
    Includes bibliographical references and index.
    ISBN 978-1-4200-6426-1 (hardcover : alk. paper)
    1. Mathematical statistics--Longitudinal studies. I. Nunez-Anton, Vicente A. II. Title.
III. Series.

QA276.Z475 2010
519.5--dc22   2009024574

Visit the Taylor & Francis Web site at

To our wives, Bridget and Pilar,
and our children, Nathan, Joshua, Bethany, Anna, Abby, Vicente, and Irene

Contents

Preface

This book describes a class of models for longitudinal data called antedependence models, and some important statistical inference procedures associated with them. Known also as transition models, antedependence models postulate that certain conditional independencies exist among the observations, which are related to their time ordering. In the important case of normally distributed observations, these independencies imply that certain partial correlations are null. Antedependence models are particularly useful for modeling longitudinal data that exhibit serial correlation, i.e., correlation that decays as the elapsed time between measurements increases. The most well-known members of the class are stationary autoregressive models, but antedependence models can be much more general. For example, parsimonious antedependence models exist for which the observation variances are heterogeneous and/or the correlations between observations lagged the same distance apart change over time.

We wrote this book because it is our belief that antedependence models for longitudinal data are underappreciated and underutilized, in proportion to their usefulness. All books on the parametric modeling of longitudinal data of which we are aware feature either marginal models, which directly specify the joint distribution of the observations, or random coefficient models, in which each observation is regarded as a function of covariates with regression coefficients that vary from one subject to the next, according to some probability distribution. In those books, only a few pages or at best a short chapter are devoted to antedependence models. Such brief treatments do not do justice to these models and their associated inference problems, and they force one who wants to learn about them, in sufficient detail to exploit their structure in data analysis, to search for widely scattered journal articles on the subject, which use a variety of jargon and notation. For antedependence models to realize their full potential, we believe that this body of work needs to be brought together in one place, presented systematically, and illustrated with numerous examples. That is what this book attempts to do.

The book is organized as follows. After an introduction to the subject in Chapter 1, most of the remainder of the book divides naturally into two parts. Chapters 2 and 3 are devoted to a description of antedependence models and their properties, with Chapter 2 focusing upon unstructured antedependence,

where nothing more than the aforementioned conditions on the partial correlations are assumed, and Chapter 3 is concerned with structured antedependence models, in which additional parametric assumptions are made. Chapters
4 through 8 present inference procedures for the models. Chapter 4 considers informal model identification via simple summary statistics and graphical
methods. Chapters 5 through 7 consider formal likelihood-based procedures
for normal antedependence models: maximum likelihood and residual maximum likelihood estimation of parameters (Chapter 5); likelihood ratio tests
and penalized likelihood model selection criteria for the model's covariance
structure (Chapter 6); and mean structure (Chapter 7). Chapter 8 summarizes
the illustrative examples presented earlier in the book in a (hopefully) coherent fashion and, using the same examples, compares the performance of antedependence models to other models commonly used for longitudinal data.
Chapter 9 takes up some related topics and extensions.

Because the topic of antedependence models is considerably narrower than the
desired coverage of most courses in longitudinal data analysis, it is neither our
intent nor our expectation that the book will serve as a primary text for such a
course. Accordingly, we have not included any chapter exercises. However, the
book can be used as a supplemental text for such a course, or as the primary
text for a special topics course.

The technical level varies throughout the book. Chapters 3 and 4 contain no
theorems or proofs, and much of both chapters can be read easily by any-
one who has been exposed to autoregressive time series models or has had
an upper-level undergraduate/master's level course in multivariate analysis. In
Chapter 2 and Chapters 5 through 8, a theorem/proof format is used to some
extent, and readers will certainly benefit from more advanced training in multivariate analysis and the theory of linear models. Even here, however, it is our
hope that the frequent appearance of examples will make the main ideas, if not
the technical details, comprehensible to nearly every reader.

Although software is necessary, of course, to actually implement an analysis of
data from the antedependence perspective, we make little mention of it in the
book. Unfortunately, scant software exists for this purpose, so we have written
much of what was needed ourselves, in either R or FORTRAN. Some relevant
R functions are available for download from Dale Zimmerman's Web page:

www.stat.uiowa.edu/~dzimmer

There are many people we would like to thank for assisting us in various ways
throughout this endeavor. First and foremost, we thank Rob Calver and Shashi
Kumar of CRC Press, the former for his continual encouragement and the latter for technical support. We are indebted to Florence Jaffrézic for providing
the fruit fly mortality data and for her valuable help and willingness to meet

with us to discuss methodological issues related to these data, and to Scott Pletcher for explaining additional aspects of these data to us. Thanks also go to Jie Li, Jorge Virto, and Ignacio Díaz-Emparanza for some data editing and programming assistance with R and Gretl. Special thanks go to Miguel Angel García-Pérez for providing access to the computer that allowed us to fit some of the models in this book. We are grateful to Kung-Sik Chan and an anonymous reviewer for providing comments on some parts of the book. We also gratefully acknowledge support from Ministerio Español de Educación y Ciencia, FEDER, Universidad del País Vasco (UPV/EHU), and Departamento de Educación del Gobierno Vasco (EPV/EHU Econometrics Research Group) under research grants MTM2004-00341, MTM2007-60112, and IT-334-07.

DALE ZIMMERMAN   Iowa City, Iowa, USA
VICENTE NÚÑEZ-ANTÓN   Bilbao, Spain

CHAPTER 1

Introduction

Antedependence models are useful, albeit underutilized, generalizations of well-known stationary autoregressive models for longitudinal data. Like stationary autoregressive models, antedependence models specify parsimonious parametric forms for the conditional mean and variance of each observation, given all the observations preceding it (as well as any observed covariates) from the same subject. Antedependence models differ, however, by allowing these parametric forms to change over the course of the longitudinal study. This makes them much more flexible than their stationary autoregressive counterparts and hence, as a result, they are often able to fit longitudinal data exhibiting nonstationary characteristics (e.g., increasing variances or same-lag correlations that change over time) quite well. In this introductory chapter, we motivate these models and show where they sit in the broad spectrum of statistical models used for longitudinal data analysis. We begin by describing some important features common to many longitudinal data sets, especially the tendency for observations from the same subject to be correlated. We then briefly review how various classical methods for continuous longitudinal data analysis and a more modern parametric modeling approach deal with these correlations. Next, antedependence models are described very briefly, and an example data set is used to motivate the use of a first-order antedependence model. The remainder of the chapter outlines the scope of the book and describes several longitudinal data sets that will be used throughout the book to illustrate methodology.

1.1 Longitudinal data

Longitudinal data, or repeated measures data, consist of repeated observations of a given characteristic on multiple observational units generically called *subjects*. The characteristic of interest, or *response*, may be categorical, discrete, or continuous; furthermore, the response may be univariate or multivariate (itself consisting of several component characteristics). Unless noted otherwise, however, we will assume that the response is continuous and univariate, as most

1

available results on antedependence models pertain to this type of response. We defer consideration of antedependence models for categorical/discrete and multivariate longitudinal data to the last chapter.

The response is usually measured at discrete points in time, but alternatively it could be measured at points in one-dimensional space. Examples of the former include the weights of cattle measured at weekly intervals from weaning to finishing, or the scores on a scholastic aptitude examination taken annually by students in a particular school district; exemplifying the latter are diversity indices of pollen grains found at various depths of an ice core sample, or the lengths of time ("split times") runners need to complete each consecutive 10-kilometer section of a 100-kilometer race. Henceforth, for ease of exposition we shall use terminology appropriate for a response measured at points in time, but our discourse applies equally well, with a suitable change in terminology, to the one-dimensional spatial context.

In addition to the response variable, one or more covariates may be observed on subjects. The covariates may be time-dependent (such as the time of measurement itself, or the subject's weight or health status at each time of measurement), time-independent (such as gender), or a mixture of both. If the longitudinal study is a comparative experiment in which treatments are randomized to units, then subjects may be accorded the additional status of experimental units and one or more of the covariates is then a nominal variable indicating the treatment group membership of each subject. Usually this treatment group covariate is time-independent, but it can be time-dependent as in the case of crossover experiments.

Because the response is measured at multiple times on multiple subjects, longitudinal data are replicated in two "directions." However, there is an important difference in the dependence structure in the two directions. The subjects are usually not inter-related in any meaningful way, with the important consequence that observations from different subjects may reasonably be regarded as independent. The same cannot be said of observations taken at different times from the same subject. In fact, measurements of the response from the same subject tend to be considerably more alike than measurements from different subjects. Consider, for example, the data listed in Table 1.1, which come from a longitudinal study of cattle growth reported by Kenward (1987). The data are weights (in kg) of 30 cattle receiving a treatment for intestinal parasites, which were recorded on 11 occasions; more details pertaining to these data are given in Section 1.7.1, and we subsequently refer to these data as the Treatment A cattle growth data. Weights of an additional thirty cattle that received a different treatment (Treatment B) were recorded on the same occasions, and we will also consider these data later. Figure 1.1 displays the Treatment A cattle growth data graphically as a profile plot, i.e., a plot of subjects' responses versus time, with successive measurements from the same subject connected by line segments.

Table 1.1 *Weights (in kg) of cattle from the growth study of Kenward (1987): Treatment A cattle only.*

	Weeks from start of experiment										
Cow	0	2	4	6	8	10	12	14	16	18	19
1	233	224	245	258	271	287	287	287	290	293	297
2	231	238	260	273	290	300	311	313	317	321	326
3	232	237	245	265	285	298	304	319	317	334	329
4	239	246	268	288	308	309	327	324	327	336	341
5	215	216	239	264	282	299	307	321	328	332	337
6	236	226	242	255	263	277	290	299	300	308	310
7	219	229	246	265	279	292	299	299	298	300	290
8	231	245	270	292	302	321	322	334	323	337	337
9	230	228	243	255	272	276	277	289	289	300	303
10	232	240	247	263	275	286	294	302	308	319	326
11	234	237	259	289	311	324	342	347	355	368	368
12	237	235	258	263	282	304	318	327	336	349	353
13	229	234	254	276	294	315	323	341	346	352	357
14	220	227	248	273	290	308	322	326	330	342	343
15	232	241	255	276	293	309	310	330	326	329	330
16	210	225	242	260	272	277	273	295	292	305	306
17	229	241	252	265	274	285	303	308	315	328	328
18	204	198	217	233	251	258	272	283	279	295	298
19	220	221	236	260	274	295	300	301	310	318	316
20	233	234	250	268	280	298	308	319	318	336	333
21	234	234	254	274	294	306	318	334	343	349	350
22	200	207	217	238	252	267	284	282	282	284	288
23	220	213	229	252	254	273	293	289	294	292	298
24	225	239	254	269	289	308	313	324	327	347	344
25	236	245	257	271	294	307	317	327	328	328	325
26	231	231	237	261	274	285	291	301	307	315	320
27	208	211	238	254	267	287	306	312	320	337	338
28	232	248	261	285	292	307	312	323	318	328	329
29	233	241	252	273	301	316	332	336	339	348	345
30	221	219	231	251	270	272	287	294	292	292	299

Although there is some crossing of individual cattles' growth trajectories over time, the heaviest cattle at the outset of the study tend to remain among the heaviest for the study's duration, and similarly the lightest cattle tend to remain among the lightest. This persistence within subjects, sometimes called "tracking," manifests as sizable positive correlations among same-subject measurements, as seen in the sample correlations of these data (Table 1.2, below the main diagonal). Furthermore, these correlations, like those of many other longitudinal data sets, decrease as the elapsed time between measurements increases — a phenomenon known as *serial correlation*. This is not so easily discerned from the profile plot, but manifests clearly in the sample correlation matrix as a more-or-less monotonic attenuation of correlations within columns (or rows) as one moves away from the main diagonal.

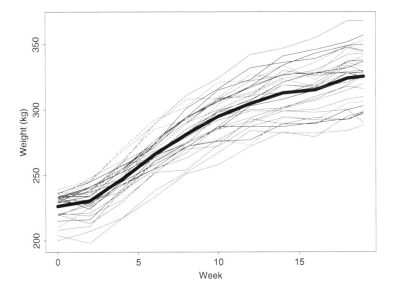

Figure 1.1 *Profile plot for the Treatment A cattle growth data. The thicker line indicates the overall mean profile.*

The inferential objectives of a longitudinal data analysis typically include describing how the response is related to time and other covariates and describing how this relationship is affected by treatments or other factors of classification. Plainly, these objectives are focused primarily upon the mean structure of the data. Unfortunately, the presence of within-subject correlation complicates the achievement of these objectives. The prospects for informative statistical analysis are far from hopeless, however, for the independent replication across

Table 1.2 *Sample variances, along the main diagonal, and correlations, below the main diagonal, of the Treatment A cattle growth data.*

106										
.82	155									
.76	.91	165								
.66	.84	.93	185							
.64	.80	.88	.94	243						
.59	.74	.85	.91	.94	284					
.52	.63	.75	.83	.87	.93	307				
.53	.67	.77	.84	.89	.94	.93	341			
.52	.60	.71	.77	.84	.90	.93	.97	389		
.48	.58	.70	.73	.80	.87	.88	.94	.96	470	
.48	.55	.68	.71	.77	.83	.86	.92	.96	.98	445

subjects opens up the possibility of estimating the parameters of non-trivial models for both the mean structure *and* the within-subject correlation structure or, more broadly, for the entire covariance structure of the data. In this way a full accounting of the covariance structure may be taken when making inferences about the mean structure.

The parametric modeling approach to which we have just alluded is the current "state of the art" of continuous longitudinal data analysis, antedependence models being one of several large classes of available models. Before describing this approach and antedependence models in more detail, we set the stage by briefly reviewing how several rather more classical methods of continuous longitudinal data analysis deal with within-subject correlation.

1.2 Classical methods of analysis

The methods of longitudinal data analysis discussed in this section have existed for a long time. Excellent summaries of them are given by Crowder and Hand (1990), Diggle et al. (2002), Davis (2002), and Weiss (2005), to which we refer the reader for further details.

The earliest methods of longitudinal data analysis either ignored within-subject correlation altogether or attempted to circumvent any difficulties it might pose to inference. In one of the most rudimentary of approaches, the data from each time point are analyzed separately. For example, if the data are grouped according to treatments or some other factor of classification, a separate analysis of variance (ANOVA) may be carried out using only the data at each time point. This "time-by-time ANOVA" approach completely ignores within-subject correlations and suffers from several additional weaknesses; for example, it fails to characterize how the response varies over time or how group effects manifest and evolve over time. Another overly simplistic approach is to perform separate comparisons among responses, or groups of responses, for each pair of measurement times. Multiple paired t tests or multiple two-factor ANOVAs, with treatment and time as the two factors (the latter being dichotomous), might be used for this purpose. This approach again ignores within-subject correlation and requires the investigator to somehow piece together all the conclusions from the individual, correlated tests into a coherent story.

A somewhat more satisfactory classical approach, known variously as the "summary-statistic approach," "response feature analysis," or "derived variable analysis," is to reduce the vector of multiple measurements on each subject to a single measurement that characterizes how the response is related to time (for example, a least squares slope coefficient from a regression of response on time for that subject), and then use various standard methods (e.g., ANOVA) to study how treatments or other factors affect this summary statistic. By reducing all measurements on a given subject to a single measurement, this approach circumvents difficulties associated with accounting for within-subject correlation and yields a set of derived observations that are often reasonably regarded as independent. However, some information on the relationship between response and time inevitably is lost.

Another classical method for the analysis of longitudinal data is known as the "repeated measures ANOVA." In this approach, an ANOVA is performed as if the data were from a split-plot experiment, with time of measurement as the split-plot factor. The ANOVA yields F-tests for the hypotheses of no time effects and no group-by-time interaction effects, but as a consequence of variance heterogeneity and within-subject correlation these tests are generally not valid. The tests are valid if and only if a condition on the covariance structure known as sphericity is satisfied (Mauchly, 1940). One type of covariance structure that satisfies the sphericity condition is compound symmetry, in which the variances of responses are constant over time and all correlations are equal, regardless of the amount of time that elapses between measurements. Although sphericity is slightly more general than compound symmetry, it is still very restrictive and in practice it is not often well-satisfied by the data. Various modifications to the repeated measures ANOVA have been devised to yield F tests that are approximately valid regardless of the within-subject covariance structure (Greenhouse and Geisser, 1959; Huynh and Feldt, 1976), but their use is waning due to the development of better methods of analysis.

Historically, the first approach to the analysis of longitudinal data that accounted directly for within-subject correlation was the "general multivariate regression approach." This approach regards the vector of responses on a given subject as a multivariate observational unit, and it assumes that vectors from different subjects are independent and distributed according to some family of distributions (usually multivariate normal). The mean vectors of these distributions may be taken to be common across subjects but otherwise arbitrary — the so-called "saturated" mean structure — or may be permitted to depend functionally on values of observed covariates, including time itself. The covariance matrix is regarded as an unknown parameter to be estimated and typically is assumed to be homogeneous either across all subjects or across subjects within specified groups (e.g., treatment groups). No additional structure (e.g., serial correlation or homogeneous variances) is imposed upon the covariance matrix beyond that required for positive definiteness, however. The covariance matrix is estimated by the ordinary sample covariance matrix in the saturated case, or by the sample covariance matrix of residuals from the fitted mean structure in the more general case. This estimate plays an important role in various inferential methods directed towards the data's mean structure, e.g., Hotelling's T^2 and multivariate analysis of variance.

Although the general multivariate regression approach is completely flexible with regard to the covariance structure imposed, unfortunately it has several shortcomings. First, it is not always applicable, for it requires that the data be *balanced* (or *rectangular*), i.e., that the measurement times be common across all subjects, with no measurements missing. Thus, it cannot be used, without substantial modification, when data from some subjects are missing (incomplete). Missingness is commonplace for longitudinal data, as a result of such things as staggered entry, failure to appear for an appointment, or early withdrawal ("dropout") of subjects from the study. Nor can the general multivariate regression approach be used when the number of measurement times exceeds the number of subjects, for in this event the sample covariance matrix is singular and hence not positive definite (Dykstra, 1970). Most importantly, even when the general multivariate approach is applicable it may be quite inefficient due to the large number of parameters in the covariance structure that must be estimated. For balanced data with n measurement times, the covariance matrix has $n(n+1)/2$ distinct parameters. If the number of subjects is not substantially larger than n, much of the information content of the data is in some sense "spent" on estimating the covariance matrix, leaving less for estimating the mean structure.

1.3 Parametric modeling

Relatively recently, a parametric modeling approach to longitudinal data analysis has gained standing that, while similar to the general multivariate approach in some respects, does not suffer from its shortcomings. Although it involves modeling both the data's mean structure and its covariance structure, the real novelty of the approach is its parsimonious modeling of the latter and it is upon this aspect that we focus most of our attention. Parsimonious modeling of the covariance structure generally results in more efficient estimation of the data's mean structure and more appropriate estimates of the standard errors of the estimated mean structure, in comparison to the general multivariate approach (Diggle, 1988). Moreover, it can deal effectively with unbalanced and missing data, and it can be employed even when the number of measurement times is large relative to the number of subjects, provided that the assumed model is sufficiently parsimonious. This approach, with several parsimonious model options, has been coded into widely available software, for example PROC MIXED of SAS and the lme function of S-Plus.

In order to describe the parametric modeling approach in more detail, we define some notation. Let N denote the number of subjects, let

$$\mathbf{Y}_s = (Y_{s1}, Y_{s2}, \ldots, Y_{sn_s})^T$$

be the vector of n_s measurements of the response (in chronological order) on subject s and let

$$\mathbf{t}_s = (t_{s1}, t_{s2}, \ldots, t_{sn_s})^T$$

be the corresponding vector of measurement times ($s = 1, \ldots, N$). Note that the data are balanced if and only if $\mathbf{t}_1 = \mathbf{t}_2 = \cdots = \mathbf{t}_N$. If the data are indeed balanced, then we may dispense with the subscript "s" on the number of measurement times, representing it more simply by n. Any covariates associated with Y_{si}, possibly including but not limited to the time of measurement, t_{si}, will be collected in a vector denoted by $\mathbf{x}_{si}$.

It is assumed that the $\mathbf{Y}_s$'s are independent random vectors, each with its own mean vector $\boldsymbol{\mu}_s = (\mu_{si})$ and covariance matrix $\boldsymbol{\Sigma}_s = (\sigma_{sij})$. Thus, subject-specific mean vectors are allowed, due to possible differences in measured co-variates and measurement times across subjects, but the elements within these vectors may be related through their functional dependence on relatively few parameters. Similarly, elements in the N covariance matrices may be functionally related, both within and across subjects, via relatively few parameters. Indeed, if the model imposes no such functional relationships, there are too many parameters to be estimated from the data.

Within the parametric modeling paradigm, three types of models can be distinguished, as discussed by Diggle et al. (2002). The first type, *marginal* models, specify parametric functions for the elements of the mean vectors and covariance matrices directly, i.e.,

$$\mu_{si} = \mu(\mathbf{x}_{si}; \boldsymbol{\beta}), \quad \sigma_{sij} = \sigma(t_{si}, t_{sj}; \boldsymbol{\theta}). \tag{1.1}$$

Here, $\boldsymbol{\beta}$ and $\boldsymbol{\theta}$ are functionally independent vectors of fixed parameters belonging to specified parameter spaces. Usually the parameter space for $\boldsymbol{\beta}$ is unrestricted, but the parameter space for $\boldsymbol{\theta}$ must be such that all subjects' covariance matrices are positive definite. In (1.1) the elements of the covariance matrices are expressed as functions of measurement times only, but this could be extended to permit dependence on other covariates, if desired. Correlations being more readily interpretable than covariances, an alternative marginal formulation models the variances and correlations $\{\rho_{sij}\}$, so that the model for σ_{sij} given above is replaced by

$$\sigma_{sii} = \sigma(t_{si}; \boldsymbol{\theta}_1), \quad \rho_{sij} = \rho(t_{si}, t_{sj}; \boldsymbol{\theta}_2).$$

Here, $\boldsymbol{\theta}_1$ and $\boldsymbol{\theta}_2$ belong to specified parameter spaces that yield positive definite covariance matrices, and they are functionally independent so that variances may be modeled separately from correlations. An extremely parsimonious example of a marginally specified covariance structure of this type is the compound symmetry model, in which

$$\sigma_{sii} \equiv \sigma^2, \quad \rho_{sij} \equiv \rho \quad \left(\sigma^2 > 0, \; -\frac{1}{\max n_s} < \rho < 1\right). \quad (1.2)$$

Another extremely parsimonious example is the stationary continuous-time first-order autoregressive covariance model , in which

$$\sigma_{sii} \equiv \sigma^2, \quad \rho_{sij} = \rho^{|t_{si}-t_{sj}|} \quad (\sigma^2 > 0, \, 0 \le \rho < 1). \quad (1.3)$$

Here and throughout the book, we use the term "stationary" in conjunction with a longitudinal covariance structure to mean that the variances are equal across time and the correlations depend on the times of measurement only through the absolute value of their differences. Such a covariance structure, together with a constant mean structure, ensures that the corresponding random process is weakly stationary.

The second class of models, *random coefficient* models, are models for the conditional mean and variance of each measurement, given the covariates and a set of subject-specific random regression coefficient vectors $\mathbf{b}_1, \ldots, \mathbf{b}_N$. For example, we might specify that

$$E(Y_{si}|\mathbf{x}_{si}, \mathbf{b}_s) = \mathbf{x}_{si}^T \mathbf{b}_s, \quad \operatorname{cov}(Y_{si}, Y_{sj}|\mathbf{x}_{si}, \mathbf{x}_{sj}, \mathbf{b}_s) = \sigma^2$$

(where "cov," here and throughout the book, is short for "covariance"), and that $\mathbf{b}_1, \ldots, \mathbf{b}_N$ are independent and identically distributed normal vectors with mean vector $\boldsymbol{\beta}$ and covariance matrix $\mathbf{G}$, whose elements are unknown parameters belonging to specified parameter spaces. Although the observations are conditionally uncorrelated in this formulation, marginally they are correlated within subjects as a result of the shared realized value of $\mathbf{b}_s$ among all measurements from subject s. Thus the covariance structure of the observations is specified rather more indirectly than for a marginal model, but a marginal model is implied nonetheless.

The third class of models, *transition* models, specify parametric forms for the

the conditional mean and variance of each Y_{si} given the covariates $\mathbf{x}_{si}$ and all preceding measurements, $\{Y_{s,i-1}, \ldots, Y_{s1}\}$, on the sth subject. A commonly used model of this type for balanced, equally-spaced data is the stationary first-order autoregressive, or AR(1), model

$$
\begin{aligned}
Y_{s1} - \mu(\mathbf{x}_{s1}; \boldsymbol{\beta}) &= \epsilon_{s1}, \\
Y_{si} - \mu(\mathbf{x}_{si}; \boldsymbol{\beta}) &= \phi[Y_{s,i-1} - \mu(\mathbf{x}_{s,i-1}; \boldsymbol{\beta})] + \epsilon_{si}, \\
& \qquad i = 2, \ldots, n,
\end{aligned}
\tag{1.4}
$$

where the "autoregressive coefficient" ϕ satisfies $-1 < \phi < 1$ and the "innovations" $\{\epsilon_{si}\}$ are independent zero-mean normal random variables with $\mathrm{var}(\epsilon_{s1}) = \delta/(1 - \phi^2)$ and $\mathrm{var}(\epsilon_{si}) = \delta$ for $i = 2, \ldots, n$ and all s. Note that transition models, like random coefficient models, imply certain marginal models. For example, the model obtained by marginalizing the first-order autoregressive model (1.4) and restricting its parameter space for ϕ to $[0, 1)$ coincides with model (1.3) when the data are balanced and measurement times are equally spaced.

Whatever the class of parametric models used for (continuous) longitudinal data, the standard approach to parameter estimation and other kinds of statistical inference is to assume that the joint distribution of the observations (or some transformation thereof) is multivariate normal and use the method of maximum likelihood or a common variation of it known as restricted, or residual, maximum likelihood (REML).

1.4 Antedependence models, in brief

For the AR(1) model in particular and for stationary autoregressive models in general, the innovation variances and the autoregressive coefficients corresponding to a given lag are constant over time (apart possibly from some "start-up values," such as the variance of ϵ_{s1} in the AR(1) model). Consequently, the marginal variances of responses are constant over time and the marginal correlations between measurements equidistant in time are equal. Often in practice, however, longitudinal data do not satisfy these stationarity assumptions. Nevertheless, we may still use a transition model if we are willing to consider a more general class of transition models known as antedependence models. A first-order normal antedependence model (in its most general form) for balanced data is given by

$$
\begin{aligned}
Y_{s1} - \mu(\mathbf{x}_{s1}; \boldsymbol{\beta}) &= \epsilon_{s1}, \\
Y_{si} - \mu(\mathbf{x}_{si}; \boldsymbol{\beta}) &= \phi_{i-1}[Y_{s,i-1} - \mu(\mathbf{x}_{s,i-1}; \boldsymbol{\beta})] + \epsilon_{si}, \\
& \qquad i = 2, \ldots, n,
\end{aligned}
\tag{1.5}
$$

where $\epsilon_{s1}, \ldots, \epsilon_{sn}$ are independent normal random variables with mean zero and variances given by $\mathrm{var}(\epsilon_{si}) = \delta_i > 0$, and the autoregressive coefficients $\phi_1, \ldots, \phi_{n-1}$ are unconstrained. Plainly, this model generalizes the AR(1) model (1.4) by allowing the innovation variances and lag-one autoregressive coefficients to vary over time. This, in turn, allows the marginal variances and lag-one correlations among responses to vary over time as well. Correlations beyond lag one are also allowed to vary over time, but in a way that is completely determined by the lag-one correlations. In fact, under this model the correlations $\{\rho_{sij}\}$ satisfy

$$\rho_{sij} = \prod_{m=j}^{i-1} \rho_{m+1,m}, \quad i > j,$$

where $\rho_{m+1,m}$ is the lag-one-ahead correlation at time m, which is assumed to be common across subjects and to satisfy $-1 < \rho_{m+1,m} < 1$ for $m = 1, \ldots, n-1$. Thus, correlations among observations lagged two or more measurement times apart are completely determined as the product of the lag-one correlations corresponding to each intervening pair of consecutive observations. (This result, the relationship between the marginal variances and the $\rho_{m+1,m}$'s on the one hand and the innovation variances and lag-one autoregressive coefficients on the other, and many other results for first-order antedependence models will be established in Chapter 2.)

Higher-order antedependence models generalize higher-order stationary autoregressive models similarly, by allowing variances and higher-order same-lag correlations to vary over time in more flexible ways.

Because antedependence models do not impose any stationarity assumptions, they can provide for a considerably better fit than stationary autoregressive models to many longitudinal data sets exhibiting serial correlation. The example of the next section illustrates this point. Moreover, maximum likelihood and REML estimates of the parameters of an important class of such models known as *unstructured* antedependence models, of which (1.5) is the first-order case, exist in closed form when the data are balanced and the mean structure depends only on time-independent covariates (e.g., the saturated case).

1.5 A motivating example

Consider again the Treatment A cattle growth data, in particular their estimated within-subject covariance structure (Table 1.2), which is computed from the residuals from an estimated saturated mean structure. We noted previously that the correlations are all positive and decrease more-or-less monotonically within columns of the correlation matrix. Two other features of the covariance structure are also worth noting. First, the variances (Table 1.2, main diagonal elements) are not homogeneous, but instead tend to increase over time; in fact, the variance quadruples (approximately) from the beginning of the study to its end. Second, same-lag correlations (correlations within a given subdiagonal below the main diagonal) are not constant, but instead tend to increase somewhat over time.

How might this covariance structure be modeled parametrically? Because the data are balanced, one option would be the general multivariate approach. However, in light of the discernible behavior of the variances and correlations, such an approach would appear to be substantially overparameterized. The compound symmetry model is an altogether unsuitable alternative, owing to the widely disparate sample variances and clear evidence of serial correlation in the data. Furthermore, sphericity is rejected unequivocally (p-value $< 1.0 \times 10^{-8}$). We might briefly entertain stationary autoregressive models or other stationary time series models, but these models do not comport with the nonstationarity manifested in the sample variances and same-lag (within-subdiagonal) sample correlations.

Instead of attempting to model the nonstationarity exhibited by these data, we could try to reduce or eliminate it by transforming the response variable — a standard ploy for reducing the functional dependence of the variance on the mean in applied regression modeling — and then use a stationary autoregressive model for the transformed data. Several power transformations (including logarithms) of the cattle weights were attempted, and the one that best stabilized the variance was the inverse square root transformation. Table 1.3 shows the sample variances and correlations of the transformed data. Although an inverse square-root transformation renders the variances of these data relatively stationary, the nonstationary behavior of the correlations persists after transformation; in fact, the correlations change very little. It appears, unfortunately, that we cannot "transform away" the nonstationarity of the variances and correlations simultaneously.

However, the first-order antedependence model (1.5) introduced in the previous section is flexible enough to accommodate the nonstationarity exhibited by both the variances and correlations of these data. Table 1.4 displays the REML estimates of the marginal variances and correlations of the first-order antedependence model, based on the data measured in the original scale. Comparison with Table 1.2 suggests that this model fits remarkably well. In fact, the variances and lag-one correlations fit perfectly, which, as we will see later, is a universal property of REML estimation for this model. Thus a comparison of the lag-two and higher correlations is actually more relevant, and we see that the correspondence of these correlations in the two tables is quite good. In particular, the fitted first-order antedependence model is able to track the increase in the data's same-lag correlations over time.

Table 1.3 *Sample variances, along the main diagonal, and correlations, below the main diagonal, of the inverse-square-root transformed Treatment A cattle growth data. Variances are in units of* 10^{-8}.

247										
.83	342									
.78	.91	290								
.68	.86	.93	253							
.65	.82	.89	.94	279						
.60	.75	.85	.92	.94	287					
.53	.63	.75	.82	.86	.93	276				
.54	.68	.78	.84	.89	.94	.93	282			
.53	.62	.73	.79	.84	.91	.94	.97	316		
.49	.60	.72	.75	.80	.87	.88	.95	.96	356	
.49	.57	.69	.72	.77	.83	.85	.92	.95	.98	330

Table 1.4 *REML variance estimates, along the main diagonal, and REML correlation estimates, below the main diagonal, for the first-order antedependence model fitted to the Treatment A cattle growth data.*

106										
.82	155									
.75	.91	165								
.69	.84	.93	185							
.65	.79	.87	.94	243						
.61	.74	.82	.89	.94	284					
.57	.69	.76	.83	.88	.93	307				
.53	.65	.71	.77	.82	.87	.93	341			
.52	.63	.69	.75	.79	.84	.90	.97	389		
.50	.60	.66	.72	.76	.81	.87	.93	.96	470	
.49	.59	.65	.71	.75	.80	.86	.92	.95	.98	445

The first-order antedependence model is not the only nonstationary alternative to an AR(1) model for these data. Another possibility is to model the data's correlation structure as that of a stationary AR(1) model, but let the variances be arbitrary (but positive). This model, known as the heterogeneous AR(1) or ARH(1) model, is another first-order antedependence model, which is more parsimonious than the general case but not as parsimonious as an AR(1). Table 1.5 displays REML estimates of the marginal variances and correlations under this model. Note that the variance estimates coincide with those from the fitted first-order antedependence model. Note also that this model is not sufficiently flexible to track the heterogeneity among the same-lag correlations. We will see later that other antedependence models exist that are superior, for these data, to both the first-order antedependence model of the previous section and the heterogeneous AR(1) model.

Table 1.5 *REML variance estimates, along the main diagonal, and REML correlation estimates, below the main diagonal, for the heterogeneous AR(1) model fitted to the Treatment A cattle growth data.*

106										
.94	155									
.89	.94	165								
.83	.89	.94	185							
.78	.83	.89	.94	243						
.73	.78	.83	.89	.94	284					
.69	.73	.78	.83	.89	.94	307				
.65	.69	.73	.78	.83	.89	.94	341			
.61	.65	.69	.73	.78	.83	.89	.94	389		
.58	.61	.65	.69	.73	.78	.83	.89	.94	470	
.54	.58	.61	.65	.69	.73	.78	.83	.89	.94	445

1.6 Overview of the book

This book describes antedependence models, their properties, and some important statistical inference procedures associated with them. Descriptions of the models and their properties are the topics of Chapters 2 and 3. Chapter 2 defines and develops antedependence in its most general form, known as *un-structured* antedependence, and the related notion of partial antecorrelation. It also establishes several equivalent characterizations and parameterizations of unstructured antedependence for normal variables. Chapter 3 is concerned with *structured* antedependence models, which retain the essential conditional independence properties of unstructured antedependence models but are more parsimonious. Beginning with Chapter 4, the book turns towards inference for the models within the classical longitudinal sampling framework. Informal model identification, using relevant summary statistics and exploratory graphical methods, is the topic of Chapter 4 itself. More formal, likelihood-based inference procedures for normal antedependence models are derived and presented in Chapters 5 through 7. Chapter 5 deals with maximum likelihood and residual maximum likelihood estimation of parameters, Chapter 6 with likelihood ratio tests and penalized likelihood model selection criteria for the model's covariance structure, and Chapter 7 with such tests and criteria for the model's mean structure. Numerous examples are provided in Chapters 4 through 7 to illustrate the methodology and to demonstrate the advantages of a longitudinal data analysis based on an antedependence model. Chapter 8 summarizes these illustrative examples and uses them to compare the performance of antedependence models to some other parametric models commonly used for longitudinal data.

Throughout all of the chapters summarized above, we limit our focus in some ways, for the sake of simplicity. For example, we consider only longitudinal data that are univariate and continuous, and we suppose that their means are modeled only as linear functions of unknown parameters. Extensions of models and inference procedures to multivariate or discrete data and to nonlinear mean structure, along with some other topics, are described briefly in Chapter 9.

1.7 Four featured data sets

We conclude this chapter by introducing four longitudinal data sets, including the cattle growth data featured in previous sections, that will be used extensively in this book to illustrate various methods of statistical inference associated with antedependence models. All of these data sets arise from studies within the biological sciences, broadly defined. They exhibit a wide range of characteristics that can occur with longitudinal data, including observational and experimental data, equally and unequally spaced measurement times, time-independent and time-dependent covariates, saturated and unsaturated mean structure, treatment groups and no grouping factors, homogeneous and heterogeneous within-group covariance matrices, and complete and missing data (monotone and non-monotone).

1.7.1 Cattle growth data

The cattle growth data (Kenward, 1987) come from a comparative experiment in which cattle receiving two treatments, generically labeled A and B, for intestinal parasites were weighed 11 times over a 133-day period. Thirty animals received Treatment A and thirty received Treatment B. The first 10 measurements on each animal were made at two-week intervals and the final measurement was made one week after the tenth measurement. Measurement times were common across animals and no observations were missing. Although measurement times were not quite equally spaced (due to the shorter interval between the tenth and eleventh measurements), the data are balanced. We already displayed the data (Table 1.1) and the profile plot (Figure 1.1) for the cattle receiving Treatment A; the data and profile plot for the cattle receiving Treatment B are given in Table 1.6 and Figure 1.2, respectively. The main objective of the analysis is to characterize how cattle growth is affected by the treatments. In particular, the experimenter wishes to know if there is a difference in growth between treatment groups, and if so, the time of measurement at which it first occurs.

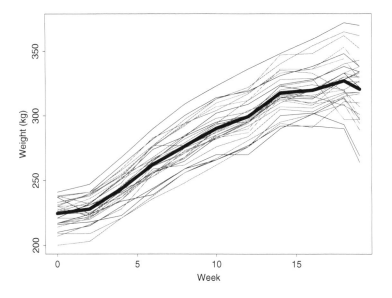

Figure 1.2 *Profile plot for the Treatment B cattle growth data. The thicker line indicates the overall mean profile.*

1.7.2 100-km race data

The 100-km race data, kindly provided by Ian Jollife of the University of Kent, consist of the "split" times for each of 80 competitors in each 10-km section

Table 1.6 *Weights (in kg) of cattle from the growth study of Kenward (1987): Treatment B cattle only.*

	Weeks from start of experiment										
Cow	0	2	4	6	8	10	12	14	16	18	19
1	210	215	230	244	259	266	277	292	292	290	264
2	230	240	258	277	277	293	300	323	327	340	343
3	226	233	248	277	297	313	322	340	354	365	362
4	233	239	253	277	292	310	318	333	336	353	338
5	238	241	262	282	300	314	319	331	338	348	338
6	225	228	237	261	271	288	300	316	319	333	330
7	224	225	239	257	268	290	304	313	310	318	318
8	237	241	255	276	293	307	312	336	336	344	328
9	237	224	234	239	256	266	276	300	302	293	269
10	233	239	259	283	294	313	320	347	348	362	352
11	217	222	235	256	267	285	295	317	315	308	301
12	228	223	246	266	277	287	300	312	308	328	333
13	241	247	268	290	309	323	336	348	359	372	370
14	221	221	240	253	273	282	292	307	306	317	318
15	217	220	235	259	262	276	284	305	303	315	317
16	214	221	237	256	271	283	287	314	316	320	298
17	224	231	241	256	265	283	295	314	313	328	334
18	200	203	221	236	248	262	276	294	291	311	310
19	238	232	252	268	285	298	303	320	324	320	327
20	230	222	243	253	268	284	290	316	314	330	330
21	217	224	242	265	284	302	309	324	328	338	334
22	209	209	221	238	256	267	281	295	301	309	289
23	224	227	245	267	279	294	312	328	329	297	297
24	230	231	244	261	272	283	294	318	320	333	338
25	216	218	223	243	259	270	270	290	301	314	297
26	231	239	254	276	294	304	317	335	333	319	307
27	207	216	228	255	275	285	296	314	319	330	330
28	227	236	251	264	276	287	297	315	309	313	294
29	221	232	251	274	284	295	300	323	319	333	322
30	233	238	254	266	282	294	295	310	320	327	326

of a 100-km race held in 1984 in the United Kingdom. The data include, in addition to the split times, the ages of all but four of the competitors. The 10-km sections, which in this situation play the role usually played by measurement times, are evenly spaced (obviously). Every competitor finished the race, hence the data are balanced. Tables 1.7 and 1.8 provide the data, and Figure 1.3 is a profile plot of the data. Everitt (1994a, 1994b) performs some additional graphical and exploratory analyses of these data. The analysis objective here is to find a parsimonious model that adequately describes how competitor performance on each 10-km section is related to the section number ($i = 1, 2, \ldots, 10$), age, and performance on previous sections.

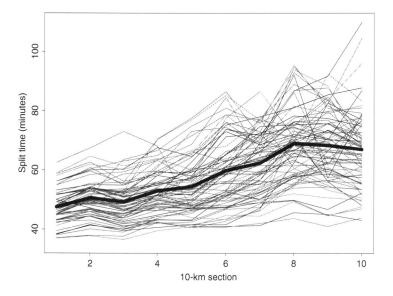

Figure 1.3 *Profile plot for the 100-km race data. The thicker line indicates the overall mean profile.*

1.7.3 Speech recognition data

The speech recognition data (Tyler et al., 1988) consist of scores (percentage of correct responses) on a sentence test administered under audition-only conditions to groups of human subjects wearing one of two types of cochlear implants, referred to here as A and B. Implants were surgically implanted five to six weeks prior to being electrically connected to an external speech processor.

Table 1.7: *100-km race data set, first 40 competitors.*

						Section					
Subject	Age	1	2	3	4	5	6	7	8	9	10
1	39	37.0	37.8	36.6	39.6	41.0	41.0	41.3	45.7	45.1	43.1
2	39	39.5	42.2	40.0	42.3	40.6	40.8	42.0	43.7	41.0	43.9
3	—	37.1	38.0	37.7	42.4	41.6	43.5	48.7	49.7	44.8	47.0
4	36	37.0	37.8	36.6	39.6	41.0	44.8	44.5	49.4	44.6	47.7
5	34	42.2	44.5	41.9	43.4	43.0	47.2	49.1	49.9	46.8	52.3
6	46	43.0	44.6	41.2	42.1	42.5	46.8	47.5	55.8	56.6	58.6
7	35	43.2	44.4	41.0	43.4	43.0	47.2	52.4	57.3	54.4	53.5
8	47	43.2	46.7	44.8	47.5	47.4	47.7	49.9	52.1	50.7	50.0
9	30	38.5	41.4	40.1	43.2	43.2	51.5	56.7	71.5	56.2	48.2
10	—	42.5	43.1	40.6	44.5	45.4	52.3	59.7	59.3	55.0	49.6
11	48	38.0	40.1	39.1	43.8	46.6	51.9	59.2	63.5	57.6	58.4
12	39	46.0	50.4	46.8	47.4	44.1	43.4	46.3	55.0	64.9	56.2
13	32	44.8	46.0	43.1	46.5	46.3	49.0	52.5	58.4	60.9	55.2
14	43	44.8	46.0	43.1	46.5	46.3	49.0	52.5	58.4	60.9	55.2
15	35	47.0	49.4	46.8	48.6	47.8	50.8	50.3	54.0	54.4	53.6
16	47	45.0	46.7	45.3	49.9	47.8	51.2	54.1	58.7	53.3	50.7
17	38	45.0	46.7	43.8	48.0	47.2	47.5	51.7	57.3	60.4	55.6
18	25	43.1	44.5	41.0	42.5	40.6	42.8	46.5	73.2	70.8	63.4
19	49	45.2	46.9	45.5	48.8	50.1	51.2	56.4	55.2	56.6	53.5
20	50	43.0	46.1	44.7	47.4	47.1	46.8	54.6	60.4	68.0	51.6
21	24	38.3	41.6	39.6	40.7	41.6	41.6	47.2	62.4	82.7	77.1
22	51	45.0	47.1	45.3	49.1	46.8	47.4	50.3	55.1	66.4	64.6
23	41	43.2	46.1	45.2	48.4	49.9	49.6	52.7	58.1	62.8	62.6
24	41	41.2	44.6	43.8	48.4	48.8	53.4	58.9	68.6	59.1	53.4
25	47	49.2	48.8	48.7	51.8	48.2	52.8	50.2	58.0	58.7	57.5
26	43	48.0	52.9	49.6	50.1	48.1	48.1	49.1	54.6	62.7	64.0
27	41	46.0	49.9	47.7	50.4	52.9	51.4	55.6	57.8	59.7	55.8
28	45	46.1	46.0	42.2	44.4	46.0	49.0	53.3	66.7	72.9	67.6
29	39	45.1	49.7	46.5	46.5	49.3	58.8	58.7	64.7	64.0	63.4
30	42	48.0	52.9	49.6	50.1	48.4	50.0	58.5	62.9	60.1	60.1
31	55	49.2	54.5	51.3	56.1	53.9	53.2	53.4	58.8	62.4	59.4
32	42	47.0	49.4	46.8	49.7	50.3	55.5	59.8	67.1	64.2	70.4
33	39	48.2	54.1	51.2	53.5	54.8	55.7	55.2	65.7	62.3	62.3
34	37	46.5	50.8	48.0	51.4	50.0	58.6	61.6	61.5	61.9	75.4
35	33	47.3	51.2	49.5	52.6	57.9	58.6	66.4	70.6	56.4	55.6
36	37	48.2	53.9	50.9	54.0	52.4	59.3	77.5	60.6	55.8	61.4
37	34	43.3	45.2	42.7	44.9	47.3	52.9	69.3	92.2	57.3	79.1
38	35	52.0	53.0	50.0	51.6	55.4	56.3	56.7	68.4	66.9	65.4
39	37	49.2	54.5	50.8	53.6	53.4	56.0	62.3	65.8	66.1	65.6
40	24	49.3	52.8	51.1	53.8	52.4	59.3	63.2	73.7	62.3	62.3

Table 1.8:   *100-km race data set, last 40 competitors.*

					Section					
Subject	Age	1	2	3	4	5	6	7	8	9	10
41	28	47.2	51.3	49.5	52.6	51.6	60.1	64.4	66.5	66.6	76.9
42	35	49.2	48.8	49.2	54.2	60.8	60.4	64.0	69.9	66.1	65.1
43	39	45.2	50.2	48.7	53.6	53.5	60.3	59.2	71.4	75.9	71.8
44	50	45.3	51.0	46.9	50.0	51.0	59.7	78.2	68.9	69.7	72.8
45	47	49.2	48.3	46.2	51.6	51.9	61.1	71.8	74.6	70.3	69.9
46	52	49.2	48.7	48.8	52.6	57.9	65.3	71.7	64.1	70.7	68.2
47	22	46.0	49.2	47.5	51.5	54.1	61.4	66.5	76.5	77.0	68.6
48	47	46.5	51.1	49.9	56.1	53.6	58.2	66.3	76.2	70.6	76.3
49	49	51.6	54.0	52.1	55.1	57.4	61.0	63.4	70.2	73.4	67.9
50	—	47.2	50.0	48.1	50.8	55.3	62.6	70.5	76.1	72.7	72.8
51	44	45.0	50.3	48.8	53.6	54.4	58.9	67.6	77.7	79.9	81.1
52	57	48.0	52.9	49.6	50.1	53.5	65.6	72.8	74.1	72.6	78.1
53	36	53.2	55.1	55.0	59.3	59.4	63.2	66.1	66.7	73.7	68.3
54	45	62.5	67.5	73.1	68.2	47.1	51.9	58.3	68.5	64.4	62.1
55	—	49.2	48.8	53.4	56.1	59.8	65.2	72.8	71.4	79.8	70.5
56	46	49.2	48.6	47.5	51.8	57.7	63.5	63.5	69.5	92.6	83.4
57	38	51.6	53.7	49.2	58.3	56.4	65.3	74.8	75.4	75.8	69.2
58	49	58.7	62.7	56.3	58.6	66.3	62.9	67.4	71.4	69.6	60.8
59	39	49.2	53.3	53.7	54.8	59.3	73.9	70.8	86.3	61.8	78.7
60	38	59.0	64.6	61.4	64.0	60.2	64.0	66.2	69.5	69.3	64.9
61	44	50.1	53.9	52.7	59.8	58.2	71.4	72.3	78.4	77.5	74.9
62	43	55.0	58.5	59.4	63.4	57.0	66.4	67.7	68.7	75.9	77.7
63	62	47.2	52.1	51.7	61.0	73.2	74.5	69.2	76.5	75.6	70.9
64	31	51.7	54.0	53.0	55.6	56.0	62.9	76.2	81.1	85.5	87.8
65	36	50.0	47.3	44.1	51.7	62.8	75.3	78.1	81.2	85.5	87.8
66	28	56.2	59.7	55.6	58.2	64.4	76.1	68.4	75.3	84.8	70.5
67	46	56.2	59.7	55.6	58.2	64.4	76.1	68.4	75.3	84.8	70.5
68	34	46.5	51.7	52.3	61.7	66.8	68.1	76.9	74.9	83.7	96.1
69	62	56.2	60.0	60.4	67.7	64.7	73.1	68.7	72.1	70.3	86.9
70	39	49.2	53.0	52.5	55.5	57.1	77.7	86.6	71.2	82.5	104.6
71	43	51.6	54.2	58.7	59.8	65.4	76.2	73.1	93.0	83.7	74.3
72	46	58.1	62.0	60.2	63.7	65.7	78.8	69.4	81.7	79.2	81.9
73	20	48.2	54.0	52.5	55.5	60.1	73.9	71.8	86.8	91.9	110.0
74	49	55.0	60.9	55.0	63.5	68.8	84.6	64.8	95.0	81.8	75.4
75	35	56.2	60.4	63.1	65.0	71.7	78.8	77.0	89.0	83.5	61.0
76	42	48.0	52.9	52.8	70.5	77.1	85.3	76.9	93.9	88.4	68.2
77	20	46.5	51.0	63.6	66.7	75.0	81.0	76.0	95.4	80.9	79.3
78	30	46.5	51.0	63.6	66.7	75.0	81.0	76.0	95.4	80.9	79.3
79	27	52.2	55.5	55.9	70.6	77.7	86.6	71.6	86.9	87.8	71.2
80	34	50.5	55.4	64.1	66.3	75.6	86.6	71.6	87.3	89.2	73.4

Subjects were profoundly, bilaterally deaf, thus preconnection baseline values for the sentence test were all zero. Twenty subjects received implant A and 21 received implant B. Measurements were scheduled at 1, 9, 18, and 30 months after connection. There was some variation in actual follow-up times, however, so these times were not exact. Moreover, some subjects did not show up for one or more of their scheduled follow-ups, so some data are missing. Table 1.9 gives the data, and Figure 1.4 displays profile plots of the data for each type of implant. Our interest centers on describing how the audiologic performance of individuals receiving each type of implant depends on the elapsed time since implantation. More specifically, we wish to know, primarily, how the mean profiles of the two implant types compare to one another, and secondarily, whether a subject's audiologic performance tends to become more consistent over time.

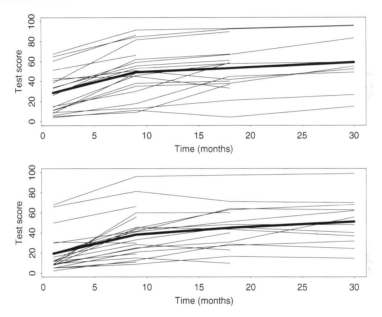

Figure 1.4 *Profile plots for the speech recognition data: Top panel, cochlear implant type A; bottom panel, cochlear implant type B. The thicker lines indicate the overall mean profiles for each implant type.*

1.7.4 Fruit fly mortality data

The fruit fly mortality data were kindly provided to us by J. Curtsinger, A. Khazaeli, and F. Jaffrézic of the INRA Quantitative and Applied Genetics

Table 1.9: *Speech recognition data set.*

Subject	Implant type	Months after connection 1	9	18	30
1	A	28.57	53.00	57.83	59.22
2	A	9.00	13.00	21.00	26.50
3	A	60.37	86.41	——	——
4	A	33.87	55.50	61.06	——
5	A	42.86	44.93	33.00	——
6	A	26.04	61.98	67.28	——
7	A	15.00	30.00	58.53	——
8	A	33.00	59.00	66.80	83.20
9	A	11.29	38.02	40.00	——
10	A	11.00	35.10	37.79	54.80
11	A	40.55	50.69	41.70	52.07
12	A	3.90	11.06	4.15	14.90
13	A	6.00	17.74	44.70	48.85
14	A	64.75	84.50	92.40	95.39
15	A	38.25	81.57	89.63	——
16	A	67.50	91.47	92.86	96.00
17	A	14.29	45.62	58.00	——
18	A	5.00	9.00	37.00	——
19	A	51.15	66.13	——	——
20	A	8.00	48.16	——	——
21	B	11.75	19.00	——	——
22	B	5.00	11.00	——	——
23	B	8.76	24.42	40.00	——
24	B	5.00	20.79	27.42	31.80
25	B	2.30	12.67	28.80	24.42
26	B	12.90	28.34	23.00	——
27	B	19.23	45.50	43.32	36.80
28	B	68.00	96.08	97.47	99.00
29	B	20.28	41.01	51.15	61.98
30	B	16.00	33.00	45.39	40.09
31	B	65.90	81.30	71.20	70.00
32	B	5.00	8.76	16.59	14.75
33	B	9.22	14.98	9.68	——
34	B	11.29	44.47	62.90	68.20
35	B	30.88	29.72	——	——
36	B	29.72	41.40	64.00	62.67
37	B	8.00	43.55	48.16	48.00
38	B	8.76	60.00	60.00	——
39	B	11.55	55.81	——	——
40	B	49.77	66.27	——	——
41	B	8.00	25.00	30.88	55.53

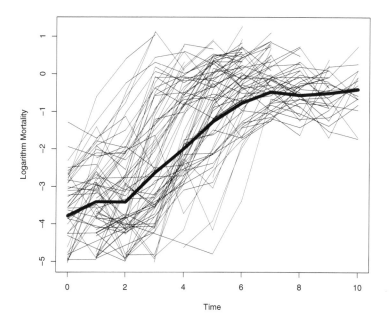

Figure 1.5 *Profile plot for the fruit fly mortality data. The thicker line indicates the overall mean profile.*

(Centre de Researche de Jouy, Jouy-en-Josas, France). The data are age-specific measurements of mortality for 112 cohorts of *Drosophila melanogaster*, the common fruit fly. The cohorts were derived from 56 recombinant inbred lines, each line replicated twice. Cohorts consisted of roughly an equivalent number (500 to 1000) of flies. Every day, dead flies were retrieved from the cage holding each cohort and counted, but these counts were pooled into 11 5-day intervals for analysis. Raw mortality rates were measured as $-\log\left(\frac{N(t+1)}{N(t)}\right)$, where $N(t)$ is the number of flies in the cohort living at the beginning of time interval t $(t = 0, 1, \ldots, 10)$. To make the responses more normally distributed, these raw mortality rates were log-transformed; thus, ultimately the response variable for analysis was $\log\left\{-\log\left(\frac{N(t+1)}{N(t)}\right)\right\}$. For reasons unknown to us, approximately 22% of the data is missing, and the missingness is not monotone. Tables 1.10 through 1.12 list the data, and Figure 1.5 is the profile plot. Our analysis objective is to find a parsimonious model that adequately describes how mortality, averaged over recombinant inbred lines, changes with age (time) and how mortality at any given age is related to mortality at previous ages.

Table 1.10:   *Fruit fly mortality data set, first 50 cohorts.*

				Five day period from start of study						
1	2	3	4	5	6	7	8	9	10	11
−2.83	−2.96	−3.14	−1.27	−0.68	−0.63	−1.70	−0.09	0.33	−0.37	—
−3.49	−4.86	—	−3.45	−4.81	−4.11	−1.77	−0.83	−0.50	0.02	−0.72
−3.79	−3.25	−4.14	−4.82	−3.70	−2.01	−1.13	−0.56	0.12	−0.17	−1.70
−3.77	−4.16	−4.84	−3.73	−2.37	−2.27	−0.53	−0.02	−0.44	−0.80	0.23
−3.44	−4.11	−4.79	−4.78	−3.15	−1.00	−1.35	−0.54	−0.73	−0.17	0.10
−3.77	−4.85	−4.84	−2.60	−1.85	−0.92	−0.43	−0.98	−0.02	−0.53	−0.67
−3.39	−3.65	−3.62	−3.60	−3.57	−3.25	−2.98	−2.21	−1.22	−0.81	−0.70
−3.57	−2.60	−3.47	−2.49	−0.08	−0.18	−0.51	−0.50	0.09	—	—
−3.27	−2.89	−4.80	−4.10	−1.87	0.40	0.71	0.09	—	—	—
−4.82	—	−4.11	−1.11	0.32	—	—	—	—	—	—
—	−4.25	−3.83	−1.35	0.15	0.88	0.09	—	—	—	—
—	—	—	—	−4.63	−3.92	−1.97	−0.78	−0.67	−0.79	−0.30
−4.96	—	—	−4.95	—	—	−3.84	−3.82	−2.37	−2.04	−1.64
—	−3.67	−4.05	−3.33	−2.58	−2.38	−1.42	−0.97	−1.20	−1.16	−0.65
—	−4.89	−4.18	−3.24	−2.39	−1.49	−1.71	−1.19	−1.42	−0.96	−0.09
−3.88	−4.96	−4.26	−4.24	−4.92	−4.22	−4.21	−2.12	−1.31	−1.33	−1.07
−4.90	−4.89	—	—	—	−4.89	—	−2.19	−0.76	−0.37	−0.57
−2.58	−2.11	−2.12	0.19	−0.24	0.28	−0.37	—	—	—	—
−3.80	−4.88	−4.87	−3.24	−2.39	−1.23	−1.98	−0.41	−0.44	−0.51	0.88
−4.99	—	−4.99	—	−3.36	−1.77	−1.81	−1.13	−0.82	0.19	0.03
−4.06	−4.74	−4.73	−4.72	−4.71	−2.59	−1.71	−0.43	−0.82	−0.22	0.38
−3.45	−3.00	−1.86	−0.89	−0.57	−0.87	−2.06	−1.36	−1.64	−0.58	0.33
−2.70	−1.41	−0.82	−0.13	−1.05	−0.60	−1.61	−0.21	−1.25	−0.90	−0.37
—	—	—	−5.01	−2.00	0.15	−0.13	−0.12	−0.58	−1.25	0.09
−4.25	—	—	−1.98	−0.85	0.47	−0.37	−0.02	−0.90	—	—
—	—	—	—	—	−2.76	−0.27	0.19	0.70	−0.37	—
−4.96	−4.26	−4.94	−3.31	−4.20	−1.86	−0.38	−0.24	−0.72	−0.55	−0.53
−4.95	−3.83	−2.69	−1.77	−0.62	−0.34	0.12	0.83	—	—	—
−4.68	−3.27	−3.93	−3.21	−2.03	−1.98	−1.41	−0.47	0.04	−0.24	0.48
−3.63	—	−4.02	−4.00	−1.90	−1.21	−0.70	−0.58	−0.02	−0.21	—
−3.58	−3.56	−4.92	−3.29	−2.35	−0.73	0.92	—	—	—	—
—	—	−4.82	−4.82	−2.70	−2.12	−1.36	−0.83	−1.06	−1.38	−0.82
−4.30	−4.28	−3.86	−4.25	−2.96	−1.75	−1.28	−0.54	−0.80	−1.33	−0.70
−3.41	−3.67	−4.05	−2.92	−1.98	−1.91	−1.20	−1.50	−1.35	−0.65	0.01
−4.36	—	—	−3.94	−3.92	−2.32	−1.00	0.31	−0.49	−0.48	0.23
−3.36	—	—	—	−4.25	−2.83	−1.97	−0.37	−0.15	0.21	−0.17
—	−3.87	−3.85	−3.53	−2.21	−0.98	−0.38	−0.30	−1.25	−0.27	−0.58
—	−4.83	—	−2.85	−2.53	−2.15	−1.31	−1.16	−0.77	−0.83	−0.96
−3.57	—	−3.54	−3.10	−3.05	−1.98	−1.90	−1.42	−1.81	−2.29	−1.74
−3.23	−3.19	−2.08	−1.35	−0.57	0.36	0.53	−0.37	—	—	—
—	−2.84	−1.97	−0.09	−0.09	−0.37	0.23	—	−0.37	—	—
−4.27	−3.14	−3.80	−1.93	−0.82	−0.10	0.70	−0.37	−0.37	—	—
—	−4.23	−3.81	−0.92	−0.03	0.46	0.67	—	—	—	—
−3.07	−2.76	−1.75	−0.37	−0.03	0.53	0.33	—	—	—	—
−3.56	−2.56	−1.04	0.06	−0.19	−0.12	0.23	—	—	—	—
−2.90	−1.74	−0.76	1.12	0.09	—	—	—	—	—	—
−3.46	−2.48	−0.48	1.11	—	—	—	—	—	—	—
−4.99	−2.88	—	−4.93	−1.80	−0.64	−0.06	0.79	—	—	—
−3.86	—	−4.25	−3.53	−2.80	−1.20	−0.43	−0.11	0.19	0.09	—
−4.99	−3.59	−3.34	−3.30	−1.76	−1.63	−1.01	0.33	0.73	—	—

Table 1.11:   *Fruit fly mortality data set, next 50 cohorts.*

				Five day period from start of study						
1	2	3	4	5	6	7	8	9	10	11
–3.56	–2.96	–3.25	–2.31	–0.76	–1.18	–0.68	0.52	—	—	—
–4.77	—	–2.79	–1.85	–0.84	–0.54	–0.73	–0.17	0.09	–0.90	—
–5.03	–2.92	–4.97	–1.66	–0.80	–0.25	–0.98	–1.07	–1.64	0.23	—
—	–4.90	–3.79	–3.47	–2.87	–2.01	–0.64	–0.10	–0.09	–0.09	—
–4.01	–3.06	–2.10	–1.01	0.79	0.67	—	—	—	—	—
–2.31	–1.26	–0.25	0.31	0.01	0.33	—	—	—	—	—
–2.83	–0.59	0.40	1.04	—	—	—	—	—	—	—
—	–4.74	–2.92	–3.57	–2.30	–0.81	–0.16	0.50	–0.67	–0.90	—
–2.91	–2.01	–2.29	–1.48	–0.50	0.00	0.58	—	—	—	—
–3.01	–3.12	–2.20	–0.96	0.20	0.91	—	—	—	—	—
–1.94	–2.19	–2.07	–1.36	–0.35	0.27	0.83	—	—	—	—
–4.95	–4.24	–4.92	—	–2.23	–1.09	–0.50	–0.19	–0.30	0.19	–0.90
–4.23	–4.21	–4.20	—	–2.05	–0.92	–1.16	–0.44	–2.25	–0.77	–0.45
–2.78	–3.20	–3.68	–2.94	–3.60	–1.34	–0.15	–0.09	–0.17	0.09	–0.37
—	–4.16	–3.73	–4.12	–2.36	–0.71	–0.04	–0.15	–0.50	–1.70	–0.09
–4.80	—	—	–4.09	–4.08	–2.42	–1.89	–0.85	–0.46	–0.65	–0.97
–4.88	—	—	—	—	–1.75	–0.04	0.51	0.33	–0.37	—
—	–3.72	—	—	—	–4.80	–3.39	–0.67	–0.87	–1.53	–1.73
–3.36	–3.33	–3.52	—	–4.19	–3.25	–1.76	–1.46	–1.63	–0.70	–1.19
–4.61	–4.60	—	–3.19	–3.85	–2.29	–1.96	–1.40	–0.97	–0.19	–0.28
—	—	–4.92	–4.22	–2.93	–0.89	–0.63	–0.37	–0.09	–0.67	–1.70
–4.26	–4.24	–4.92	–3.81	–2.93	–0.91	0.37	1.10	—	—	—
–4.23	–3.29	–3.25	–4.14	–2.59	–1.36	–0.45	–0.34	0.30	0.09	–0.37
—	—	–3.79	–4.87	–4.17	–1.95	–0.25	0.84	0.48	—	—
—	–4.32	—	—	—	–2.48	–0.93	–0.17	–0.59	–0.43	–0.37
–4.92	–3.81	–4.89	–4.89	–4.18	–2.64	–2.10	–1.58	–0.59	–0.74	–0.61
–3.17	–3.13	–4.20	–2.91	–3.01	–1.75	–1.45	–0.58	–0.09	–0.21	0.33
–3.82	–2.80	–4.84	–3.21	—	–2.35	–1.36	–0.52	–0.10	0.14	0.58
—	—	–4.92	—	–3.10	–2.89	–1.44	–1.26	–0.39	0.19	0.26
—	–4.90	–4.20	–3.48	–2.41	–2.14	–1.66	–1.02	–0.67	–0.62	–0.17
–3.85	–2.60	–4.17	–3.74	–2.21	–0.42	0.75	—	—	—	—
–2.65	–3.81	–2.93	–0.49	0.61	0.26	—	—	—	—	—
–4.93	–4.92	–4.92	–2.68	–1.47	–1.46	–0.59	–0.85	–0.27	–0.58	0.33
—	–4.92	—	–4.22	–2.01	–0.18	0.43	0.88	—	—	—
—	—	–3.38	–2.93	–2.15	–2.12	–1.27	–0.78	–0.33	–0.09	0.73
—	—	–4.28	–3.56	–1.86	–1.59	–0.35	–0.98	–0.41	–0.45	0.09
–4.93	–1.91	–1.49	–0.38	–0.83	0.50	0.58	—	—	—	—
–2.65	–1.79	–1.97	–0.58	–0.50	–0.71	–0.29	–0.53	—	–0.09	—
–2.77	–1.92	–2.55	–0.23	0.75	—	—	—	—	—	—
–3.25	–3.73	–4.82	—	–4.81	–2.83	–1.01	–0.98	–0.50	–0.37	0.48
–3.11	–3.07	–4.14	–3.19	–2.54	–1.18	–0.81	–0.48	0.33	—	—
–2.70	–1.83	–0.61	0.44	0.58	–0.37	—	—	—	—	—
–3.80	–3.48	–4.15	–3.02	–1.10	0.34	–0.17	0.41	—	—	—
–4.31	—	–4.99	–3.59	–2.27	–3.04	–1.09	0.69	—	—	—
–3.78	–2.34	–3.14	–3.32	–3.29	–2.02	–1.25	–0.96	–0.88	–1.54	–0.55
–4.96	—	–4.95	–3.55	–3.11	–2.76	–1.31	–0.16	–0.37	0.05	–0.17
–3.86	–3.55	–3.52	–4.89	–1.38	–0.89	–0.57	–0.44	–1.25	0.09	–1.50
–1.29	–1.71	–1.13	–2.58	–1.06	–0.29	0.04	—	—	—	—
–2.50	–2.41	–3.45	–1.21	–1.01	0.33	0.33	—	—	—	—
–4.61	–2.78	—	—	–2.72	–0.83	0.50	0.88	—	—	—

Table 1.12:   *Fruit fly mortality data set, last 12 cohorts.*

				Five day period from start of study						
1	2	3	4	5	6	7	8	9	10	11
−4.28	−4.27	−4.26	−4.24	−3.30	−2.12	−0.85	−0.10	−0.23	0.03	−0.67
−3.43	−2.68	−3.10	−2.57	−0.43	0.50	0.48	—	—	—	—
—	−2.99	−2.05	−1.64	−3.73	−2.58	−1.76	−1.25	−0.25	−0.24	−0.67
−4.95	−4.95	−2.97	−3.49	−2.63	−1.21	−0.16	0.22	0.26	−0.90	—
—	−3.31	−2.75	−0.07	−0.57	0.37	−0.37	—	—	—	—
−4.17	−3.74	−3.01	−4.78	−4.07	−1.37	−0.82	−0.63	−1.47	−0.77	−0.63
—	−3.85	−4.23	−3.81	−2.79	−1.23	0.02	−0.45	−1.12	0.38	—
−3.82	−4.90	−3.08	−3.22	−2.04	0.04	0.44	0.73	—	—	—
−2.70	−1.40	−2.96	−2.06	−0.55	0.05	0.16	0.33	—	—	—
−3.72	−4.10	−4.09	−4.07	−1.51	0.01	1.26	—	—	—	—
—	−2.24	−1.93	−2.51	−2.43	−4.17	−1.89	−0.73	−0.20	0.09	−0.67
−3.86	−3.43	−2.69	−0.56	−0.13	0.05	—	—	—	—	—

CHAPTER 2

Unstructured Antedependence Models

This chapter introduces antedependence models in their most general form, which are called *unstructured* antedependence models because no additional structure beyond that needed for antedependence is imposed upon the model. We begin with definitions of antedependence and partial antecorrelation, which are equivalent in the important case of normal random variables. We then establish a number of equivalent characterizations of the covariance structure of partially antecorrelated variables, as well as some properties of determinants and traces involving such a covariance structure. These results will, in later chapters, be very important for deriving inferential procedures for antedependence models. The results are specialized to the important first-order case, and then generalized to the so-called variable-order case. Finally, relationships between these models and some other, more well known conditional independence models are described.

Those readers who have a certain squeamishness insofar as the "innards" of matrices are concerned may not regard this chapter as their favorite. For, the derivation of important properties of antedependence models requires that the observations' covariance matrix and other related matrices be dissected into various submatrices and individual elements, and these pieces are assigned notation that, in order to be sufficiently precise, is unavoidably cumbersome.

2.1 Antedependent random variables

Consider n random variables $Y_1, Y_2, \ldots, Y_n$ whose indices are ordered in a meaningful way, for example serially in time (chronologically) or along a transect in two-dimensional or three-dimensional space. In the following formal definition of antedependence and subsequently, we use the symbol $\perp$ to denote the independence of two random variables or two sets of random variables;

for example, we write $Y_1 \perp (Y_2, Y_3)$ to indicate that Y_1 and (Y_2, Y_3) are independent. Furthermore, we write $Y_1 \perp Y_2 | Y_3$ to indicate that Y_1 and Y_2 are conditionally independent, given any value of Y_3.

Definition 2.1. Index-ordered random variables $Y_1, \ldots, Y_n$ are said to be antedependent of order p, or AD(p), if

$$Y_i \perp (Y_{i-p-q-1}, Y_{i-p-q-2}, \ldots, Y_1) | (Y_{i-1}, Y_{i-2}, \ldots, Y_{i-p-q}),$$
$$i = p + q + 2, p + q + 3, \ldots, n; \quad q = 0, 1, \ldots, n - p - 2;$$

that is, if each variable, given at least p immediately preceding variables, is independent of all further preceding variables.

Definition 2.1 of antedependence is Gabriel's original definition (Gabriel, 1962), but there are several equivalent, somewhat less cumbersome, definitions that we will consider subsequently. To illustrate the definition, consider $n = 4$ random variables Y_1, Y_2, Y_3, Y_4. These variables are AD(2) if $Y_4 \perp Y_1 | (Y_2, Y_3)$, and they are AD(1) if

$$Y_3 \perp Y_1 | Y_2, \quad Y_4 \perp (Y_1, Y_2) | Y_3, \quad Y_4 \perp Y_1 | (Y_2, Y_3). \qquad (2.1)$$

The order, p, of antedependence may be any integer between 0 and $n - 1$, inclusive. The extreme cases $p = 0$ and $p = n - 1$ are equivalent to complete independence and arbitrary (general multivariate) dependence structure, respectively. It is clear from their definition that AD(p) variables are nested; that is,

$$AD(0) \subset AD(1) \subset AD(2) \subset \cdots \subset AD(n - 1). \qquad (2.2)$$

Thus antedependence models of ever-increasing order partition the universe of all dependence structures. Put another way, every set of random variables with ordered indices is antedependent of some order.

Since antedependence is defined with respect to a particular ordering of indices, a collection of variables that are AD(p) with respect to one ordering may not be AD(p) with respect to another ordering. The cases $p = 0$ and $p = n - 1$ are exceptions; for these cases antedependence holds with respect to any ordering. Of course, for longitudinal data only one ordering, namely chronological ordering, would seem to be relevant.

Owing to the large number of individual conditional independence conditions that can be generated by the phrases "at least" and "all further" in Definition 2.1, it appears that there could be some redundancies among the required set of conditions. This is indeed the case. Consider, for example, the conditions (2.1) required of four variables to be AD(1), and suppose that the joint probability density function of these variables exists. If the second of the three conditions in (2.1) holds, then we have, using obvious notation for the various conditional densities,

$$
\begin{aligned}
f_{4|2,3}(Y_4|Y_2,Y_3) &= \int f_{4|1,2,3}(Y_4|Y_1,Y_2,Y_3)f_{1|2,3}(Y_1|Y_2,Y_3)\,dY_1 \\
&= \int f_{4|3}(Y_4|Y_3)f_{1|2,3}(Y_1|Y_2,Y_3)\,dY_1 \\
&= f_{4|3}(Y_4|Y_3) \\
&= f_{4|1,2,3}(Y_4|Y_1,Y_2,Y_3),
\end{aligned}
$$

and thus the third condition in (2.1) is redundant. Alternatively, if the first and third conditions in (2.1) hold, and if $Y_4 \perp Y_2|Y_3$, which is part but not all of the second condition, then

$$
\begin{aligned}
f_{1,4|3}(Y_1,Y_4|Y_3) &= \int f_{1,4|2,3}(Y_1,Y_4|Y_2,Y_3)f_{2|3}(Y_2|Y_3)\,dY_2 \\
&= \int f_{1|2,3}(Y_1|Y_2,Y_3)f_{4|2,3}(Y_4|Y_2,Y_3)f_{2|3}(Y_2|Y_3)\,dY_2 \\
&= \int f_{1|2,3}(Y_1|Y_2,Y_3)f_{4|3}(Y_4|Y_3)f_{2|3}(Y_2|Y_3)\,dY_2 \\
&= f_{4|3}(Y_4|Y_3)f_{1|3}(Y_1|Y_3),
\end{aligned}
$$

which establishes that the remaining part of the second condition holds, i.e., $Y_4 \perp Y_1|Y_3$. Thus it is possible to define AD(1) variables using fewer conditional independence conditions than those originally specified by Gabriel. Indeed, the same sorts of arguments can be used with arbitrary n and p to establish the following two simpler, but equivalent, definitions of AD(p) variables.

Definition 2.2. Index-ordered random variables $Y_1, \ldots, Y_n$ are said to be antedependent of order p, or AD(p), if either of the following are true:

(a) $Y_i \perp (Y_{i-p-1}, Y_{i-p-2}, \ldots, Y_1)|(Y_{i-1}, Y_{i-2}, \ldots, Y_{i-p})$, for $i = p+2, p+3, \ldots, n$; that is, if each variable, given exactly p immediately preceding variables, is independent of all further preceding variables.

(b) $Y_i \perp Y_{i-p-q-1}|(Y_{i-1}, Y_{i-2}, \ldots, Y_{i-p-q})$, for $i = p+q+2, p+q+3, \ldots, n$; $q = 0, 1, \ldots, n-p-2$; that is, if each pair of variables with indices differing by more than p units are independent, given all the intervening variables between the pair.

The number of pairwise conditional independence conditions in either case of Definition 2.2 is $(n-p)(n-p-1)/2$, and these conditions contain no redundancies. Note that Definition 2.2(b) reveals that if the index-ordered variables $Y_1, \ldots, Y_n$ are AD(p), then so are $Y_n, Y_{n-1}, \ldots, Y_1$; that is, the variables in reverse order are also AD(p).

It is quite easy to construct examples of AD(p) random variables. For example, let $\epsilon_1, \ldots, \epsilon_n$ be independent random variables, and define

$$Y_i = \sum_{k=1}^{i} \epsilon_k, \quad i = 1, \ldots, n. \tag{2.3}$$

It can be shown that $Y_i \perp Y_{i+2} | Y_{i+1}$ for all i; for example, if the probability density functions of these variables exist, then for $i = 1, 2, \ldots, n-2$,

$$
\begin{aligned}
f_{i,i+2|i+1}(Y_i, Y_{i+2}|Y_{i+1}) &= f_{i+2|i,i+1}(Y_{i+2}|Y_i, Y_{i+1}) f_{i|i+1}(Y_i|Y_{i+1}) \\
&= f_{i+2|i+1}(Y_{i+2}|Y_{i+1}) f_{i|i+1}(Y_i|Y_{i+1}).
\end{aligned}
$$

Thus, $\{Y_i : i = 1, \ldots, n\}$ are AD(1). Similarly, if cumulative sums are taken of arbitrary AD(p) variables, i.e., if the ϵ_k's in (2.3) are AD(p) variables, then the Y_i's are AD($p+1$) variables. We will consider these and other types of antedependent variables in more detail in Chapter 3.

The term "antedependence," like its original definition, is due to Gabriel (1962). Note that the Latin root "ante" means "before" or "preceding," so the term is quite apt. The same concept has been called the general Markov property by other writers (e.g., Feller, 1968, p. 419), and it would also be reasonable to call it "banded precision matrix structure" or "generalized autoregressive structure," for reasons that will become clear in the next section. However, none of these alternatives are as catchy as the original. There is one unfortunate aspect of the term, however: when it is spoken, hearers often misunderstand it as "anti-dependence" and think it may be either an antonym of dependence, a type of negative dependence, or who knows what. The confusion is usually cleared up by spelling the word for the listener.

2.2 Antecorrelation and partial antecorrelation

Consider again n index-ordered random variables $Y_1, Y_2, \ldots, Y_n$, but now suppose that their variances exist. In this event, the conditional independence properties defining AD(p) random variables imply certain corresponding properties of the conditional covariances and conditional correlations among the variables. Let $\sigma_{ij|B}$ and $\rho_{ij|B}$ denote the conditional covariance and conditional correlation, respectively, between Y_i and Y_j, given all variables whose indices belong to a set B. If the variables are AD(p), then it follows from Definition 2.2 that both of the following are true:

$$
\begin{aligned}
\rho_{ij|j-1,j-2,\ldots,j-p} &= 0, \quad i = 1, 2, \ldots, j-p-1, \\
\rho_{i-p-q-1,i|i-1,i-2,\ldots,i-p-q} &= 0, \quad q = 0, 1, \ldots, n-p-2.
\end{aligned}
$$

The same type of result holds, of course, for the conditional covariances. However, the converse is not generally true; that is, if either equivalent property of the conditional correlations holds, this is not sufficient to imply that the variables are AD(p). The converse does hold, however, if $Y_1, \ldots, Y_n$ are jointly normally distributed, as a consequence of the normality of all conditional distributions and the equivalence of uncorrelatedness and independence for normal variables.

The property of the conditional correlations given above is of interest in its own right, even though it is not generally equivalent to antedependence. It appears that no term has previously been given to this property, but it seems reasonable to call it "antecorrelation" of order p, or AC(p). As noted above, AC(p) and AD(p) are equivalent for jointly normally distributed variables.

A property of a covariance structure ostensibly very similar to antecorrelation of order p is one for which the corresponding *partial* correlations, rather than conditional correlations, are equal to zero. We call this property partial antecorrelation of order p, or PAC(p), and we will give a formal definition of it shortly. First, however, we digress briefly to review some notation, terminology, and results associated with partial correlations.

Let $\mathbf{Y} = (Y_1, \ldots, Y_n)^T$ be a random vector with positive definite covariance matrix $\mathbf{\Sigma}$. Let $\{i : j\}$ denote any sequence of consecutive integers $i, i + 1, \ldots, j$, and let A and B be any nonempty disjoint subsets of $\{1 : n\}$. Suppose that we form the two vectors of variables, $\mathbf{Y}_A$ and $\mathbf{Y}_B$, consisting of those elements of $\mathbf{Y}$ whose indices belong to A and B, respectively. Let $\boldsymbol{\mu}_A = E(\mathbf{Y}_A)$, $\boldsymbol{\mu}_B = E(\mathbf{Y}_B)$, $\mathbf{\Sigma}_{AA} = \text{var}(\mathbf{Y}_A)$, $\mathbf{\Sigma}_{BB} = \text{var}(\mathbf{Y}_B)$, $\mathbf{\Sigma}_{AB} = \text{cov}(\mathbf{Y}_A, \mathbf{Y}_B)$, and $\mathbf{\Sigma}_{BA} = \mathbf{\Sigma}_{AB}^T$. Then the *regression of* $\mathbf{Y}_A$ *on* $\mathbf{Y}_B$ is the random vector given by

$$\hat{\mathbf{Y}}_{A \cdot B} = \boldsymbol{\mu}_A + \mathbf{\Sigma}_{AB} \mathbf{\Sigma}_{BB}^{-1} (\mathbf{Y}_B - \boldsymbol{\mu}_B).$$

The *residual* vector corresponding to this regression is given by

$$\mathbf{Y}_A - \hat{\mathbf{Y}}_{A \cdot B}.$$

Observe that the ith element of $\hat{\mathbf{Y}}_{A \cdot B}$ is a linear function of the elements of $\mathbf{Y}_B$.

The following theorem gives some well-known results about the regression and residual vectors that will be useful to us.

Theorem 2.1. *Let Y_{Ai} denote the ith element of $\mathbf{Y}_A$. In the regression of $\mathbf{Y}_A$ on $\mathbf{Y}_B$:*

(a) *the ith element of $\hat{\mathbf{Y}}_{A \cdot B}$ minimizes the mean squared error $E[Y_{Ai} - (c + \mathbf{k}^T \mathbf{Y}_B)]^2$ among all linear functions $c + \mathbf{k}^T \mathbf{Y}_B$;*

(b) *the ith element of $\hat{\mathbf{Y}}_{A \cdot B}$ maximizes the correlation between Y_{Ai} and $c + \mathbf{k}^T \mathbf{Y}_B$ among all linear functions $c + \mathbf{k}^T \mathbf{Y}_B$, and the maximized value of this correlation is given by*

$$R_{i \cdot B} = (\boldsymbol{\sigma}_{iB}^T \mathbf{\Sigma}_{BB}^{-1} \boldsymbol{\sigma}_{iB} / \sigma_{ii})^{1/2}, \qquad (2.4)$$

where $\sigma_{iB} = cov(\mathbf{Y}_B, Y_{Ai})$ and $\sigma_{ii} = var(Y_{Ai})$;

(c) $cov(\mathbf{Y}_A - \hat{\mathbf{Y}}_{A \cdot B}, \mathbf{Y}_B) = \mathbf{0}$;

(d) $\boldsymbol{\Sigma}_{AA \cdot B} \equiv var(\mathbf{Y}_A - \hat{\mathbf{Y}}_{A \cdot B}) = \boldsymbol{\Sigma}_{AA} - \boldsymbol{\Sigma}_{AB}\boldsymbol{\Sigma}_{BB}^{-1}\boldsymbol{\Sigma}_{BA}$;

(e) *if the joint distribution of $\mathbf{Y}_A$ and $\mathbf{Y}_B$ is multivariate normal, then $\hat{\mathbf{Y}}_{A \cdot B}$ is the conditional mean, and $\boldsymbol{\Sigma}_{AA \cdot B}$ is the conditional variance, of $\mathbf{Y}_A$ given $\mathbf{Y}_B$.*

Proof. Proofs of parts (a), (b), (c), and (e) of the theorem may be found in Anderson (1984, pp. 36–40). For part (d), we have

$$
\begin{aligned}
var(\mathbf{Y}_A - \hat{\mathbf{Y}}_{A \cdot B}) &= var(\mathbf{Y}_A) + var(\hat{\mathbf{Y}}_{A \cdot B}) - cov(\mathbf{Y}_A, \hat{\mathbf{Y}}_{A \cdot B}) \\
&\quad - cov(\mathbf{Y}_A, \hat{\mathbf{Y}}_{A \cdot B})^T \\
&= \boldsymbol{\Sigma}_{AA} + \boldsymbol{\Sigma}_{AB}\boldsymbol{\Sigma}_{BB}^{-1}\boldsymbol{\Sigma}_{BB}\boldsymbol{\Sigma}_{BB}^{-1}\boldsymbol{\Sigma}_{BA} - \boldsymbol{\Sigma}_{AB}\boldsymbol{\Sigma}_{BB}^{-1}\boldsymbol{\Sigma}_{BA} \\
&\quad - (\boldsymbol{\Sigma}_{AB}\boldsymbol{\Sigma}_{BB}^{-1}\boldsymbol{\Sigma}_{BA})^T \\
&= \boldsymbol{\Sigma}_{AA} - \boldsymbol{\Sigma}_{AB}\boldsymbol{\Sigma}_{BB}^{-1}\boldsymbol{\Sigma}_{BA}.
\end{aligned}
$$

□

The quantity $R_{i \cdot B}$ defined in part (b) of Theorem 2.1 is called the *multiple correlation coefficient between Y_{Ai} and $\mathbf{Y}_B$*, and the matrix $\boldsymbol{\Sigma}_{AA \cdot B}$ defined in part (d) of the theorem is called the *partial covariance matrix of $\mathbf{Y}_A$, adjusted for (or partialling out) $\mathbf{Y}_B$*. Denote the (i, j)th element of $\boldsymbol{\Sigma}_{AA \cdot B}$ by $\sigma_{ij \cdot B}$; this is called the *partial covariance between Y_{Ai} and Y_{Aj}* (or the *partial variance of Y_{Ai} if $i = j$*) *adjusted for the variables in $\mathbf{Y}_B$*. The *partial correlation coefficient between Y_{Ai} and Y_{Aj} adjusted for the variables in $\mathbf{Y}_B$* is then given by

$$
\rho_{ij \cdot B} = \frac{\sigma_{ij \cdot B}}{(\sigma_{ii \cdot B}\sigma_{jj \cdot B})^{1/2}}.
$$

We can broaden this definition of a partial correlation coefficient to include the ordinary correlation coefficient as follows: if B is empty, then $\rho_{ij \cdot B} = \rho_{ij}$.

Now we are in a position to define partial antecorrelation of order p.

Definition 2.3. *Index-ordered random variables $Y_1, \dots, Y_n$ are said to be partially antecorrelated of order p, or PAC(p), if*

$$
\rho_{i, i+p+q+1 \cdot \{i+1:i+p+q\}} = 0
$$

for all $q \in \{0 : n - p - 2\}$ and $i \in \{1 : n - p - q - 1\}$; that is, if the partial correlation between each pair of variables with indices differing by more than p units, adjusted for all intervening variables between the pair, is equal to zero.

PAC(p) neither implies nor is implied by AD(p), but they coincide when Y_1, $\dots, Y_n$ are jointly normally distributed. [In this case, of course, they also

coincide with AC(p), for then $\Sigma_{i,i+p+q+1 \cdot \{i+1:i+p+q\}}$ coincides with the conditional covariance matrix of $(Y_i, Y_{i+p+q+1})^T$ given $(Y_{i+1}, \ldots, Y_{i+p+q})^T$.]
For this reason, while we continue to use the PAC(p) terminology for most of the remainder of this chapter, our discourse applies equally well to normal AD(p) variables.

We call the characterization of PAC(p) random variables provided by Definition 2.3 the *intervenor-adjusted partial correlation characterization*, or simply the *intervenor-adjusted characterization*. For this characterization, the natural parameterization of a PAC(p) covariance structure is in terms of the quantities

$$\{\sigma_{11}, \sigma_{22}, \ldots, \sigma_{nn}, \rho_{12}, \rho_{23}, \ldots, \rho_{n-1,n}, \rho_{13 \cdot 2}, \ldots, \rho_{n-2,n \cdot n-1}, \ldots,$$

$$\rho_{1,p+1 \cdot \{2:p\}}, \ldots, \rho_{n-p,n \cdot \{n-p+1:n-1\}}\}, \tag{2.5}$$

but partial covariances could be used in place of the partial correlations if desired. These parameters are in one-to-one correspondence with the elements of the covariance matrix; indeed, either set of parameters can be exchanged for the other through the use of the well-known recursion formula for partial correlations, i.e.,

$$\rho_{i,i+k \cdot \{i+1:i+j\}}$$

$$= \frac{\rho_{i,i+k \cdot \{i+1:i+j-1\}} - \rho_{i,i+j \cdot \{i+1:i+j-1\}} \rho_{i+k,i+j \cdot \{i+1:i+j-1\}}}{[(1 - \rho_{i,i+j \cdot \{i+1:i+j-1\}}^2)(1 - \rho_{i+k,i+j \cdot \{i+1:i+j-1\}}^2)]^{1/2}},$$

$$i = 1, \ldots, n-1; \; k = 2, \ldots, n-i; \; j = 1, \ldots, k-1.$$

We write this as

$$\Xi = H(\Sigma), \tag{2.6}$$

where H is a one-to-one matrix-valued function mapping the marginal covariance matrix to the matrix $\Xi = (\xi_{ij})$ of parameters of the intervenor-adjusted characterization. Here $\xi_{ii} = \sigma_{ii}$ for $i = 1, \ldots, n$ and $\xi_{ji} = \xi_{ij} = \rho_{ij \cdot \{i+1:j-1\}}$ for $i < j$.

The number of distinct parameters in (2.5) is easily seen to be

$$n + (n-1) + \cdots + (n-p) = \frac{(2n-p)(p+1)}{2}. \tag{2.7}$$

Thus, unless $p = n - 1$, the PAC(p) covariance structure can be parameterized with fewer parameters than an arbitrary (general multivariate) covariance matrix for n variables, which has $n(n+1)/2$ parameters. Note also that as n increases, the number of parameters in a PAC(p) covariance structure increases linearly, in contrast to quadratic growth for the general case.

Although the parameters listed in (2.5) are distinct in the sense that none of them can be expressed as a function of the others, they are not unconstrained, for they must be such that Σ is positive definite. For each marginal variance, the constraint is merely $\sigma_{ii} > 0$, and the ordinary correlations between consecutive variables need only satisfy $-1 < \rho_{i,i+1} < 1$ for all i. For the partial correlations adjusted for at least one intervenor, however, the constraints are interdependent and much more complicated. Thus, it is not generally possible to completely specify the parameter space for a PAC(p) covariance structure in terms of simple inequalities on each of the individual parameters listed in (2.5).

2.3 Equivalent characterizations

The intervenor-adjusted characterization of PAC(p) random variables described in the previous section is rather indirect in terms of the implied marginal covariance structure of the variables. For the purpose of formulating particular PAC models or normal AD models, it would be preferable to have a more direct characterization. It turns out that the covariance structure of PAC(p) random variables can be characterized alternatively in several equivalent ways, each quite interesting in its own right, and each with important consequences for model formulation. In this section we establish these equivalences. In particular, we derive properties of the precision matrix, i.e., the inverse of the covariance matrix $\boldsymbol{\Sigma}$, for PAC(p) variables, and we also derive properties of the modified Cholesky decomposition of the precision matrix. These lead to the *precision matrix characterization* and *autoregressive characterization* of PAC(p) variables. Finally, we determine what these characterizations imply about the structure of $\boldsymbol{\Sigma}$ itself, yielding the *marginal characterization* of PAC(p) variables.

2.3.1 Precision matrix characterization

We begin the process of deriving a precision matrix characterization of PAC(p) variables by giving a result that expresses a partial covariance from a regression of two variables on $q+1$ variables in terms of partial variances and covariances from a regression of the same two variables on a subset of q of the $q + 1$ variables.

Lemma 2.1. *Let* $Y_1, \ldots, Y_n$ *be random variables with a positive definite covariance matrix* $\boldsymbol{\Sigma}$. *Let* B *be any subset of* $\{1 : n\}$ *consisting of at most* $n - 3$ *elements, and suppose that* i, j, k *are elements of* $\{1 : n\}$ *not in* B *and such that* $i \neq k$ *and* $j \neq k$. *Then*

$$\sigma_{ij \cdot B, k} = \sigma_{ij \cdot B} - \frac{\sigma_{ik \cdot B} \sigma_{jk \cdot B}}{\sigma_{kk \cdot B}}.$$

Proof. The lemma follows immediately upon putting $n_1 = 2$ and $n_2 = 1$ in Corollary A.1.1.3. □

Lemma 2.1 may be used to establish a useful result about partial covariances between PAC(p) variables lagged sufficiently far apart when those variables are regressed on certain subsets of the remaining variables, as developed by the next two lemmas.

Lemma 2.2. *If $Y_1, \ldots, Y_n$ are PAC(p) random variables with positive definite covariance matrix Σ, then $\sigma_{ij \cdot B} = 0$, where $i \neq j$ and B is any set of p or more consecutive indices between i and j.*

Proof. Without loss of generality suppose that $j > i$. If $j = i + p + 1$ the result follows easily from Definition 2.3, so suppose that $B = \{i + 1 : i + p\}$ where $i + p + 1 < j$. By Lemma 2.1, we have

$$\sigma_{ij \cdot B} - \sigma_{ij \cdot B, i+p+1} \quad \propto \quad \sigma_{i,i+p+1 \cdot B} \sigma_{i+p+1,j \cdot B},$$

$$\sigma_{ij \cdot B, i+p+1} - \sigma_{ij \cdot B, i+p+1, i+p+2} \quad \propto \quad \sigma_{i,i+p+2 \cdot B, i+p+1}$$
$$\times \sigma_{i+p+2,j \cdot B, i+p+1},$$

$$\vdots$$

$$\sigma_{ij \cdot B, \{i+p+1:j-3\}} - \sigma_{ij \cdot B, \{i+p+1:j-2\}} \quad \propto \quad \sigma_{i,j-2 \cdot B, \{i+p+1:j-3\}}$$
$$\times \sigma_{j-2,j \cdot B, \{i+p+1:j-3\}},$$

$$\sigma_{ij \cdot B, \{i+p+1:j-2\}} - \sigma_{ij \cdot B, \{i+p+1:j-1\}} \quad \propto \quad \sigma_{i,j-1 \cdot B, \{i+p+1:j-2\}}$$
$$\times \sigma_{j-1,j \cdot B, \{i+p+1:j-2\}}.$$

Observe that the partial covariance immediately to the right of the proportionality symbol in each relation is equal to zero since the variables are PAC(p), and, for the same reason, so is the partial covariance immediately to the left of the proportionality symbol in the last relation. Thus $\sigma_{ij \cdot B, \{i+p+1:j-2\}} = 0$ and this zero "floats up," initially to the second partial covariance in the penultimate relation and then successively to the second partial covariance in each relation, including the first relation. Thus $\sigma_{ij \cdot B} = 0$ for this particular B. The same argument is easily extended to $B = \{i+k : i+k+p+m\}$ where k and m are positive and nonnegative integers, respectively, such that $i+k+p+m \leq j$. $\square$

For an $n \times n$ matrix $\mathbf{A} = (a_{ij})$, define the pth *subdiagonal* $(p = 1, \ldots, n-1)$ to be the set $\{a_{ij} : i - j = p\}$ and the pth *superdiagonal* $(p = 1, \ldots, n-1)$ to be the set $\{a_{ij} : j - i = p\}$. Together, the pth subdiagonal and pth superdiagonal of $\mathbf{A}$ are called the pth *off-diagonals* of $\mathbf{A}$. Clearly, if $\mathbf{A}$ is symmetric, then its pth subdiagonal and pth superdiagonal coincide, hence specifying either one also specifies the other. We shall find it convenient to denote the set of indices of elements on off-diagonals $p+1, p+2, \ldots, n$ of an $n \times n$ symmetric matrix by $I_p^n = \{(i,j) : i \in \{1 : n\}, j \in \{1 : n\}, |i - j| > p\}$.

Lemma 2.3. *If $Y_1, \ldots, Y_n$ are PAC(p) random variables with positive definite covariance matrix Σ, then $\sigma_{ij \cdot \{i,j\}^c} = 0$ for all $(i, j) \in I_p^n$.*

Proof. Without loss of generality suppose that $j - i > p$. Since $\sigma_{1n \cdot \{2:n-1\}} = 0$ trivially, we suppose further that $j < n$. By Lemma 2.1, we have

$$\sigma_{ij \cdot \{i+1:j-1\}} - \sigma_{ij \cdot \{i+1:j-1\}, j+1} \quad \propto \quad \sigma_{i,j+1 \cdot \{i+1:j-1\}}$$
$$\times \, \sigma_{j,j+1 \cdot \{i+1:j-1\}},$$

$$\sigma_{ij \cdot \{i+1:j-1\}, j+1} - \sigma_{ij \cdot \{i+1:j-1\}, j+1, j+2} \quad \propto \quad \sigma_{i,j+2 \cdot \{i+1:j-1\}, j+1}$$
$$\times \, \sigma_{j,j+2 \cdot \{i+1:j-1\}, j+1},$$

$$\vdots$$

$$\sigma_{ij \cdot \{i+1:j-1\}, \{j+1:n-1\}} - \sigma_{ij \cdot \{i+1:j-1\}, \{j+1:n\}} \quad \propto \quad \sigma_{in \cdot \{i+1:j-1\}, \{j+1:n-1\}}$$
$$\times \, \sigma_{jn \cdot \{i+1:j-1\}, \{j+1:n-1\}}.$$

Observe that the partial covariance immediately to the right of the proportionality symbol in each relation is equal to zero by Lemma 2.2, and so is the first partial covariance in the first relation. Thus this zero "trickles down" to each successive relation including the last, implying that $\sigma_{ij \cdot \{i+1:j-1\}, \{j+1:n\}} = 0$. The same type of argument is easily extended for partialling out those variables with indices less than i to yield $\sigma_{ij \cdot \{1:i-1\}, \{i+1:j-1\}, \{j+1:n\}} = 0$ for all $(i, j) \in I_p^n$. $\square$

Some additional notation needed for the next lemma in particular, but also for various results in this and other sections, is as follows. Let $\Sigma_{k:m}$ denote the submatrix of any positive definite matrix Σ consisting of its rows $k, k + 1, \ldots, m$ and columns $k, k + 1, \ldots, m$; let $\sigma_{k:m,l}$ denote the vector consisting of elements in rows $k, k + 1, \ldots, m$ and column l of Σ; and let $\sigma_{k:m}^{ij}$ denote the (i, j)th element of $\Sigma_{k:m}^{-1}$. We write σ^{ij} for $\sigma_{1:n}^{ij}$.

Lemma 2.4. *Let $Y_1, \ldots, Y_n$ be random variables with positive definite covariance matrix Σ. Then $\sigma^{ij} = 0$ for all $(i, j) \in I_p^n$ if and only if $\sigma_{k:m}^{ij} = 0$ for all $(i, j) \in I_p^{m-k+1}$ and all k and m.*

Proof. Putting $k = 1$ and $m = n$ into the second condition yields the first condition. To show that the first condition implies the second condition, it suffices to establish that the first condition implies that $\sigma_{1:n-1}^{ij} = 0$ and $\sigma_{2:n}^{ij} = 0$ for all $(i, j) \in I_p^n$, as the second condition can then be obtained by repeated use of these results for successively larger k and successively smaller m.

We show first that if $\sigma^{ij} = 0$ for all $(i, j) \in I_p^n$, then $\sigma_{1:n-1}^{ij} = 0$ for all $(i, j) \in I_p^{n-1}$. Partition $\Sigma \, (= \Sigma_{1:n})$ and $\Sigma^{-1} \, (= \Sigma^{1:n})$ as follows:

$$\Sigma = \begin{pmatrix} \Sigma_{1:n-1} & \sigma_{1:n-1,n} \\ \sigma_{1:n-1,n}^T & \sigma_{nn} \end{pmatrix}, \quad \Sigma^{-1} = \begin{pmatrix} \Sigma^{1:n-1} & \sigma^{1:n-1,n} \\ (\sigma^{1:n-1,n})^T & \sigma^{nn} \end{pmatrix}.$$

Then, by Theorem A.1.1(a),

$$
\begin{aligned}
\mathbf{\Sigma}^{1:n-1} &= \mathbf{\Sigma}_{1:n-1}^{-1} + \left(\sigma_{nn} - \boldsymbol{\sigma}_{1:n-1,n}^{T}\mathbf{\Sigma}_{1:n-1}^{-1}\boldsymbol{\sigma}_{1:n-1,n}\right)^{-1} \\
&\quad \times \mathbf{\Sigma}_{1:n-1}^{-1}\boldsymbol{\sigma}_{1:n-1,n}\boldsymbol{\sigma}_{1:n-1,n}^{T}\mathbf{\Sigma}_{1:n-1}^{-1} \\
&= \mathbf{\Sigma}_{1:n-1}^{-1} + \left\{(\sigma^{nn})^{2}(\sigma_{nn} - \boldsymbol{\sigma}_{1:n-1,n}^{T}\mathbf{\Sigma}_{1:n-1}^{-1}\boldsymbol{\sigma}_{1:n-1,n})\right\}^{-1} \\
&\quad \times \boldsymbol{\sigma}^{1:n-1,n}(\boldsymbol{\sigma}^{1:n-1,n})^{T} \quad\quad\quad\quad\quad\quad (2.8)
\end{aligned}
$$

where the second equality follows from Theorem A.1.1(e). Now if $\sigma_{1:n}^{ij} = 0$ for all $(i,j) \in I_p^n$, then all elements on off-diagonals $p+1,\ldots,n$ of $\mathbf{\Sigma}^{1:n-1}$ equal zero and all elements of $\boldsymbol{\sigma}^{1:n-1,n}$ that lie on off-diagonals $p+1,\ldots,n$ of $\mathbf{\Sigma}_{1:n}^{-1}$ equal zero. Hence, by (2.8), all elements of $\mathbf{\Sigma}_{1:n-1}^{-1}$ on off-diagonals $p+1,\ldots,n-1$ must likewise equal zero, i.e., $\sigma_{1:n-1}^{ij} = 0$ for all $(i,j) \in I_p^{n-1}$. That the first condition of Lemma 2.4 implies that $\sigma_{2:n}^{ij} = 0$ for all $(i,j) \in I_p^{n-1}$ can be shown similarly, based on partitioning $\mathbf{\Sigma}$ and $\mathbf{\Sigma}^{-1}$ alternatively as

$$
\mathbf{\Sigma} = \begin{pmatrix} \sigma_{11} & \boldsymbol{\sigma}_{2:n,1}^{T} \\ \boldsymbol{\sigma}_{2:n,1} & \mathbf{\Sigma}_{2:n} \end{pmatrix}, \quad \mathbf{\Sigma}^{-1} = \begin{pmatrix} \sigma^{11} & (\boldsymbol{\sigma}^{2:n,1})^{T} \\ \boldsymbol{\sigma}^{2:n,1} & \mathbf{\Sigma}^{2:n} \end{pmatrix}.
$$

$\square$

Lemma 2.5. *Let $Y_1,\ldots,Y_n$ be random variables with positive definite covariance matrix $\mathbf{\Sigma}$. Then, the partial covariance between Y_i and Y_j ($i \neq j$) adjusted for all remaining variables in $Y_1,\ldots,Y_n$ is given by*

$$
\sigma_{ij\cdot\{i,j\}^C} = -\sigma^{ij}(\sigma_{ii\cdot\{i,j\}^C}\sigma_{jj\cdot\{i,j\}^C})^{1/2}/(\sigma^{ii}\sigma^{jj})^{1/2}.
$$

Proof. The lemma follows immediately from Corollary A.1.1.2 upon permutation of indices. $\square$

We are now ready to give the precision matrix characterization of PAC(p) random variables.

Theorem 2.2. *Random variables $Y_1,\ldots,Y_n$ with positive definite covariance matrix $\mathbf{\Sigma}$ are PAC(p) if and only if the elements of $\mathbf{\Sigma}^{-1}$ satisfy $\sigma^{ij} = 0$ for all $(i,j) \in I_p^n$.*

Proof. Suppose first that $Y_1,\ldots,Y_n$ are PAC(p) random variables, and suppose that $(i,j) \in I_p^n$ with $i < j$. Then Lemma 2.3 yields $\sigma_{ij\cdot\{i,j\}^C} = 0$. It follows from Lemma 2.5 and the symmetry of $\mathbf{\Sigma}^{-1}$ that $\sigma^{ij} = 0$ for all $(i,j) \in I_p^n$.

Conversely, suppose that $\sigma^{ij} = 0$ for all $(i,j) \in I_p^n$. Then by Lemma 2.4, for all k and m, $\sigma_{k:m}^{ij} = 0$ for all $(i,j) \in I_p^n$, and in particular, $\sigma_{k:m}^{km} = 0$ whenever $m - k > p$. But this implies, by Lemma 2.5, that $\sigma_{km\cdot k+1,\ldots,m-1} = 0$ for all $(m,k) \in I_p^n$. Hence $Y_1,\ldots,Y_n$ are PAC(p) random variables. $\square$

Theorem 2.2 tells us that the zeroes in the precision matrix of PAC(p) variables have a distinctive banded structure. For example, in the case of six PAC(2) random variables, the precision matrix is given by

$$\Sigma^{-1} = \begin{pmatrix} \sigma^{11} & \sigma^{21} & \sigma^{31} & 0 & 0 & 0 \\ \sigma^{21} & \sigma^{22} & \sigma^{32} & \sigma^{42} & 0 & 0 \\ \sigma^{31} & \sigma^{32} & \sigma^{33} & \sigma^{43} & \sigma^{53} & 0 \\ 0 & \sigma^{42} & \sigma^{43} & \sigma^{44} & \sigma^{54} & \sigma^{64} \\ 0 & 0 & \sigma^{53} & \sigma^{54} & \sigma^{55} & \sigma^{65} \\ 0 & 0 & 0 & \sigma^{64} & \sigma^{65} & \sigma^{66} \end{pmatrix}.$$

Thus, we see that the covariance structure of these six variables may be parameterized by $\sigma^{11}, \sigma^{22}, \ldots, \sigma^{66}, \sigma^{21}, \ldots, \sigma^{64}$. More generally, Theorem 2.2 reveals that the covariance structure of n PAC(p) variables may be parameterized by those elements of Σ^{-1} with indices belonging to the complement of I_p^n in $\{1:n\} \times \{1:n\}$, i.e., by

$$\{\sigma^{11}, \sigma^{22}, \ldots, \sigma^{nn}, \sigma^{21}, \sigma^{32}, \ldots, \sigma^{n,n-1}, \sigma^{31}, \ldots, \sigma^{p+1,1} \ldots, \sigma^{n,n-p}\}.$$
$$(2.9)$$

The number of distinct parameters in this parameterization is easily seen to be

$$n + (n-1) + \cdots + (n-p) = \frac{(2n-p)(p+1)}{2},$$

as it must be, of course, since the mapping from the parameters in (2.5) to those in (2.9) is one-to-one.

There is a one-to-one correspondence between the parameters in (2.9) and those in the following list:

$$\{1/\sigma^{11}, 1/\sigma^{22}, \ldots, 1/\sigma^{nn}, \pi_{21}, \pi_{32}, \ldots, \pi_{n,n-1}, \pi_{31}, \ldots, \pi_{p+1,1} \ldots, \pi_{n,n-p}\}$$
$$(2.10)$$

where

$$\pi_{ij} = -\frac{\sigma^{ij}}{(\sigma^{ii}\sigma^{jj})^{1/2}}.$$

Now, by Lemma 2.5, π_{ij} is the partial correlation coefficient between Y_i and Y_j adjusted for the $n-2$ other variables. Furthermore, by Theorems 2.1(d) and A.1.1(a), $1/\sigma^{ii}$ is the partial variance of Y_i adjusted for the $n-1$ other variables. Thus, an alternative, arguably more interpretable version of the precision matrix characterization results from reparameterizing the elements of the precision matrix in terms of partial variances $[\sigma_{ii \cdot \{i\}^C} = 1/\sigma^{ii}]$ and partial correlations $[\rho_{ij \cdot \{i,j\}^C} = \pi_{ij}]$, and specifying that

$$\rho_{ij \cdot \{i,j\}^C} = 0 \quad \text{if } |i - j| > p.$$

Although the parameters listed in (2.9) are distinct in the sense that none of them can be expressed as a function of the others, they are not unconstrained, for they must be such that Σ^{-1} (or equivalently Σ) is positive definite. For each diagonal element of $\mathbf{\Sigma}^{-1}$, the constraint is merely $\sigma^{ii} > 0$. For each non-trivial off-diagonal element of $\mathbf{\Sigma}^{-1}$, however, the constraint involves complicated, nonlinear functions of the remaining non-trivial, off-diagonal elements. Hence it is not generally possible to completely specify the parameter space for the precision matrix formulation in terms of simple inequalities on individual parameters. The alternative parameterization in terms of partial variances and partial correlations, given by (2.10), offers no particular advantage in this regard.

2.3.2 Autoregressive characterization

Another characterization of PAC(p) random variables may be obtained by considering residuals from a particular sequence of regressions for index-ordered random variables with positive definite covariance matrix. In this sequence, each variable is regressed on all of its predecessors in the ordered list. That is, we regress Y_2 on Y_1, Y_3 on Y_1 and Y_2, and so forth. Let $\mu_i = E(Y_i)$, $\boldsymbol{\mu}_{1:i} = (\mu_1, \ldots, \mu_i)^T$, $\boldsymbol{\mu} = \boldsymbol{\mu}_{1:n}$, $\mathbf{Y}_{1:i} = (Y_1, \ldots, Y_i)^T$, and $\mathbf{Y} = \mathbf{Y}_{1:n}$. Also, put $\epsilon_1 = Y_1 - \mu_1$, and for $i = 2, \ldots, n$ let ϵ_i denote the residual from the regression of Y_i on its predecessors, i.e.,

$$\epsilon_1 = Y_1 - \mu_1, \tag{2.11}$$

$$\epsilon_2 = (Y_2 - \mu_2) - (\sigma_{12}/\sigma_{11})(Y_1 - \mu_1), \tag{2.12}$$

$$\epsilon_3 = (Y_3 - \mu_3) - \boldsymbol{\sigma}_{1:2,3}^T \mathbf{\Sigma}_{1:2}^{-1}(\mathbf{Y}_{1:2} - \boldsymbol{\mu}_{1:2}), \tag{2.13}$$

$$\vdots$$

$$\epsilon_n = (Y_n - \mu_n) - \boldsymbol{\sigma}_{1:n-1,n}^T \mathbf{\Sigma}_{1:n-1}^{-1}(\mathbf{Y}_{1:n-1} - \boldsymbol{\mu}_{1:n-1}). \tag{2.14}$$

These equations may be written in matrix form as

$$\boldsymbol{\epsilon} = \mathbf{T}(\mathbf{Y} - \boldsymbol{\mu}), \tag{2.15}$$

where $\mathbf{T}$ is a lower triangular matrix with ones along the main diagonal — a so-called unit lower triangular matrix — and elements $\{t_{ij}\}$ below the main diagonal, where t_{ij} is the jth element of $-\boldsymbol{\sigma}_{1:i-1,i}^T \mathbf{\Sigma}_{1:i-1}^{-1}$. That is,

$$\mathbf{T} = \begin{pmatrix} 1 & 0 & 0 & \cdots & 0 & 0 & 0 \\ t_{21} & 1 & 0 & \cdots & 0 & 0 & 0 \\ t_{31} & t_{32} & 1 & \cdots & 0 & 0 & 0 \\ \vdots & & & & & & \\ t_{n-1,1} & t_{n-1,2} & t_{n-1,3} & \cdots & t_{n-1,n-2} & 1 & 0 \\ t_{n1} & t_{n2} & t_{n3} & \cdots & t_{n,n-2} & t_{n,n-1} & 1 \end{pmatrix}.$$

It follows from (2.15) that $\text{var}(\boldsymbol{\epsilon}) = \mathbf{T}\mathbf{\Sigma}\mathbf{T}^T$ and, since $\mathbf{T}$ is nonsingular, that this covariance matrix is positive definite. Moreover, the ϵ_i's are uncorrelated as a consequence of Theorem 2.1(c). Therefore, we have that

$$\mathbf{T}\mathbf{\Sigma}\mathbf{T}^T = \mathbf{D}$$

or equivalently that

$$\mathbf{\Sigma} = \mathbf{T}^{-1}\mathbf{D}(\mathbf{T}^T)^{-1} \tag{2.16}$$

or

$$\mathbf{\Sigma}^{-1} = \mathbf{T}^T\mathbf{D}^{-1}\mathbf{T}, \tag{2.17}$$

where $\mathbf{D}$ is an $n \times n$ diagonal matrix with positive main diagonal elements δ_i. Furthermore, $\delta_1 = \text{var}(\epsilon_1) = \text{var}(Y_1)$ and $\delta_i = \text{var}(\epsilon_i) = \text{var}(Y_i - \hat{Y}_{i\cdot 1:i-1})$ for $i = 2, \ldots, n$. As a consequence of Theorem 2.1(d) we have

$$\delta_i = \sigma_{ii\cdot\{1:i-1\}} = \sigma_{ii} - \boldsymbol{\sigma}_{1:i-1,i}^T\mathbf{\Sigma}_{1:i-1}^{-1}\boldsymbol{\sigma}_{1:i-1,i}, \quad i = 2, \ldots, n. \tag{2.18}$$

The decomposition of $\mathbf{\Sigma}^{-1}$ given by (2.17) is known variously as its $U'DU$ decomposition (Harville, 1997), square-root-free Cholesky decomposition (Tanabe and Sagae, 1992), or *modified Cholesky decomposition* (Pourahmadi, 1999); we will use the latter term. This decomposition is known to be unique; see Theorem A.1.2. Thus, the mapping from $\mathbf{\Sigma}$ to $(\mathbf{T}, \mathbf{D})$ is one-to-one, and consequently $\mathbf{\Sigma}$ may be parameterized by the $n(n+1)/2$ non-trivial elements of $\mathbf{T}$ and $\mathbf{D}$, i.e., by

$$\delta_1, \ldots, \delta_n, t_{21}, t_{32}, \ldots, t_{n,n-1}, t_{31}, t_{42}, \ldots, t_{n1}.$$

Equivalently, $\mathbf{\Sigma}$ may be parameterized by

$$\delta_1, \ldots, \delta_n, \phi_{21}, \phi_{32}, \ldots, \phi_{n,n-1}, \phi_{31}, \phi_{42}, \ldots, \phi_{n1},$$

where $\phi_{ij} = -t_{ij}$. Moreover, the parameters in this last list have useful interpretations: ϕ_{ij} is the coefficient corresponding to Y_j in the regression of Y_i on its predecessors, and δ_i is the variance of the residual from that same regression. We borrow terminology from time series analysis and refer to the ϕ_{ij}'s as *autoregressive coefficients*, and to the δ_i's as *innovation variances* (the ϵ_i's being the *innovations*). Using this notation, and observing that

$$(\phi_{i1}, \phi_{i2}, \ldots, \phi_{i,i-1}) = \boldsymbol{\sigma}_{1:i-1,i}^T\mathbf{\Sigma}_{1:i-1}^{-1} \tag{2.19}$$

we see that the ith equation in the sequential regression (2.11) through (2.14) may be rewritten as

$$Y_i - \mu_i = \sum_{k=1}^{i-1} \phi_{i,i-k}(Y_{i-k} - \mu_{i-k}) + \epsilon_i, \quad i = 1, \ldots, n \tag{2.20}$$

where $\delta_i = \text{var}(\epsilon_i)$. (Here and subsequently, if a sum's upper limit of summation is smaller than its lower limit of summation, we take the sum to equal 0.)

Our development to this point in this section applies to any positive definite co-variance structure, not merely that of PAC(p) variables. However, for PAC(p) variables it turns out that certain subdiagonal elements of $\mathbf{T}$, or equivalently certain autoregressive coefficients, are equal to zero, as described by the following theorem.

Theorem 2.3. *Random variables $Y_1, \ldots, Y_n$ with positive definite covariance matrix Σ are PAC(p) if and only if ϕ_{ij}, the coefficient corresponding to Y_j in the regression of Y_i on $Y_1, Y_2, \ldots, Y_{i-1}$, is equal to 0 for all $(i, j) \in I_p^n$ with $i > j$.*

Proof. According to (2.19), ϕ_{ij} is given by the jth element of $\boldsymbol{\sigma}_{1:i-1,i}^T \Sigma_{1:i-1}^{-1}$. By Corollary A.1.1.1,

$$\boldsymbol{\sigma}_{1:i-1,i}^T \Sigma_{1:i-1}^{-1} = (-\sigma_{1:i}^{i1}/\sigma_{1:i}^{ii}, -\sigma_{1:i}^{i2}/\sigma_{1:i}^{ii}, \ldots, -\sigma_{1:i}^{i,i-1}/\sigma_{1:i}^{ii}).$$

Now, if $Y_1, \ldots, Y_n$ are PAC(p) variables, then $\sigma_{1:i}^{ij} = 0$ for all $j = 1, \ldots, i - p - 1$ by Lemma 2.4. Thus ϕ_{ij} is equal to zero for all $(i, j) \in I_p^n$ with $i > j$.

Conversely, suppose that the coefficient corresponding to Y_j in the regression of Y_i on $Y_1, \ldots, Y_{i-1}$ is equal to zero for all $(i, j) \in I_p^n$ with $i > j$. Then the subdiagonal elements of $\mathbf{T}$ in (2.15) equal 0 for all $(i, j) \in I_p^n$ with $i > j$. Consider now an element σ^{ij} of Σ^{-1} for which $(i, j) \in I_p^n$ and $i > j$, and recall from expression (2.17) that $\Sigma^{-1} = \mathbf{T}^T \mathbf{D}^{-1} \mathbf{T}$. By direct matrix multiplication and recalling that $\phi_{ij} = -t_{ij}$, we obtain

$$\sigma^{ij} = \frac{t_{ij}}{\delta_i} + \sum_{k=i+1}^{n} \frac{t_{ki} t_{kj}}{\delta_k}$$

$$= -\frac{\phi_{ij}}{\delta_i} + \sum_{k=i+1}^{n} \frac{\phi_{ki} \phi_{kj}}{\delta_k}.$$

If $(i, j) \in I_p^n$ then $\phi_{kj} = 0$ for $k = i, \ldots, n$ and thus $\sigma^{ij} = 0$. It follows from Theorem 2.2 and the symmetry of Σ^{-1} that $Y_1, \ldots, Y_n$ are PAC(p) variables. $\square$

The upshot of Theorem 2.3 is that for PAC(p) random variables, the zeros in the unit lower triangular matrix $\mathbf{T}$ of the precision matrix's modified Cholesky decomposition, like those in the precision matrix itself, have a banded structure; that is, those subdiagonals beyond the pth consist of all zeros. For example, the unit lower triangular matrix corresponding to five PAC(2) random variables is

$$\mathbf{T} = \begin{pmatrix} 1 & 0 & 0 & 0 & 0 \\ -\phi_{21} & 1 & 0 & 0 & 0 \\ -\phi_{31} & -\phi_{32} & 1 & 0 & 0 \\ 0 & -\phi_{42} & -\phi_{43} & 1 & 0 \\ 0 & 0 & -\phi_{53} & -\phi_{54} & 1 \end{pmatrix},$$

and we see that the variables' covariance structure may be parameterized in this case by $\delta_1, \ldots, \delta_5$ and $\phi_{21}, \phi_{32}, \ldots, \phi_{53}$. In general, the covariance structure of n PAC(p) random variables may be parameterized by

$$\delta_1, \ldots, \delta_n, \phi_{21}, \phi_{32}, \phi_{42}, \ldots, \phi_{n,n-1}, \phi_{31}, \ldots, \phi_{n,n-p}.$$

Upon counting the elements in this list, we find that the autoregressive formulation of a PAC(p) covariance structure has $(2n-p)(p+1)/2$ distinct parameters, the same (necessarily) as the number of parameters in the intervenor-adjusted and precision matrix formulations. However, there is an important difference between the parameters for this formulation and those of the previous two: whereas the parameters of the intervenor-adjusted and precision matrix formulations must satisfy positive definiteness constraints, which are "messy" in terms of what they require of at least some of the individual parameters, the autoregressive parameters $\{\phi_{ij}\}$ are completely unconstrained and each of the innovation variances $\{\delta_i\}$ need only be positive. This makes the autoregressive formulation of models for PAC(p) [and normal AD(p)] variables much easier to deal with in practice, as we will see later.

It also follows immediately from Theorem 2.3 that for PAC(p) variables, equations (2.20) may be written using fewer terms. More specifically, we may write

$$Y_i - \mu_i = \sum_{k=1}^{p_i} \phi_{i,i-k}(Y_{i-k} - \mu_{i-k}) + \epsilon_i, \quad i = 1, \ldots, n \qquad (2.21)$$

where $p_i = \min(p, i-1)$. Furthermore, for PAC(p) variables, the general expressions (2.19) and (2.18) for the autoregressive coefficients $\{\phi_{ij}\}$ and innovation variances $\{\delta_i\}$, respectively, may be reexpressed in terms of smaller vectors and matrices. Specifically, we have

$$(\phi_{i,i-p_i}, \phi_{i,i-p_i+1}, \ldots, \phi_{i,i-1}) = \boldsymbol{\sigma}_{i-p_i:i-1,i}^T \boldsymbol{\Sigma}_{i-p_i:i-1}^{-1}, \qquad (2.22)$$

$$\delta_i = \begin{cases} \sigma_{11}, & \text{for } i = 1 \\ \sigma_{ii} - \boldsymbol{\sigma}_{i-p_i:i-1,i}^T \boldsymbol{\Sigma}_{i-p_i:i-1}^{-1} \boldsymbol{\sigma}_{i-p_i:i-1,i}, & \text{for } i = 2, \ldots, n. \end{cases} \qquad (2.23)$$

In the proof of Theorem 2.3, (2.17) was used to show that the elements of PAC(p) variables' precision matrix $\boldsymbol{\Sigma}^{-1}$ with indices in I_p^n were equal to zero. In fact, further use of (2.17) yields explicit expressions for every element of $\boldsymbol{\Sigma}^{-1}$ in terms of the parameters of the autoregressive formulation. By direct matrix multiplication we obtain

$$\sigma^{ii} = \begin{cases} \dfrac{1}{\delta_i} + \displaystyle\sum_{k=i+1}^{\min(i+p,n)} \dfrac{\phi_{ki}^2}{\delta_k}, & \text{for } i = 1, \ldots, n-1 \\ \dfrac{1}{\delta_n}, & \text{for } i = n, \end{cases} \qquad (2.24)$$

and

$$
\sigma^{ij} = \begin{cases}
0, & \text{for } j = 1, \ldots, i - p - 1 \\
-\dfrac{\phi_{i,i-p}}{\delta_i}, & \text{for } j = i - p \\
-\dfrac{\phi_{ij}}{\delta_i} + \displaystyle\sum_{k=i+1}^{\min(i+p-1,n)} \dfrac{\phi_{ki}\phi_{kj}}{\delta_k}, & \text{for } j = i - p + 1, \ldots, i - 1.
\end{cases}
$$

$$(2.25)$$

2.3.3 Marginal characterization

The three characterizations of the covariance structure of PAC(p) random variables that we have described so far are elegant and tidy, as each specifies that certain quantities (intervenor-adjusted partial correlations, off-diagonal elements of Σ^{-1}, or subdiagonal elements of $\mathbf{T}$) involved in a one-to-one relationship with the covariance matrix Σ are zero. However, they are all somewhat indirect in the sense that they do not directly specify properties of the elements $\{\sigma_{ij}\}$ of Σ itself. In this section we determine these properties. Although they turn out to be generally more cumbersome than properties of the parameters of the other characterizations, they are nevertheless interesting and useful.

We begin by showing how expressions may be obtained for the elements of a PAC(p) covariance matrix Σ in explicit terms of the autoregressive coefficients and innovation variances. Assume that $p > 0$; there are no non-trivial autoregressive relationships between the variables otherwise. Recall from (2.21) that the ith equation in the sequence of regressions (2.11) through (2.14) may be written as

$$
Y_i - \mu_i = \sum_{k=1}^{p_i} \phi_{i,i-k}(Y_{i-k} - \mu_{i-k}) + \epsilon_i, \quad i = 1, \ldots, n, \tag{2.26}
$$

where $p_i = \min(p, i - 1)$. Multiplying both sides of (2.26) by $(Y_{i-j} - \mu_{i-j})$ and taking expectations yields

$$
\sigma_{i,i-j} = \sum_{k=1}^{p_i} \phi_{i,i-k}\sigma_{i-k,i-j} + \delta_i I_{\{j=0\}}, \quad i = 1, \ldots, n; \ j = 0, \ldots, i - 1. \tag{2.27}
$$

This set of equations allows all elements of Σ, beginning with σ_{11}, to be obtained recursively from previously obtained elements of Σ and parameters of the autoregressive formulation. Furthermore, if desired, the previously obtained elements of Σ may then be eliminated so as to express each element of

Σ completely in terms of the parameters of the autoregressive formulation. To illustrate, suppose that $p = 2$. Then we have

$$
\begin{aligned}
\sigma_{11} &= \delta_1, \\
\sigma_{21} &= \phi_{21}\sigma_{11} = \phi_{21}\delta_1, \\
\sigma_{22} &= \phi_{21}\sigma_{12} + \delta_2 = \phi_{21}^2\delta_1 + \delta_2, \\
\sigma_{31} &= \phi_{32}\sigma_{21} + \phi_{31}\sigma_{11} = (\phi_{32}\phi_{21} + \phi_{31})\delta_1, \\
\sigma_{32} &= \phi_{32}\sigma_{22} + \phi_{31}\sigma_{12} = (\phi_{32}\phi_{21}^2 + \phi_{31}\phi_{21})\delta_1 + \phi_{32}\delta_2, \\
\sigma_{33} &= \phi_{32}\sigma_{23} + \phi_{31}\sigma_{13} + \delta_3 \\
&= (\phi_{31}^2 + 2\phi_{32}\phi_{31}\phi_{21} + \phi_{32}^2\phi_{21}^2)\delta_1 + \phi_{32}^2\delta_2 + \delta_3,
\end{aligned}
$$

and so on. Of course, these equations are merely an elementwise rendering of (2.16) for PAC(2) variables.

The recursive equations just described may also be used to obtain expressions of covariance matrix elements $\{\sigma_{ij} : (i, j) \in I_p^n\}$ in explicit terms of elements $\{\sigma_{ij} : (i, j) \notin I_p^n\}$. Let us first illustrate this for a special case and then generalize it. Consider the case $p = 2$ and $n = 5$, for which the covariance matrix is

$$
\Sigma = \begin{pmatrix}
\sigma_{11} & \sigma_{12} & \sigma_{13} & \sigma_{14} & \sigma_{15} \\
\sigma_{21} & \sigma_{22} & \sigma_{23} & \sigma_{24} & \sigma_{25} \\
\sigma_{31} & \sigma_{32} & \sigma_{33} & \sigma_{34} & \sigma_{35} \\
\sigma_{41} & \sigma_{42} & \sigma_{43} & \sigma_{44} & \sigma_{45} \\
\sigma_{51} & \sigma_{52} & \sigma_{53} & \sigma_{54} & \sigma_{55}
\end{pmatrix}.
$$

Then, by (2.27),

$$
\begin{pmatrix} \sigma_{41} \\ \sigma_{42} \\ \sigma_{43} \end{pmatrix} = \begin{pmatrix} \sigma_{21} & \sigma_{31} \\ \sigma_{22} & \sigma_{32} \\ \sigma_{23} & \sigma_{33} \end{pmatrix} \begin{pmatrix} \phi_{42} \\ \phi_{43} \end{pmatrix}. \tag{2.28}
$$

Solving the last two equations in (2.28) for the autoregressive coefficients yields

$$
\begin{pmatrix} \phi_{42} \\ \phi_{43} \end{pmatrix} = \begin{pmatrix} \sigma_{22} & \sigma_{32} \\ \sigma_{23} & \sigma_{33} \end{pmatrix}^{-1} \begin{pmatrix} \sigma_{42} \\ \sigma_{43} \end{pmatrix},
$$

and putting this solution back into the first equation in (2.28) yields

$$
\sigma_{41} = (\sigma_{21}, \sigma_{31}) \begin{pmatrix} \sigma_{22} & \sigma_{32} \\ \sigma_{23} & \sigma_{33} \end{pmatrix}^{-1} \begin{pmatrix} \sigma_{42} \\ \sigma_{43} \end{pmatrix}. \tag{2.29}
$$

In similar fashion, (2.27) also gives

$$
\begin{pmatrix} \sigma_{51} \\ \sigma_{52} \\ \sigma_{53} \\ \sigma_{54} \end{pmatrix} = \begin{pmatrix} \sigma_{31} & \sigma_{41} \\ \sigma_{32} & \sigma_{42} \\ \sigma_{33} & \sigma_{43} \\ \sigma_{34} & \sigma_{44} \end{pmatrix} \begin{pmatrix} \phi_{53} \\ \phi_{54} \end{pmatrix}. \tag{2.30}
$$

Solving the last two equations in (2.30) for the autoregressive coefficients yields

$$
\begin{pmatrix} \phi_{53} \\ \phi_{54} \end{pmatrix} = \begin{pmatrix} \sigma_{33} & \sigma_{43} \\ \sigma_{34} & \sigma_{44} \end{pmatrix}^{-1} \begin{pmatrix} \sigma_{53} \\ \sigma_{54} \end{pmatrix},
$$

and putting this solution back into the first two equations in (2.30) yields

$$
\begin{pmatrix} \sigma_{51} \\ \sigma_{52} \end{pmatrix} = \begin{pmatrix} \sigma_{31} & \sigma_{41} \\ \sigma_{32} & \sigma_{42} \end{pmatrix} \begin{pmatrix} \sigma_{33} & \sigma_{43} \\ \sigma_{34} & \sigma_{44} \end{pmatrix}^{-1} \begin{pmatrix} \sigma_{53} \\ \sigma_{54} \end{pmatrix}. \tag{2.31}
$$

Thus are the elements of a PAC(2) covariance matrix Σ that lie on off-diagonals beyond the second, namely σ_{41}, σ_{51}, and σ_{52}, expressed in terms of elements of Σ lying on the main diagonal and the first two off-diagonals.

We state the procedure for general p and n as the following theorem. The theorem may be proved by induction, but the proof is tedious and is therefore omitted.

Theorem 2.4. *Let $Y_1, \ldots, Y_n$ be PAC(p) random variables with positive definite covariance matrix Σ. For $i = p + 2, \ldots, n$, define*

$$
\zeta_i = \begin{pmatrix} \sigma_{i1} \\ \vdots \\ \sigma_{i,i-p-1} \end{pmatrix}, \quad \Upsilon_i = \begin{pmatrix} \sigma_{i-p,1} & \cdots & \sigma_{i-1,1} \\ & \vdots & \\ \sigma_{i-p,i-p-1} & \cdots & \sigma_{i-1,i-p-1} \end{pmatrix},
$$

$$
\Psi_i = \begin{pmatrix} \sigma_{i-p,i-p} & \cdots & \sigma_{i-1,i-p} \\ & \vdots & \\ \sigma_{i-p,i-1} & \cdots & \sigma_{i-1,i-1} \end{pmatrix}, \quad \eta_i = \begin{pmatrix} \sigma_{i,i-p} \\ \vdots \\ \sigma_{i,i-1} \end{pmatrix}.
$$

Then $\{\zeta_i : i = p + 2, \ldots, n\}$ comprises all those elements of Σ whose indices belong to I_p^n, and these may be obtained in terms of elements with indices not belonging to I_p^n by iteratively applying the equations

$$
\zeta_i = \Upsilon_i \Psi_i^{-1} \eta_i, \quad i = p + 2, \ldots, n. \tag{2.32}
$$

As a consequence of Theorem 2.4, it is possible to parameterize the marginal formulation of PAC(p) variables by only those elements of Σ that lie on its main diagonal and first p off-diagonals, i.e., by

$$
\sigma_{11}, \ldots, \sigma_{nn}, \sigma_{21}, \ldots, \sigma_{n,n-1}, \sigma_{31}, \ldots, \sigma_{n,n-p}.
$$

However, the constraints on these parameters required for positive definiteness of Σ are generally quite complicated, apart from those on the σ_{ii}'s. So this parameterization of a PAC(p) covariance structure, while useful theoretically, is not necessarily the most convenient one to use for estimation purposes.

2.4 Some results on determinants and traces

In this section we use the special properties of PAC(p) variables that were derived in the previous section to establish some results on the determinant and trace of those variables' covariance matrix. These results will be useful later for deriving and computing maximum likelihood estimators and likelihood ratio test statistics under a normal AD(p) model. Here we merely present the results; proofs can be found in Appendix 2.

Here and subsequently, we write the determinant and trace of a square matrix $\mathbf{A}$ as $|\mathbf{A}|$ and $\text{tr}(\mathbf{A})$, respectively.

Theorem 2.5. *Let $Y_1, \ldots, Y_n$ be PAC(p) random variables with positive definite covariance matrix $\boldsymbol{\Sigma}$.*

(a) *Let $\mathbf{T}$ and $\mathbf{D} = \text{diag}(\delta_1, \ldots, \delta_n)$ be the unit lower triangular and diagonal matrices, respectively, of the modified Cholesky decomposition of $\boldsymbol{\Sigma}^{-1}$. Then*

$$|\boldsymbol{\Sigma}| = \prod_{i=1}^{n} \delta_i.$$

(b) *Let $R^2_{i \cdot \{i-p_i:i-1\}}$ denote the multiple correlation coefficient between Y_i and $Y_{i-p_i}, \ldots, Y_{i-1}$, where $i > 1$ and $p_i = \min(i-1, p)$. Then*

$$|\boldsymbol{\Sigma}| = \prod_{i=1}^{n} \sigma_{ii} \prod_{i=2}^{n} (1 - R^2_{i \cdot \{i-p_i:i-1\}})$$

where we define $R^2_{i \cdot \{i-p_i:i-1\}} = 0$ if $p = 0$.

(c) *Let $\rho_{ij \cdot B}$ denote the partial correlation coefficient between Y_i and Y_j adjusted for $\{Y_k : k \in B\}$. Then*

$$|\boldsymbol{\Sigma}| = \prod_{i=1}^{n} \sigma_{ii} \prod_{i=2}^{n} (1 - \rho^2_{i,i-1}) \prod_{i=3}^{n} (1 - \rho^2_{i,i-2 \cdot i-1}) \cdots$$
$$\times \prod_{i=p+1}^{n} (1 - \rho^2_{i,i-p \cdot \{i-p+1:i-1\}}).$$

(d) *Let $\boldsymbol{\Sigma}_{k:m}$ denote the submatrix consisting of elements in rows $k, k+1, \ldots, m$ and columns $k, k+1, \ldots, m$ of $\boldsymbol{\Sigma}$. Then*

$$|\boldsymbol{\Sigma}| = \frac{\prod_{i=1}^{n-p} |\boldsymbol{\Sigma}_{i:i+p}|}{\prod_{i=1}^{n-p-1} |\boldsymbol{\Sigma}_{i+1:i+p}|} \tag{2.33}$$

where we define

$$\prod_{i=1}^{n-p-1} |\boldsymbol{\Sigma}_{i+1:i+p}| = 1 \quad \text{if } p = 0 \text{ or } p = n-1.$$

Parts (a), (c), and (d) of Theorem 2.5 express the determinant of the covariance matrix of PAC(p) random variables in computationally efficient terms of the parameters of its autoregressive, intervenor-adjusted, and marginal formulations, respectively. The reader may wonder whether it is also possible to give a computationally efficient expression for the determinant in terms of the parameters of the precision matrix formulation, i.e., the nonzero elements of the precision matrix. The authors are not aware of any such general expression; however, algorithms for computing the determinant of tridiagonal and pentadiagonal matrices, due to El-Mikkawy (2004) and Sogabe (2008), respectively, could be used to evaluate the determinant of $\boldsymbol{\Sigma}^{-1}$ efficiently when $Y_1, \ldots, Y_n$ are PAC(1) or PAC(2).

We conclude this section with a theorem for the trace of the product of a symmetric matrix and the precision matrix of PAC(p) variables.

Theorem 2.6. *Let $Y_1, \ldots, Y_n$ be PAC(p) random variables with positive definite covariance matrix $\boldsymbol{\Sigma}$, and let $\mathbf{A}$ be an $n \times n$ symmetric matrix. Then*

$$\text{tr}(\mathbf{A}\boldsymbol{\Sigma}^{-1}) = \sum_{i=1}^{n-p} \text{tr}[\mathbf{A}_{i:i+p}(\boldsymbol{\Sigma}_{i:i+p})^{-1}] - \sum_{i=1}^{n-p-1} \text{tr}[\mathbf{A}_{i+1:i+p}(\boldsymbol{\Sigma}_{i+1:i+p})^{-1}]$$

(2.34)

where we define

$$\sum_{i=1}^{n-p-1} \text{tr}[\mathbf{A}_{i+1:i+p}(\boldsymbol{\Sigma}_{i+1:i+p})^{-1}] = 0 \quad \text{if } p = 0 \text{ or } p = n - 1.$$

2.5 The first-order case

The results on PAC(p) random variables presented so far are valid for any integer p between 0 and n, inclusive. In this section we consider how these results specialize for the simplest non-trivial case of p, i.e., $p = 1$. This case is probably the most important one from a practical point of view, as its normal cousin, the AD(1) model, offers the opportunity to model longitudinal data exhibiting variance heterogeneity and nonstationary serial correlation very parsimoniously.

According to Definition 2.3 — our original definition of PAC(p) variables — index-ordered random variables $Y_1, \ldots, Y_n$ are PAC(1) if and only if

$$\rho_{i,i+q+2 \cdot \{i+1:i+q+1\}} = 0 \quad \text{for all } i = 1, \ldots, n - q - 2 \text{ and}$$
$$\text{all } q = 0, \ldots, n - 3. \quad (2.35)$$

In other words, the variables are PAC(1) if and only if the partial correlation between each pair of non-adjacent variables, adjusted for all intervening variables, is equal to 0. Of course, if the variables are jointly normally distributed, then they are AD(1) variables also.

It turns out that the property of the partial autocorrelations given by (2.35) confers a fascinating structure upon the marginal covariance matrix of PAC(1) variables, in particular upon the marginal correlations. Consider $\sigma_{ij \cdot m}$, for $i > m > j$. Then by (2.35) we have

$$
\begin{aligned}
0 = \sigma_{ij \cdot m} &= \sigma_{ij} - \frac{\sigma_{im}\sigma_{jm}}{\sigma_{mm}} \\
&= \rho_{ij}(\sigma_{ii}\sigma_{jj})^{1/2} - \frac{\rho_{im}(\sigma_{ii}\sigma_{mm})^{1/2}\rho_{jm}(\sigma_{jj}\sigma_{mm})^{1/2}}{\sigma_{mm}},
\end{aligned}
$$

implying that

$$
\rho_{ij} = \rho_{im}\rho_{mj} \quad \text{for } i > m > j. \tag{2.36}
$$

That is, the correlation between any two of the (ordered) variables is equal to the product of two correlations, each one being a correlation between one of the two variables and an arbitrary intervening variable. Furthermore, by repeatedly substituting for any correlation on the right-hand side of (2.36) between non-adjacent variables until all such "non-adjacent" correlations have been replaced with products of "adjacent" (lag-one) correlations, we obtain

$$
\rho_{ij} = \prod_{m=j}^{i-1} \rho_{m+1,m}. \tag{2.37}
$$

Thus, the correlation between any variables lagged two or more indices apart is equal to the product of the lag-one correlations corresponding to each intervening pair of consecutive variables, and we can write the covariance matrix as follows, putting $\rho_i = \rho_{i+1,i}$ and $\nu_{ij} = (\sigma_{ii}\sigma_{jj})^{1/2}$:

$$
\Sigma = \begin{pmatrix}
\sigma_{11} & & & & & & \\
\nu_{21}\rho_1 & \sigma_{22} & & & & symm & \\
\nu_{31}\rho_1\rho_2 & \nu_{32}\rho_2 & \sigma_{33} & & & & \\
\nu_{41}\rho_1\rho_2\rho_3 & \nu_{42}\rho_2\rho_3 & \nu_{43}\rho_3 & \sigma_{44} & & & \\
\vdots & \vdots & \vdots & \ddots & \ddots & & \\
\vdots & \vdots & \vdots & \ddots & \ddots & \sigma_{n-1,n-1} & \\
\nu_{n1}\prod_{i=1}^{n-1}\rho_i & \cdots & \cdots & \cdots & \cdots & \nu_{n,n-1}\rho_{n-1} & \sigma_{nn}
\end{pmatrix}
$$

$$
\tag{2.38}
$$

Note that this covariance matrix is completely determined by the $(2n-1)$ elements on its main diagonal and first subdiagonal, or equivalently, by the $(2n-1)$ parameters $\sigma_{11}, ..., \sigma_{nn}, \rho_1, ..., \rho_{n-1}$. In this case, the parameter constraints required for positive definiteness of the covariance matrix are easy to specify; they are $\sigma_{ii} > 0$ for $i = 1, \ldots, n$ and $-1 < \rho_i < 1$ for $i = 1, ..., n-1$.

It is interesting to note that the marginal correlations displayed within any column of (2.38) necessarily are monotone decreasing in magnitude as a function of distance from the main diagonal. However, there is no requirement of constancy (stationarity in the time series context) of correlations between variables lagged by the same number of indices. It is also worth noting that the multiplicative structure of the correlations can be derived by another route, namely by specializing the general expression (2.32) for obtaining elements on higher-order off-diagonals of Σ from those on its main diagonal and first p off-diagonals.

Expression (2.37) may be used to obtain expressions for the covariances (rather than the correlations) between variables lagged two or more indices apart in terms of the lag-one covariances, as follows:

$$\sigma_{ij} = (\sigma_{ii}\sigma_{jj})^{1/2} \prod_{m=j}^{i-1} \frac{\sigma_{m+1,m}}{(\sigma_{m+1,m+1}\sigma_{mm})^{1/2}}$$

$$= \frac{\prod_{m=j}^{i-1} \sigma_{m+1,m}}{\prod_{m=j+1}^{i-1} \sigma_{mm}}. \tag{2.39}$$

Another important feature of the covariance structure given by (2.37) through (2.39) is as follows. Suppose we strike out the jth row and jth column of Σ in (2.38), and then renumber the variances along the main diagonal as $\sigma_{11}, \sigma_{22}, \sigma_{33}, \ldots, \sigma_{n-1,n-1}$ and the lag-one correlations along the first subdiagonal as $\rho_1, \rho_2, \ldots, \rho_{n-1}$. Then the resulting $(n-1) \times (n-1)$ matrix retains the same structure as Σ in (2.38). In fact, if we strike out rows $j_1, j_2, \ldots, j_k$ and columns $j_1, j_2, \ldots, j_k$ of Σ, where $1 \leq k \leq n-1$, the resulting $(n-k) \times (n-k)$ matrix also retains the same structure. This shows that any subsequence of PAC(1) variables is also PAC(1), and also that any subsequence of normal AD(1) variables is also AD(1).

The precision matrix of PAC(1) variables also has an interesting structure. Theorem 2.2 tells us that this matrix must be tridiagonal; that is, its elements on all off-diagonals but the first must equal zero. But the theorem gives us no information about its elements on the main diagonal and first off-diagonal. However, by solving the equation $\Sigma\Sigma^{-1} = I$ for the non-trivial elements of the tridiagonal matrix Σ^{-1}, the following explicit formula for the precision matrix may be obtained relatively easily:

$$
\sigma^{ij} =
\begin{cases}
\dfrac{\sigma_{22}}{\sigma_{11}\sigma_{22} - \sigma_{12}^2} & i = j = 1 \\[2ex]
\dfrac{\sigma_{n-1,n-1}}{\sigma_{n-1,n-1}\sigma_{nn} - \sigma_{n,n-1}^2} & i = j = n \\[2ex]
\dfrac{\sigma_{i-1,i-1}\sigma_{ii}^2\sigma_{i+1,i+1} - \sigma_{i,i-1}^2\sigma_{i+1,i}^2}{\sigma_{ii}^2(\sigma_{i-1,i-1}\sigma_{ii} - \sigma_{i,i-1}^2)(\sigma_{ii}\sigma_{i+1,i+1} - \sigma_{i+1,i}^2)} & i = j \neq 1, n \\[2ex]
\dfrac{-\sigma_{i,i-1}}{\sigma_{ii}\sigma_{i-1,i-1} - \sigma_{i,i-1}^2} & i = j + 1 \\[2ex]
\sigma^{ji} & i = j - 1 \\[2ex]
0 & |i - j| > 1.
\end{cases}
$$

$$(2.40)$$

Expressed alternatively in terms of the marginal variances and lag-one correlations, this formula is as follows:

$$
\sigma^{ij} =
\begin{cases}
\dfrac{1}{\sigma_{11}(1 - \rho_1^2)} & i = j = 1 \\[2ex]
\dfrac{1}{\sigma_{nn}(1 - \rho_{n-1}^2)} & i = j = n \\[2ex]
\dfrac{1 - \rho_{i-1}^2\rho_i^2}{\sigma_{ii}(1 - \rho_{i-1}^2)(1 - \rho_i^2)} & i = j \neq 1, n \\[2ex]
\dfrac{-\rho_i}{(\sigma_{ii}\sigma_{jj})^{1/2}(1 - \rho_i^2)} & i = j + 1 \\[2ex]
\sigma^{ji} & i = j - 1 \\[2ex]
0 & |i - j| > 1.
\end{cases}
$$

These expressions for the inverse have been derived and rederived, in various guises and for various special cases, several times in the literature. Apparently the first to give the inverse explicitly was Guttman (1955). Roy and Sarhan (1956) independently also gave the inverse for a special case, and this was subsequently generalized by Greenberg and Sarhan (1959). Much later, Barrett and Feinsilver (1978) gave the same expression, referring to it as the inverse of a positive definite covariance matrix satisfying a condition they called the "triangle property," which turns out to be equivalent to (2.37); hence, any covariance matrix satisfying this property is an AD(1) covariance matrix and vice versa. There is also a substantial related literature on the inversion of a nonsingular tridiagonal matrix; see, for example, Schlegel (1970) and Mallik (2001).

The autoregressive characterization of the covariance structure of PAC(1) variables is given by the modified Cholesky decomposition of the precision matrix, i.e., $\mathbf{T}\boldsymbol{\Sigma}\mathbf{T}^T = \mathbf{D}$, where $\mathbf{D} = \text{diag}(\delta_1, \ldots, \delta_n)$ and

$$
\mathbf{T} = \begin{pmatrix}
1 & 0 & 0 & 0 & 0 & 0 \\
-\phi_1 & 1 & 0 & 0 & 0 & 0 \\
0 & -\phi_2 & 1 & 0 & 0 & 0 \\
\vdots & & \ddots & \ddots & & \\
0 & \cdots & 0 & -\phi_{n-2} & 1 & 0 \\
0 & \cdots & 0 & 0 & -\phi_{n-1} & 1
\end{pmatrix},
$$

and where, for simplicity of notation, we have put $\phi_i = \phi_{i+1,i}$, for $i = 1, \ldots, n-1$. Written as individual equations as in (2.21), we have

$$
\begin{aligned}
Y_1 - \mu_1 &= \epsilon_1, \\
Y_i - \mu_i &= \phi_{i-1}(Y_{i-1} - \mu_{i-1}) + \epsilon_i, \quad i = 2, \ldots, n,
\end{aligned}
$$

which is essentially identical to model (1.5) introduced in Chapter 1. Using (2.22) and (2.23), the parameters of this characterization, $\{\phi_i\}$ and $\{\delta_i\}$, may be expressed in terms of those of the marginal characterization as follows:

$$
\phi_i = \frac{\sigma_{i+1,i}}{\sigma_{ii}} \quad \text{for } i = 1, \ldots, n-1,
$$

$$
\delta_i = \begin{cases}
\sigma_{11} & \text{for } i = 1 \\
\sigma_{ii} - \dfrac{\sigma_{i,i-1}^2}{\sigma_{i-1,i-1}} & \text{for } i = 2, \ldots, n.
\end{cases}
$$

Furthermore, the general recursive approach for expressing the elements of $\boldsymbol{\Sigma}$ in terms of the autoregressive parameters, which was given by (2.27), simplifies for PAC(1) variables to

$$
\sigma_{i,i-j} = \phi_{i-1}\sigma_{i-1,i-j} + \delta_i I_{\{j=0\}}, \quad i = 1, \ldots, n;\ j = 0, \ldots, i-1.
$$

Written out equation by equation, we have

$$
\begin{aligned}
\sigma_{11} &= \delta_1, \\
\sigma_{21} &= \phi_1\sigma_{11} = \phi_1\delta_1, \\
\sigma_{22} &= \phi_1\sigma_{12} + \delta_2 = \phi_1^2\delta_1 + \delta_2, \\
\sigma_{31} &= \phi_2\sigma_{21} = \phi_2\phi_1\delta_1, \\
\sigma_{32} &= \phi_2\sigma_{22} = \phi_2\phi_1^2\delta_1 + \phi_2\delta_2, \\
\sigma_{33} &= \phi_2\sigma_{23} + \delta_3 = \phi_2^2\phi_1^2\delta_1 + \phi_2^2\delta_2 + \delta_3,
\end{aligned}
$$

and so on.

The general expressions for elements of the precision matrix of PAC(p) variables in terms of the autoregressive parameterization, given by (2.24) and (2.25), simplify for PAC(1) variables to

$$
\sigma^{ii} = \begin{cases}
\dfrac{1}{\delta_i} + \dfrac{\phi_i^2}{\delta_{i+1}} & \text{for } i = 1, \ldots, n-1, \\[2ex]
\dfrac{1}{\delta_n} & \text{for } i = n,
\end{cases}
$$

$$
\sigma^{ij} = \begin{cases}
0 & \text{for } i = 2, \ldots, n \text{ and } j = 1, \ldots, i-2, \\[2ex]
-\dfrac{\phi_{i-1}}{\delta_i} & \text{for } i = 2, \ldots, n \text{ and } j = i-1.
\end{cases}
$$

Four general expressions for the determinant of the covariance matrix of PAC(p) variables were given in Theorem 2.5. For PAC(1) variables, these expressions all specialize easily to

$$
|\mathbf{\Sigma}| = \prod_{i=1}^{n} \sigma_{ii} \prod_{i=1}^{n-1} (1 - \rho_i^2). \tag{2.41}
$$

Similarly, the general expression for $\mathrm{tr}(\mathbf{A}\mathbf{\Sigma}^{-1})$ given by Theorem 2.6, where $\mathbf{A}$ is any $n \times n$ symmetric matrix, specializes to

$$
\begin{aligned}
\mathrm{tr}(\mathbf{A}\mathbf{\Sigma}^{-1}) &= \sum_{i=1}^{n-1} \mathrm{tr}[\mathbf{A}_{i:i+1}(\mathbf{\Sigma}_{i:i+1})^{-1}] - \sum_{i=1}^{n-2} \mathrm{tr}[\mathbf{A}_{i+1:i+1}(\mathbf{\Sigma}_{i+1:i+1})^{-1}] \\
&= \sum_{i=1}^{n-1} \mathrm{tr}\left[\begin{pmatrix} a_{ii} & a_{i,i+1} \\ a_{i+1,i} & a_{i+1,i+1} \end{pmatrix} \begin{pmatrix} \sigma_{ii} & \sigma_{i,i+1} \\ \sigma_{i+1,i} & \sigma_{i+1,i+1} \end{pmatrix}^{-1} \right] \\
&\quad - \sum_{i=1}^{n-2} \frac{a_{i+1,i+1}}{\sigma_{i+1,i+1}} \\
&= \sum_{i=1}^{n-1} \frac{a_{ii}\sigma_{i+1,i+1} - 2a_{i,i+1}\sigma_{i,i+1} + a_{i+1,i+1}\sigma_{ii}}{\sigma_{ii}\sigma_{i+1,i+1} - \sigma_{i,i+1}^2} - \sum_{i=2}^{n-1} \frac{a_{ii}}{\sigma_{ii}}.
\end{aligned}
$$

2.6 Variable-order antedependence

Our discussion to this point has presumed that the order, p, of antedependence or partial antecorrelation is constant for all variables. It is possible, however, to define more general versions of antedependence or partial antecorrelation by allowing the order to vary among the variables. We give the following two formal definitions, the first of which is due to Macchiavelli and Arnold (1994).

Definition 2.4. *Index-ordered random variables* $Y_1, \ldots, Y_n$ *are said to be antedependent of variable order* $(p_1, p_2, \ldots, p_n)$, *or* $AD(p_1, p_2, \ldots, p_n)$, *if* Y_i, *given at least* p_i *immediately preceding variables, is independent of all further preceding variables* $(i = 1, \ldots, n)$.

Definition 2.5. *Index-ordered random variables* $Y_1, \ldots, Y_n$ *are said to be partially antecorrelated of variable order* $(p_1, p_2, \ldots, p_n)$, *or* $PAC(p_1, p_2, \ldots, p_n)$, *if the partial correlation between* Y_i *and* Y_j, *adjusted for at least* p_i *variables immediately preceding* Y_i, *is equal to zero for all* $j < i - p_i$ $(i = 1, \ldots, n)$.

Note that $p_i \leq i - 1$ necessarily (and hence $p_1 = 0$), and that if $p_i = 0$ then Y_i is either independent of, or uncorrelated with, all its predecessors, depending on whether the variables are $AD(p_1, \ldots, p_n)$ or $PAC(p_1, \ldots, p_n)$. Note also that $AD(p_1, \ldots, p_n)$ variables are nested, in the sense that

$$AD(p_1, \ldots, p_n) \subset AD(p_1 + q_1, \ldots, p_n + q_n)$$

if $q_i \geq 0$ for all i; and that

$$AD(p_1, \ldots, p_n) \subset AD(\max_i p_i). \tag{2.42}$$

As was the case for the covariance structure of constant-order PAC variables, the covariance structure of variable-order PAC variables may alternatively be characterized in several — though not quite as many — equivalent ways. The following theorem essentially gives these alternative characterizations. We give the theorem without proof; a proof can be constructed along the lines of proofs of certain lemmas and theorems in Section 2.3 pertaining to constant-order PAC variables.

Theorem 2.7. *Random variables* $Y_1, \ldots, Y_n$ *with positive definite covariance matrix* $\boldsymbol{\Sigma}$ *are* $PAC(p_1, \ldots, p_n)$ *if and only if either of the following is true:*

(i) *For all* $i = 1, \ldots, n$ *and* $j < i - p_i$, *the partial correlation between* Y_i *and* Y_j, *adjusted for all intervening variables* $Y_{j+1}, \ldots, Y_{i-1}$, *is equal to zero.*

(ii) *In the modified Cholesky decomposition of the precision matrix, i.e.,* $\boldsymbol{\Sigma}^{-1} = \mathbf{T}^T \mathbf{D}^{-1} \mathbf{T}$, *the subdiagonal elements* $-\phi_{ij}$ *of the unit lower triangular matrix* $\mathbf{T}$ *satisfy* $-\phi_{ij} = 0$ *for all* j *such that* $j < i - p_i$ $(i = 2, \ldots, n)$. *Equivalently,*

$$Y_i - \mu_i = \sum_{k=1}^{p_i} \phi_{i,i-k}(Y_{i-k} - \mu_{i-k}) + \epsilon_i, \quad i = 1, \ldots, n, \tag{2.43}$$

where $\epsilon_1, \ldots, \epsilon_n$ *are uncorrelated zero-mean random variables with* $\mathrm{var}(\epsilon_i) = \delta_i$.

We should point out that Theorem 2.7 does not give a precision matrix characterization of variable-order PAC variables. Recall that Theorem 2.2 gave such a characterization in the constant-order case; in particular, n variables with positive definite covariance matrix Σ are PAC(p) if and only if all elements of Σ^{-1} with indices in I_p^n are zero. On this basis it might be conjectured that n variables with positive definite covariance matrix Σ are PAC($p_1, \ldots, p_n$) if and only if all elements of Σ^{-1} in $I_{p_1, \ldots, p_n}$ are zero, where $I_{p_1, \ldots, p_n} = \{(i, j) : i = 2, \ldots, n; \, j < i - p_i\}$. But this conjecture is generally not true. Of course, by virtue of Theorem 2.2 and (2.42), all elements of $\Sigma^{-1} = (\sigma^{ij})$ with indices in $I_{\max p_i}^n$ are zero; furthermore, it can be shown that all elements of Σ^{-1} with indices satisfying $j < \min_{k=i,\ldots,n}(k - p_k)$ are zero. Note that this last set of indices is a subset of $I_{p_1, \ldots, p_n}$. Nevertheless, for a precision matrix satisfying this last property, the orders $(p_1, \ldots, p_n)$ are generally not uniquely determined, so this property does not yield a precision matrix characterization either.

The autoregressive formulation given by (2.43) is a particularly convenient way to represent variable-order PAC variables. As an example, if $Y_1, \ldots, Y_6$ are PAC(0,1,1,2,3,1), then they may be represented as follows (where we take $\mu_i \equiv 0$ to reduce clutter):

$$
\begin{aligned}
Y_1 &= \epsilon_1, \\
Y_2 &= \phi_{21} Y_1 + \epsilon_2, \\
Y_3 &= \phi_{32} Y_2 + \epsilon_3, \\
Y_4 &= \phi_{43} Y_3 + \phi_{42} Y_2 + \epsilon_4, \\
Y_5 &= \phi_{54} Y_4 + \phi_{53} Y_3 + \phi_{52} Y_2 + \epsilon_5, \\
Y_6 &= \phi_{65} Y_5 + \epsilon_6.
\end{aligned}
$$

Here, as in Theorem 2.7(ii), the ϵ_i's are uncorrelated, each with variance δ_i.

According to Theorem 2.7(ii), if variables are PAC($p_1, \ldots, p_n$), then subdiagonal elements of $\mathbf{T}$ with indices in $I_{p_1, \ldots, p_n}$ are equal to zero. This can be used to derive recursive equations analogous to (2.32) for obtaining expressions of covariance matrix elements $\{\sigma_{ij} : (i, j) \in I_{p_1, \ldots, p_n}\}$ in explicit terms of elements $\{\sigma_{ij} : (i, j) \notin I_{p_1, \ldots, p_n}\}$. We give this result as a theorem.

Theorem 2.8. *Let $Y_1, \ldots, Y_n$ be PAC($p_1, \ldots, p_n$) random variables with positive definite covariance matrix Σ. Furthermore, let $k_1 < k_2 < \cdots < k_m$ denote the row indices of all rows of Σ which have at least one element in $I_{p_1, \ldots, p_n}$. Then we may obtain $\{\sigma_{ij} : (i, j) \in I_{p_1, \ldots, p_n}\}$ in terms of $\{\sigma_{ij} : (i, j) \notin I_{p_1, \ldots, p_n}\}$ by iteratively applying the equations*

$$
\boldsymbol{\zeta}_i = \boldsymbol{\Upsilon}_i \boldsymbol{\Psi}_i^{-1} \boldsymbol{\eta}_i, \quad i = k_1, k_2, \ldots, k_m, \tag{2.44}
$$

where $\boldsymbol{\zeta}_i$, $\boldsymbol{\Upsilon}_i$, $\boldsymbol{\Psi}_i$, and $\boldsymbol{\eta}_i$ are defined by expressions identical to those in Theorem 2.4 except that p in those expressions is replaced by p_i.

It follows from Theorems 2.7 and 2.8 that the covariance structure of $\text{PAC}(p_1, \ldots, p_n)$ variables can be parameterized by either the marginal variances, non-trivial lag-one marginal correlations, and nontrivial intervenor-adjusted partial correlations

$$\{\rho_{ij\cdot\{j+1:i-1\}} : i = 3, \ldots, n; \ j = i - p_i, \ldots, i - 2\};$$

or the nonzero autoregressive coefficients and innovation variances

$$\{\delta_i : i = 1, \ldots, n\} \text{ and } \{\phi_{ij} : i = 1, \ldots, n; \ j = i - p_i, \ldots, i - 1\};$$

or the marginal covariance matrix elements

$$\{\sigma_{ij} : i = 1, \ldots, n; \ j = i - p_i, \ldots, i\}.$$

The number of parameters in any of these parameterizations is given by

$$n + \sum_{i=1}^{n} p_i = n + \sum_{i=2}^{n} p_i.$$

Results on the determinant of a $\text{PAC}(p_1, \ldots, p_n)$ covariance matrix, which are analogous to those given by Theorem 2.5 for a constant-order $\text{PAC}(p)$ covariance matrix, may be established in a very similar manner. We state these results, without proof, as the following theorem.

Theorem 2.9. *Let* $Y_1, \ldots, Y_n$ *be* $PAC(p_1, \ldots, p_n)$ *random variables with positive definite covariance matrix* Σ.

(a) *Let* $\mathbf{T}$ *and* $\mathbf{D} = diag(\delta_1, \ldots, \delta_n)$ *be the unit lower triangular and diagonal matrices, respectively, of the modified Cholesky decomposition of* Σ^{-1}. *Then*

$$|\Sigma| = \prod_{i=1}^{n} \delta_i.$$

(b) *Let* $R^2_{i\cdot\{i-p_i:i-1\}}$ *denote the multiple correlation coefficient between* Y_i *and* $Y_{i-p_i}, \ldots, Y_{i-1}$, *where* $i > 1$. *Then*

$$|\Sigma| = \prod_{i=1}^{n} \sigma_{ii} \prod_{i=2}^{n} (1 - R^2_{i\cdot\{i-p_i:i-1\}})$$

where we define $R^2_{i\cdot\{i-p_i:i-1\}} = 0$ *if* $p_i = 0$.

(c) *Let* $\rho_{ij\cdot B}$ *denote the partial correlation coefficient between* Y_i *and* Y_j *adjusted for* $\{Y_k : k \in B\}$. *Then*

$$|\Sigma| = \prod_{i=1}^{n} \sigma_{ii} \prod_{i=2}^{n} \prod_{j=1}^{p_i} (1 - \rho^2_{i,i-j\cdot\{i-j+1:i-1\}})$$

where we define

$$\prod_{j=1}^{p_i}(1 - \rho^2_{i,i-j\cdot\{i-j+1:i-1\}}) = 1 \quad \text{if } p_i = 0$$

and

$$\rho^2_{i,i-j\cdot\{i-j+1:i-1\}}) = \rho^2_{i,i-1} \quad \text{if } j = 1.$$

(d) *Let* $\boldsymbol{\Sigma}_{k:m}$ *denote the submatrix consisting of elements in rows* $k, k+1, \ldots, m$ *and columns* $k, k+1, \ldots, m$ *of* $\boldsymbol{\Sigma}$. *Then*

$$|\boldsymbol{\Sigma}| = \sigma_{11}\prod_{i=2}^{n}\frac{|\boldsymbol{\Sigma}_{i-p_i:i}|}{|\boldsymbol{\Sigma}_{i-p_i:i-1}|}$$

where we define

$$|\boldsymbol{\Sigma}_{i-p_i:i-1}| = 1 \quad \text{if } p_i = 0.$$

2.7 Other conditional independence models

Antedependence is a form of conditional independence for random variables ordered according to a one-dimensional index. As such, it is related to several other more well-known models of conditional independence. For example, Dempster's (1972) *covariance selection* model is a conditional independence model for jointly normal random variables which sets arbitrary elements of their precision matrix equal to 0. Thus, a pth-order antedependence model for jointly normal variables is a special type of covariance selection model, for which, as detailed by Theorem 2.2, the set of null precision matrix elements consists of those lying on off-diagonals $p+1, \ldots, n-1$. The pth-order antedependence model for jointly normal variables is also a special case of a Gaussian Markov random field (Rue and Held, 2005).

Graphical models (Whittaker, 1990; Lauritzen, 1996) are extensions of covariance selection models and Gaussian Markov random fields which do not require normality of the variables nor existence of the covariance matrix; they merely specify that some variables are independent, conditional on other variables. Because graphical models do not require that the conditional independence be related to any ordering among the variables, they are more general than antedependence models. Graphical models are given their name because the conditional independence structure of the variables may be characterized by a graph, with the consequence that elements of graph theory may be brought to bear on various probabilistic and inferential problems for these models. We aim here not to discuss graph theory in any detail, but just enough to be able to describe conditional independence graphs for antedependence models. Accordingly, a *graph* is a mathematical entity consisting of two sets: a set of *vertices* $\{1, 2, \ldots, n\}$ and a set of *edges* consisting of pairs of distinct vertices. An edge (i, j) in the edge set is said to be *undirected* if the edge set also contains (j, i); otherwise the edge is *directed*. The *directed conditional independence graph* of a set of index-ordered random variables $Y_1, \ldots, Y_n$ is a set of vertices and a set of directed edges $\{(i, j)\}$ satisfying the properties that (a) $i < j$ and (b) (i, j) is not in the edge set if and only if Y_i and Y_j are conditionally independent given all variables preceding Y_j (excluding Y_i). The directed conditional independence graph can be displayed as a picture, in which each vertex is represented by a circle and each directed edge (i, j) is represented by a line segment with an arrow pointing to vertex j. The arrow of a directed edge is used to indicate an asymmetry in the interaction between the two corresponding variables, specifically that Y_i "causes" Y_j but not vice versa. Thus antedependence models, with their notion of conditional independence among time-ordered random variables, may be represented by directed conditional independence graphs.

Figure 2.1 displays directed conditional independence graphs for AD(1), AD(2), and AD(3) variables when $n = 10$. The graphs have a distinctive chained-link structure, which occurs because of the conditional independence among pairs of variables given only one, two, or three immediate predecessors.

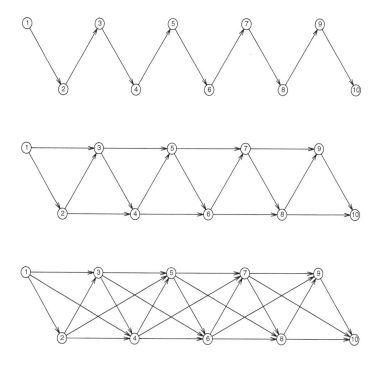

Figure 2.1 *Directed conditional independence graphs for AD(1) (top), AD(2) (middle), and AD(3) (bottom) models for 10 variables.*

CHAPTER 3

Structured Antedependence Models

We noted in the previous chapter that the unstructured antedependence (AD) covariance model is more parsimonious than the general multivariate model (provided that the order of the AD model is less than $n - 1$) while also being more flexible than stationary autoregressive models, and that consequently an unstructured AD model may be useful for longitudinal data exhibiting heterogeneous variances and nonstationary serial correlation. For many longitudinal data sets, however, an AD covariance model that is more parsimonious than an unstructured AD model, but possibly not as parsimonious as a stationary autoregressive model, may be even more useful. For example, if variances increase over time, as is common in growth studies, or if measurements equidistant in time become more highly correlated as the study progresses, then a model that incorporates these structural forms of nonstationarity is likely to be more useful. In this chapter we consider such models, which we call structured antedependence (SAD) models.

Several of the SAD models to be presented here were initially developed for use with time series data, then subsequently borrowed for use with longitudinal data. However, it is worth noting that there are some very important differences between the time series and longitudinal settings. In a typical time series setting, a response variable for a single "subject" is observed on a large number of equally spaced occasions; for example, we may observe the daily closing price of a single stock for a year, or the annual gross domestic product of a single nation for 50 years. In contrast, a longitudinal study consists of multiple subjects typically observed on relatively few occasions, which frequently are not equally spaced. Also, the notions of an "infinite past" or "infinite future," which are theoretically important for time series, often make little sense for longitudinal data and, in any case, are not needed. As a consequence, the SAD models we borrow from time series analysis will be presented in a form that allows for an abrupt start-up at time $t = 1$ rather than a form that requires an infinite past, and the asymptotic regime to which we will eventually appeal will allow the number of subjects to increase but take the number of measurement times to be fixed.

Recall that in the previous chapter we described four equivalent characterizations or formulations of the unstructured AD(p) covariance model: intervenor-adjusted, precision matrix, autoregressive, and marginal. A SAD(p) model may be obtained by imposing structure upon the parameters of any of these formulations, according to whichever seems most appropriate, useful, or interpretable in a particular situation. As we will see, classical autoregressive time series models are most easily seen as SAD models for which structure is imposed on the autoregressive formulation. The same is true for several other well-known SAD models, in part because the nonzero autoregressive coefficients of AD models are unconstrained. Nevertheless, the consequent marginal covariance structure of these models is also important and will be described to the extent reasonably possible. For a few models we will also describe the structure of the other two formulations.

Throughout, we will take the innovations in the autoregressively specified SAD models to be independent. Although the marginal covariance structure is the same in each case if we weaken this requirement to innovations that are merely uncorrelated, the response variables in the models with this weaker requirement are merely antecorrelated, not necessarily antedependent.

For the sake of clarity and economy of notation, initially we will describe each SAD model in the context of only a single subject, suppressing the subscript i indexing subjects. Hence we shall write the observations as $Y_1, \ldots, Y_n$, the measurement times as $t_1, \ldots, t_n$, and the marginal variances and correlations as $\{\sigma_{ii}\}$ and $\{\rho_{ij}\}$, respectively. Furthermore, for several of the models to be described, it will be assumed initially that the times of measurement are equally spaced, in which case we put $t_i = i$. In each such case we will subsequently describe generalizations of the model for use when measurement times are unequally spaced.

3.1 Stationary autoregressive models

Assume initially that the measurement times are equally spaced, with $t_i = i$. Then, the first-order autoregressive [AR(1)] model can be given by the following autoregressive formulation:

$$
\begin{aligned}
Y_1 - \mu_1 &= \epsilon_1, & & (3.1) \\
Y_i - \mu_i &= \phi(Y_{i-1} - \mu_{i-1}) + \epsilon_i, & i = 2, \ldots, n,
\end{aligned}
$$

where $-1 < \phi < 1$ and the innovations $\{\epsilon_i\}$ are independent zero-mean random variables with variances $\text{var}(\epsilon_1) = \delta$ and $\text{var}(\epsilon_i) = \delta(1 - \phi^2)$ for

$i = 2, \ldots, n$. There are thus two unknown parameters in the covariance structure of the model: ϕ and δ. Note that (3.1) starts at time $i = 1$ and ends at time
$i = n$; there is no need to appeal to notions of an infinite past or future, as is
common for time series.

The marginal covariance structure corresponding to (3.1) is easily shown to be
given by

$$\sigma_{ii} = \delta, \qquad \rho_{ij} = \phi^{|i-j|}. \tag{3.2}$$

Thus, for the AR(1) model the marginal variances are constant and the marginal
correlations are a function of only the elapsed time between measurements, and
consequently the model is stationary. Furthermore, the correlations decrease
(in modulus) exponentially as elapsed time increases.

A precision matrix formulation of the AR(1) covariance structure is also easily
obtained, using for example the AD(1) inverse formula (2.40) and the relationship between the elements of the precision matrix and the conditional variances
and correlations (where the conditioning is on all other observations). We find
that

$$\sigma_{ii|rest} = \begin{cases} \delta(1 - \phi^2) & \text{for } i = 1, n, \\ \delta(1 - \phi^2)/(1 + \phi^2) & \text{for } i = 2, \ldots, n-1, \end{cases}$$

$$\rho_{i,i-1|rest} = \begin{cases} \phi(1 + \phi^2)^{-1/2} & \text{for } i = 2, n, \\ \phi(1 + \phi^2)^{-1} & \text{for } i = 3, \ldots, n-1, \end{cases}$$

and of course $\rho_{i,i-j|rest} = 0$ whenever $j > 1$.

The AR(1) model can be generalized to a pth order autoregressive [AR(p)]
model, specified incompletely as

$$Y_i - \mu_i = \sum_{j=1}^{p} \phi_j (Y_{i-j} - \mu_{i-j}) + \epsilon_i, \quad i = p+1, \ldots, n. \tag{3.3}$$

Here p is an integer greater than or equal to one, the ϕ_j's satisfy certain constraints, and the ϵ_i's are independent random variables with zero means. To
complete the formulation, we take $Y_i - \mu_i$, for $i \leq p$, to be given by an autoregressive equation of order $i - 1$, with autoregressive parameters and innovation
variances chosen to achieve stationarity, i.e., to yield equal variances among re-
sponses and correlations between Y_i and Y_j that depend only on $|i - j|$. This
is always possible; for example, for the AR(2) model, we may take

$$\begin{aligned} Y_1 - \mu_1 &= \epsilon_1, \\ Y_2 - \mu_2 &= \phi_0(Y_1 - \mu_1) + \epsilon_2, \\ Y_i - \mu_i &= \phi_1(Y_{i-1} - \mu_{i-1}) + \phi_2(Y_{i-2} - \mu_{i-2}) + \epsilon_i, \qquad i = 3, \ldots, n, \end{aligned}$$

where $\phi_0 = \phi_1/(1 - \phi_2)$,

$$
\begin{aligned}
\text{var}(\epsilon_1) &= \delta, \\
\text{var}(\epsilon_2) &= \delta(1 - \phi_0^2), \\
\text{var}(\epsilon_i) &= \delta(1 - \phi_1^2 - \phi_2^2 - 2\phi_1\phi_2\phi_0), \qquad i = 3, \ldots, n.
\end{aligned}
$$

This yields marginal variances $\sigma_{ii} = \delta$ for all i, and marginal correlations given by

$$
\begin{aligned}
\rho_{i,i-1} &= \phi_1/(1 - \phi_2), \\
\rho_{i,i-2} &= (\phi_2 - \phi_2^2 + \phi_1^2)/(1 - \phi_2), \\
\rho_{i,i-j} &= \phi_1\rho_{i,i-j+1} + \phi_2\rho_{i,i-j+2}, \\
&\qquad i = 4, \ldots, n; \; j = 3, \ldots, i - 1.
\end{aligned}
$$

Note that the relationship between the parameters of the autoregressive formulation and the marginal variances and correlations is considerably more complicated for an AR(2) model than for an AR(1) model. This relationship becomes still more complicated as p increases, but with sufficient effort it can always be determined explicitly.

Like unstructured antedependence models, stationary autoregressive models are nested, i.e., AR(1) $\subset$ AR(2) $\subset \cdots \subset$ AR($n - 1$). Furthermore, upon comparison of the AD(p) model given by (2.21) (plus normality) with the AR(p) model just given, it is evident that the latter model is a special case of the former. More precisely, an AR(p) model is a special case of the unstructured AD(p) model in which, using notation from (2.21):

(a) $\phi_{i,i-j} = \phi_j$ for $i = p + 1, \ldots, n$ and $j = 1, \ldots, p$;
(b) the p roots of the AR(p) characteristic equation

$$
1 - \phi_1 x - \phi_2 x^2 - \cdots - \phi_p x^p = 0
$$

all exceed unity in modulus;
(c) $\delta_{p+1} = \delta_{p+2} = \cdots = \delta_n > 0$;
(d) the "start-up" parameter values $\{\phi_{i,i-j}: i = 2, \ldots, p; j = 1, \ldots, i - 1\}$ and $\delta_1, \delta_2, \ldots, \delta_p$ are chosen so that marginal variances are equal and marginal correlations depend only on $|i - j|$.

The number of parameters in the covariance structure of an AR(p) model is $p + 1$. Recalling from Chapter 2 that the number of covariance structure parameters in an unstructured AD(p) model is $(2n - p)(p + 1)/2$, we see that the number of covariance structure parameters of the AR(p) model, in contrast to that of the AD(p) model, does not increase with the number of measurement times.

The stationary autoregressive models presented so far are sometimes called discrete-time AR(p) models, owing to their assumption of equal spacing between observations. If the measurement times are not equally spaced the model may be applied to the time-ordered observations anyway, but in this case the covariance structure is no longer stationary (since the marginal correlations are no longer a function of elapsed time). Indeed, for such an application the nature of the model's nonstationarity is dictated by the spacings between successive measurements, which usually makes little practical sense. To retain the functional dependence of marginal correlations on elapsed time when measurement times are irregularly spaced, we must consider stationary continuous-time AR(p) processes. Such a process is defined as a stationary solution to a stochastic differential equation and is therefore rather more complicated than its discrete-time counterpart. In particular, the marginal variances and correlations generally are complicated functions of the parameters of the stochastic differential equation. However, the first-order continuous-time autoregressive process is an exception to this; its marginal variances and correlations are given by the very simple expressions

$$\sigma_{ii} = \sigma^2, \quad \rho_{ij} = \phi^{|t_i - t_j|}, \quad i, j = 1, \ldots, n, \tag{3.4}$$

where $0 \leq \phi < 1$. Note that the marginal variances are constant and the correlations decrease exponentially as a function of elapsed time. Observe also that the correlations in (3.4) and those in (3.2) coincide, provided that the measurements are equally spaced and ϕ is nonnegative. Unfortunately, this relationship does not extend to higher-order models; that is, the marginal covariance structure of equally-spaced observations of a second-order (or higher) continuous-time AR process does not coincide with that of the discrete-time AR process of the same order. Further details on continuous-time autoregressive processes are beyond the scope of this book, but the interested reader may consult Jones (1981), Jones and Ackerson (1990), and Belcher, Hampton, and Tunnicliffe Wilson (1994).

3.2 Heterogeneous autoregressive models

For the stationary autoregressive models considered in the previous section, the longitudinal observations marginally have equal variances and correlations that are functions of elapsed time. Heterogeneous extensions of AR(p) models, denoted by ARH(p), retain the same marginal correlation structure but allow the marginal variances to depend on time. Although it is possible to specify such a model by autoregressive equations, the autoregressive coefficients in these equations must necessarily be rather complicated functions of the innovation variances in order to preserve the AR(p) marginal correlation structure. It is therefore much simpler to specify an ARH(p) by its arbitrary positive marginal variances $\{\sigma_{ii}\}$ and its AR(p) marginal correlations.

Clearly, $\text{ARH}(p) \subset \text{AD}(p)$. If the functional dependence of marginal variances on time is completely general, then the number of parameters in the covariance structure is $n + p + 1$, which increases with the number of measurement times. More parsimonious heterogeneous models result from taking the variance to be a function of time, for example a log-linear function

$$\log \sigma_{ii} = \mathbf{v}_i^T \psi \quad \text{or} \quad \sigma_{ii} = \exp(\mathbf{v}_i^T \psi), \quad i = 1, \ldots, n, \qquad (3.5)$$

where $\mathbf{v}_i$ is a vector of functions of time t_i and ψ is a vector of unknown parameters. Verbyla (1993) considers modeling variances in this fashion when observations are independent, but there is no reason why the same model could not also be used for variances when observations are dependent. The advantage of a log-linear model for the variances, relative to a polynomial or some other linear model, is that its parameters are unconstrained.

3.3 Integrated autoregressive models

Again, assume initially that the measurement times are equally spaced. An integrated autoregressive model of orders p and d, or $\text{ARI}(p, d)$ model, generalizes a stationary autoregressive model by postulating that the dth-order differences among consecutive measurements, rather than the measurements themselves, follow an $\text{AR}(p)$ model. The simplest case is the $\text{ARI}(0,1)$, or random walk, model given by

$$(Y_i - \mu_i) - (Y_{i-1} - \mu_{i-1}) = \epsilon_i, \quad i = 2, \ldots, n, \qquad (3.6)$$

where $Y_1 - \mu_1 = \epsilon_1$ and $\{\epsilon_i : i = 1, \ldots, n\}$ are independent zero-mean random variables with common variance δ. For this process,

$$\sigma_{ii} = i\delta, \quad i = 1, \ldots, n,$$

and

$$\rho_{ij} = \sqrt{j/i}, \quad i \geq j = 1, \ldots, n.$$

Thus, the marginal variances increase (linearly) over time and the serial correlation decays with increasing lag (holding j fixed and increasing i) at an inverse square root rate. Furthermore, the marginal correlations between equidistant measurements are monotonic, increasing in a particular nonlinear fashion and approaching unity as time progresses. There is only one unknown parameter in the covariance structure, namely δ, of which the marginal variance is a linear function.

Observe that (3.6) may be written as

$$Y_i - \mu_i = Y_{i-1} - \mu_{i-1} + \epsilon_i, \quad i = 2, \ldots, n,$$

by which it follows that the $\text{ARI}(0,1)$ model is a special case of the $\text{AD}(1)$ model.

Although the marginal variances and correlations of the ARI(0,1) model vary quite strongly with time, the partial variances and partial correlations do not. Specifically, the partial variances and lag-one partial correlations are given by

$$\sigma_{ii|rest} = \begin{cases} \delta/2 & \text{for } i = 1, \ldots, n-1 \\ \delta & \text{for } i = n, \end{cases}$$

$$\rho_{i,i-1|rest} = \begin{cases} 1/2 & \text{for } i = 2, \ldots, n-1, \\ 1/\sqrt{2} & \text{for } i = n, \end{cases}$$

and of course $\rho_{i,i-j|rest} = 0$ whenever $j > 1$.

Another special case of an integrated autoregressive model is the ARI(1,1) model, given by

$$(Y_i - \mu_i) - (Y_{i-1} - \mu_{i-1}) = \phi[(Y_{i-1} - \mu_{i-1}) - (Y_{i-2} - \mu_{i-2})] + \epsilon_i,$$
$$i = 3, \ldots, n, \tag{3.7}$$

where $Y_1 - \mu_1 = \epsilon_1$, $Y_2 - \mu_2 = Y_1 - \mu_1 + \epsilon_2$, and the ϵ_i's are defined as for the ARI(0,1). Observe that (3.7) may be rewritten as

$$Y_i - \mu_i = (1 + \phi)(Y_{i-1} - \mu_{i-1}) - \phi(Y_{i-2} - \mu_{i-2}) + \epsilon_i, \quad i = 3, \ldots, n,$$

whereupon it is evident that this model is a special case of the AD(2) model. More generally, it can be shown that an ARI(p, d) model is a particular structured AD($p+d$) model having $p+1$ unknown covariance structure parameters.

If measurement times are unequally spaced, we may use continuous-time analogues of ARI models. The only case we will mention is the Wiener process, which is a continuous-time analogue of the random walk model. The marginal covariance function of a Wiener process is

$$\text{cov}(Y_i, Y_j) = \delta \min(t_i, t_j), \quad i, j = 1, \ldots, n,$$

which yields variances and correlations that coincide with those of the discrete-time random walk when the data are equally spaced.

3.4 Integrated antedependence models

Let $\{\epsilon_i : i = 1, \ldots, n\}$ be independent zero-mean random variables, and define

$$Y_i - \mu_i = \sum_{l=1}^{i} \epsilon_l, \quad i = 1, \ldots, n. \tag{3.8}$$

It follows from results shown in Section 2.1 that $Y_1, \ldots, Y_n$ are AD(1) variables, a fact which is also evident upon reexpressing (3.8) in autoregressive form, i.e.,

$$Y_i - \mu_i = Y_{i-1} - \mu_{i-1} + \epsilon_i, \quad i = 1, \ldots, n,$$

(where we put $Y_0 \equiv 0$ and $\mu_0 \equiv 0$). The marginal covariance structure of this AD(1) model is obtained as follows. Suppose that the variances of the ϵ_i's exist but are not necessarily constant across time. Denote these variances by δ_l $(l = 1, \ldots, n)$. Then

$$\mathrm{var}(Y_i) = \sum_{l=1}^{i} \delta_l,$$

$$\mathrm{cov}(Y_i, Y_j) = \sum_{l=1}^{\min(i,j)} \delta_l,$$

and thus for $i < j$,

$$\mathrm{corr}(Y_i, Y_j) = \frac{\sum_{l=1}^{i} \delta_l}{(\sum_{l=1}^{i} \delta_l \sum_{l=1}^{j} \delta_l)^{1/2}} = \left(1 + \sum_{l=i+1}^{j} \delta_l \bigg/ \sum_{l=1}^{i} \delta_l\right)^{-1/2}.$$

We see that: (a) for fixed $i < j$, the correlation between Y_i and Y_j is a decreasing function of j; (b) for fixed $j > i$, the correlation is an increasing function of i; (c) as both i and j are incremented equally the correlation may either decrease or increase. Thus, marginally the variances increase monotonically (but otherwise arbitrarily) over time and the correlations decrease monotically with lag (serial correlation) in a manner determined by the innovation variances. Note that the number of unknown parameters in the model's covariance structure is n, and that the marginal variances are linear functions of these parameters.

The reader has probably recognized that this model is an extension of the classical random walk given by (3.6). It is not an ARI model, however, so we need a new name for it and for models that are even more general. We define an *integrated antedependence model of orders p and d*, or ADI(p, d) model, as a model for which the dth-order differences among consecutive measurements follow an AD(p) model. By this definition, the model given by (3.8) is an ADI$(0,1)$ model. We obtain an ADI$(p, 1)$ model if the innovations in (3.8) are AD(p) or, in other words, if we form cumulative sums of AD(p) variables (either structured or unstructured).

It can be shown that an ADI(p, d) model, like an ARI(p, d) model, is a structured form of AD$(p + d)$ model. In fact, we have that

$$\mathrm{ARI}(p, d) \subset \mathrm{ADI}(p, d) \subset \mathrm{AD}(p + d).$$

The number of unknown covariance structure parameters in an ADI(p, d) model is $(p+1)(2n - p)/2$, which increases with the number of measurement times.

Since the terms in (3.8) or in an autoregressive formulation of any ADI(p, d) model are AD(p) but do not necessarily satisfy anything stronger, applications of the model to longitudinal data do not require that measurement times be equally spaced.

3.5 Unconstrained linear models

Pourahmadi (1999) introduced a family of SAD$(n-1)$ models in which the logarithms of the innovation variances and the same-lag autoregressive coefficients of the autoregressive formulation of the general multivariate model for n observations (i.e., the non-trivial elements of the matrices $\mathbf{T}$ and $\mathbf{D}$ of the modified Cholesky decomposition of $\boldsymbol{\Sigma}^{-1}$) are modeled as parsimonious linear functions, such as low-order polynomials, of either time of measurement or time between measurements (lag). For example, an SAD$(n-1)$ model for which the log-innovation variances and autoregressive coefficients follow polynomial models of time of orders m_1 and m_2, respectively, is as follows:

$$\log \delta_i = \sum_{l=1}^{m_1} \psi_l t_i^{l-1}, \quad i = 1, \ldots, n, \tag{3.9}$$

$$\phi_{i,i-j} = \sum_{l=1}^{m_2} \theta_l t_i^{l-1}, \tag{3.10}$$

$$i = 2, \ldots, n; \, j = 1, \ldots, i-1.$$

Another such model is one for which the log-innovation variances follow (3.9) but the autoregressive coefficients are given by a polynomial function of lag, i.e.,

$$\phi_{i,i-j} = \sum_{l=1}^{m_2} \theta_l |t_i - t_j|^{l-1}, \tag{3.11}$$

$$i = 2, \ldots, n; \, j = 1, \ldots, i-1;$$

in fact, this is the specific model used by Pourahmadi (1999). Either of these linear SAD$(n-1)$ models can be specialized to an SAD(p) model, with $p < n-1$, by requiring that (3.10) or (3.11) hold for $i = 2, \ldots, n$ and $j = 1, \ldots, \min(i-1, p)$ only, and setting autoregressive coefficients corresponding to lags greater than p equal to zero. The nonnull parameters, $\{\psi_l\}$ and $\{\theta_l\}$, in any such model are unconstrained, which gives these models a distinct advantage over other autoregressively specified SAD models with respect to parameter estimation. We will develop this point further in Chapter 5.

The marginal covariance structure corresponding to (3.9) and (3.10) depends, of course, on the values of the ψ_l's and θ_l's, but the precise nature of this dependence is not transparent. However, the marginal covariance structures for some low-order polynomial SAD(1) models of this type are displayed in Figure 3.1, which is the first of several figures we call "diagonal cross-section plots" presented in this chapter. Each such plot consists of two subplots, one directly on top of the other. In the top subplot, marginal variances (σ_{ii}) are plotted against a shifted time index ($t_i - 0.5$) and connected by line segments. In the bottom subplot, each set of same-lag marginal correlations ($\rho_{i,i-j}$ for fixed j) is plotted against a shifted time index and then connected by line segments. Thus, connected points in the subplots correspond either to elements on the main diagonal or to elements on a common subdiagonal of the covariance matrix. The time index for each subdiagonal is shifted so that the profile of that subdiagonal will comport with the perspective of an observer at the lower left corner of the matrix looking toward the opposite corner.

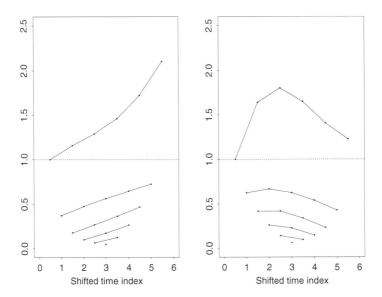

Figure 3.1 *Diagonal cross-section plots of marginal variances and correlations of an AD(1) polynomial model given by (3.9) and (3.10), with $m_1 = 1$, $m_2 = 2$, and $\psi_1 = 0$. Left panel: $\theta_1 = 0.2$, $\theta_2 = 0.1$; right panel: $\theta_1 = 1.0$, $\theta_2 = -0.1$.*

For the two specific models shown in Figure 3.1 we put $t_i = i = 1, 2, \ldots, 6$. In both models the log-innovation variances are taken to be constant ($m_1 = 1$), and that constant is $\psi_1 = 0$ (implying that the innovation variances themselves are all equal to 1.0), while the autoregressive coefficients are taken to be a linear function ($m_2 = 2$) of time. Moreover, in the first model (left panel) $\theta_1 = 0.2$ and $\theta_2 = 0.1$, so that the lag-one autoregressive coefficients are increasing (linearly) from 0.4 to 0.8 from time $t_2 = 2$ to time $t_6 = 6$. In the second model (right panel), $\theta_1 = 1.0$ and $\theta_2 = -0.1$, so that the lag-one autoregressive coefficients are decreasing (linearly) from 0.8 to 0.4 over the same time period. The plots indicate that lag-one autoregressive coefficients that increase over time result in marginal variances and correlations that do likewise, whereas lag-one autoregressive coefficients that decrease over time result in marginal correlations and variances that may increase early on but eventually decrease.

Figure 3.2 displays diagonal cross-section plots of SAD(2) models with the same measurement times and same constant log-innovation variances ($\psi_1 = 0$) as the previous two models, but with autoregressive coefficients that are functions of lag rather than time (and hence are constant over time for the same lag). In the first (left panel) of these models, $\theta_1 = 0.9$ and $\theta_2 = -0.1$, so that the autoregressive coefficients are decreasing (linearly) from 0.7 to 0.3 with increasing lag. In the second model (right panel), $\theta_1 = 0.1$ and $\theta_2 = 0.1$, so that the autoregressive coefficients are increasing (linearly) from 0.3 to 0.7 with increasing lag. In comparison to the two previous models, for which the autoregressive coefficients for lags two and higher were equal to zero, the marginal correlations for the present two models are, not surprisingly, more persistent (stronger at large lags). In fact, for the model in which the autoregressive coefficients increase with lag, the marginal correlations can be higher between an observation and a distant predecessor than between the observation and a more proximate predecessor.

3.6 Power law models

Zimmerman and Núñez-Antón (1997) introduced several families of SAD(p) models in which either the marginal correlations up to lag p, the autoregressive coefficients up to order p, or the partial correlations of variables lagged at most p variables apart, adjusted for all other variables — depending on whether additional structure is imposed upon the marginal, autoregressive, or precision matrix formulations of the unstructured AD(p) model — are given by a Box-Cox power function of time. For example, the marginally specified SAD(1) family of these models is given by

$$\begin{aligned} \sigma_{ii} &= \sigma^2 f(t_i; \psi), & i &= 1, \ldots, n, \\ \rho_{i,i-1} &= \rho^{|g(t_i;\lambda)-g(t_{i-1};\lambda)|}, & i &= 2, \ldots, n, \end{aligned} \qquad (3.12)$$

where $\sigma^2 > 0$; $f(\cdot)$ is a specified positive-valued function of time, such as the exponential function given by (3.5); ψ is a vector of relatively few parameters; $0 \leq \rho < 1$; and $g(\cdot)$ is of Box-Cox power form, i.e.,

$$g(t; \lambda) = \begin{cases} (t^\lambda - 1)/\lambda & \text{if } \lambda \neq 0 \\ \log t & \text{if } \lambda = 0. \end{cases} \qquad (3.13)$$

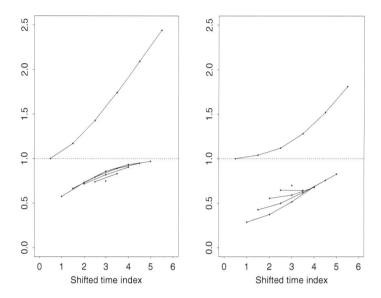

Figure 3.2 *Diagonal cross-section plots of marginal variances and correlations of an AD(1) polynomial model given by (3.9) and (3.11), with $m_1 = 1$, $m_2 = 2$, and $\psi_1 = 0$. Left panel: $\theta_1 = 0.9$, $\theta_2 = -0.1$; right panel: $\theta_1 = 0.1$, $\theta_2 = 0.1$. Marginal variances are expressed in units of log, base ten, plus one.*

Correlations corresponding to lags beyond the first, i.e., $\{\rho_{i,i-j} : i = 3, \ldots, n;\ j = 1, \ldots, i-2\}$, are taken to equal the appropriate products of the lag-one correlations, in accordance with (2.37). Note that this model does not require measurement times to be equally spaced. However, if measurement times *are* equally spaced then the lag-one correlations (and, for that matter, all same-lag correlations) are a monotone function of t: they increase if $\lambda < 1$ and decrease if $\lambda > 1$. Illustrating this is Figure 3.3, a diagonal cross-section plot of the marginal covariance structure for cases of this model with constant variance $\sigma^2 = 1.5$ and (ρ, λ) equal to either $(0.6, 0.5)$ or $(0.8, 1.5)$. When $\lambda = 1$, the same-lag correlations neither increase nor decrease, but coincide with those of the continuous-time AR(1) model. From another point of view, the power law (3.13) effects a nonlinear deformation upon the time axis, such that correlations between measurements equidistant in the deformed scale are constant: earlier portions of the time axis are "stretched out" relative to later portions if $\lambda < 1$, or "condensed" relative to later portions if $\lambda > 1$.

A natural extension of (3.12) and (3.13) to a marginally specified SAD(p)

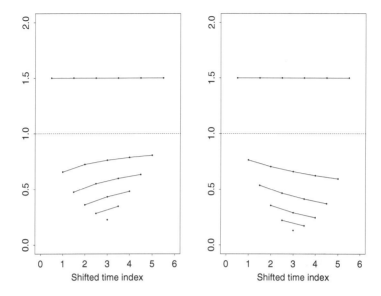

Figure 3.3 *Diagonal cross-section plots of marginal variances and correlations of the marginally formulated AD(1) power model given by (3.12) and (3.13), with constant variance. Left panel: $\rho = 0.6$, $\lambda = 0.5$; right panel: $\rho = 0.8$, $\lambda = 1.5$.*

model ($1 \le p \le n - 1$) is given by

$$
\begin{aligned}
\sigma_{ii} &= \sigma^2 f(t_i; \boldsymbol{\psi}), \quad i = 1, \ldots, n, \\
\rho_{i,i-j} &= \rho_j^{|g(t_i;\lambda_j)-g(t_{i-j};\lambda_j)|}, \\
&\quad i = j+1, \ldots, n; \ j = 1, \ldots, p,
\end{aligned}
\tag{3.14}
$$

where $0 \le \rho_j < 1$ for all j and the ρ_j's are such that $\boldsymbol{\Sigma}$ is positive definite, and all other quantities are defined as in (3.12) and (3.13). Correlations not included in (3.14) are assumed to satisfy the requirements of an AD(p) model; thus, they can be determined in terms of the parameters of (3.14) via the iterative application of equation (2.32). Again, this model does not require measurement times to be equally spaced, but when they are, it prescribes that the lag-j correlations are monotone increasing if $\lambda_j < 1$, monotone decreasing if $\lambda_j > 1$, or constant if $\lambda_j = 1$, for $j = 1, \ldots, p$.

The analogue of model (3.14) for the intervenor-adjusted AD(p) formulation adopts the same parametric model for the variances but specifies that the intervenor-adjusted partial correlations satisfy

$$\rho_{i,i-j \cdot \{i-j+1:i-1\}} = \rho_j^{|g(t_i;\lambda_j)-g(t_{i-j};\lambda_j)|}, \tag{3.15}$$
$$i = j+1,\ldots,n; \ j = 1,\ldots,p.$$

Here again $0 \le \rho_j < 1$ for all j and the ρ_j's are such that $\mathbf{\Sigma}$ is positive definite, and all other quantities are defined as in (3.12) and (3.13). Note, of course, that the first-order case of this model is merely the same as the first-order marginally specified model (3.12).

An autoregressively specified SAD(p) counterpart to (3.14) is obtained by imposing the following structure on the innovation variances and autoregressive coefficients of the autoregressive AD(p) formulation:

$$\delta_i = \delta f(t_i;\boldsymbol{\psi}), \quad i = p+1,\ldots,n,$$
$$\phi_{i,i-j} = \phi_j^{|g(t_i;\lambda_j)-g(t_{i-j};\lambda_j)|}, \tag{3.16}$$
$$i = p+1,\ldots,n; \ j = 1,\ldots,p,$$

where $\delta, \phi_1,\ldots,\phi_p$ are positive and all other quantities are defined as in (3.12) and (3.13). Note that this model does not impose any structure on the innovation variances and autoregressive coefficients corresponding to the first p measurement times. Although structure could be imposed on these "start-up" quantities as well, it has been our experience that keeping them unstructured usually results in a better-fitting model than requiring them to adhere to the structural relationship that applies to the innovation variances and autoregressive coefficients at later times.

A precision matrix specification of an SAD(p) model analogous to those given by (3.14) through (3.16) takes the elements of $\mathbf{\Sigma}^{-1}$ to satisfy

$$\sigma^{ii} = \nu f(t_i;\boldsymbol{\psi}), \quad i = 1,\ldots,n,$$
$$\pi_{i,i-j} \equiv -\frac{\sigma^{i,i-j}}{\sqrt{\sigma^{ii}\sigma^{i-j,i-j}}} = \pi_j^{|g(t_i;\lambda_j)-g(t_{i-j};\lambda_j)|}, \tag{3.17}$$
$$i = j+1,\ldots,n; \ j = 1,\ldots,p,$$

where we recall that $\sigma^{i,i-j}$ is the $(i,i-j)$th element of $\mathbf{\Sigma}^{-1}$; $\nu > 0$; $\pi_1,\ldots,\pi_p$ are positive and such that $\mathbf{\Sigma}^{-1}$ is positive definite; and all other quantities are again defined as in (3.12) and (3.13). Recall from Section 2.3.1 that $\pi_{i,i-j}$ is the partial correlation coefficient between Y_i and Y_{i-j} adjusted for the rest of the measurements, and σ^{ii} is the reciprocal of the partial variance of Y_i adjusted for the rest of the measurements.

If measurement times are equally spaced, then (3.15), (3.16), and (3.17) prescribe monotonicity for the intervenor-adjusted partial correlations, autoregressive coefficients, or ordinary partial correlations, respectively, up to order p. Note, however, that the monotonicity of either set of quantities, by itself, neither implies nor is implied by the monotonicity of the marginal variances and/or the same-lag correlations. Indeed, the range of possible marginal covariance structures engendered by (3.15) through (3.17) is not obvious. Figures 3.4 and 3.5 are diagonal cross-section plots of marginal covariance structures corresponding to some first-order cases of (3.16) and (3.17); the first-order case of (3.15) coincides with (3.12), as noted previously. For all cases of both models, $t_i = i = 1, 2, \ldots, 6$, $f(t) \equiv 1$, and $\delta = \nu = 1$.

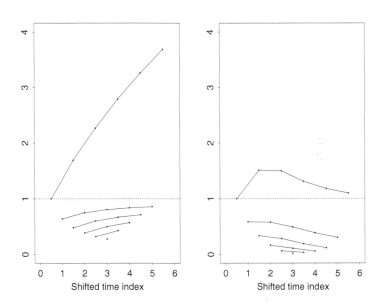

Figure 3.4 *Diagonal cross-section plots of marginal variances and correlations of the autoregressively-formulated AD(1) power model given by (3.16) and (3.13), with $\delta = 1$. Left panel: $\phi = 0.8$, $\lambda = 0.5$; right panel: $\phi = 0.8$, $\lambda = 2.0$.*

Figure 3.4 suggests that the effect that exponentiating the lag-one autoregressive coefficients by the power-transformed time scale has on the marginal covariance structure is similar to the effect of exponentiating the lag-one correlations themselves by the transformed time scale; that is, marginal same-lag correlations increase with elapsed time if $\lambda < 1$, and decrease if $\lambda > 1$. The same behavior of the marginal correlations is also seen when the lag-one partial correlations of the precision matrix formulation are exponentiated by the power-transformed time scale; however, the marginal variances do not vary monotonically for these models.

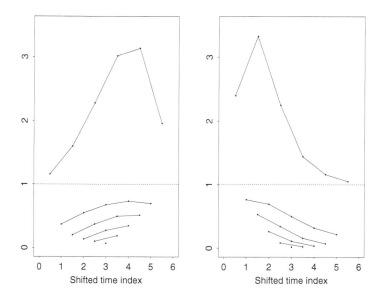

Figure 3.5 *Diagonal cross-section plots of marginal variances and correlations of the precision matrix formulation of AD(1) power model given by (3.17) and (3.13), with $\nu = 1$. Left panel: $\pi = 0.25$, $\lambda = 0.5$; right panel: $\pi = 0.75$, $\lambda = 2.0$.*

3.7 Variable-order SAD models

Each of the SAD models described in this chapter is of constant order. However, several of them may be extended easily to a variable-order SAD model. Such an extension necessarily is nonstationary, even if the constant-order SAD model from which it is extended is stationary.

As one example, consider an SAD(0,1,1,1,2,2) model for which the innovation variances and lag-one autoregressive coefficients are constant over time and the lag-two autoregressive coefficients are constant over the last two times (being equal to zero prior to that). That is,

$$
Y_i - \mu_i = \begin{cases} \epsilon_1 & \text{for } i = 1, \\ \phi_1(Y_{i-1} - \mu_{i-1}) + \epsilon_i & \text{for } i = 2, 3, 4, \\ \phi_1(Y_{i-1} - \mu_{i-1}) + \phi_2(Y_{i-2} - \mu_{i-2}) + \epsilon_i & \text{for } i = 5, 6, \end{cases}
$$
(3.18)

where the innovations are independent zero-mean random variables with constant variance δ. Figure 3.6 (left panel) is the diagonal cross-section plot of the marginal covariance structure for the case of (3.18) in which $\phi_1 = 0.8$,

$\phi_2 = 0.4$ and $\delta = 1.0$. The right panel of the same figure is the analogous plot for the SAD(0,1,2,2,1,1) model

$$Y_i - \mu_i = \begin{cases} \epsilon_1 & \text{for } i = 1, \\ \phi_1(Y_{i-1} - \mu_{i-1}) + \epsilon_i & \text{for } i = 2, 5, 6, \\ \phi_1(Y_{i-1} - \mu_{i-1}) + \phi_2(Y_{i-2} - \mu_{i-2}) + \epsilon_i & \text{for } i = 3, 4, \end{cases}$$
$$(3.19)$$

with the same parameter values as the SAD(0,1,1,1,2,2) model. The left plot shows that same-lag correlations for the first model increase monotonically, but unevenly, over time. The right plot reveals that marginal variances and correlations in the second model are not monotonic, as they increase for a while and then decrease.

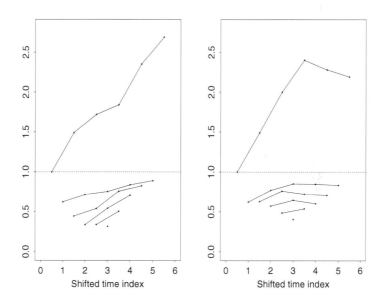

Figure 3.6 *Diagonal cross-section plots of marginal variances and correlations of variable-order SAD models. Left panel: SAD(0,1,1,1,2,2) model given by (3.18), with $\phi_1 = 0.8$, $\phi_2 = 0.4$ and $\delta = 1.0$; right panel: SAD(0,1,2,2,1,1) model given by (3.19) with same parameters as the previous model. Marginal variances are expressed in units of natural log plus one.*

3.8 Nonlinear stationary autoregressive models

All of the autoregressively specified SAD covariance models described to this point have been linear, i.e., past observations enter linearly into the equation describing the dependence of an observation on past observations. Linear autoregressively specified SAD models are a broad and useful class, but there are some types of behavior, for example clustered volatility (periods where observations' variances are large followed by periods where these variances are small), that they may not be able to model parsimoniously. In the past twenty years or so, a large number of nonlinear time series models have been developed for modeling such "chaotic" behavior. Some of these are SAD models. We give several of these SAD models below, but this is by no means a complete list. In all of the models listed, the ϵ_i's are independent zero-mean normal random variables with equal variances, and we assume that the measurements are equally spaced. Furthermore, for simplicity we omit start-up conditions and parameter constraints required for stationarity.

1. The pth-order exponential autoregressive (EXPAR) model of Haggan and Ozaki (1981):

$$Y_i - \mu_i = \sum_{j=1}^{p} [\phi_j + \pi_j \exp(-\theta(Y_{i-1} - \mu_{i-1})^2)](Y_{i-j} - \mu_{i-j}) + \epsilon_i.$$

2. The pth-order threshold autoregressive (TAR) model of Tong (1990):

$$Y_i - \mu_i = \sum_{j=1}^{p} \phi_j^{(m)} (Y_{i-j} - \mu_{i-j}) + \sqrt{\delta^{(m)}} \epsilon_i$$

$$\text{if } Y_{i-d} - \mu_{i-d} \in \Omega_m, \quad m = 1, \ldots, M$$

   where $d \in \{1, \ldots, p\}$ and $\{\Omega_m : m = 1, \ldots, M\}$ is a partition of the real line, and $\phi_j^{(m)}$ (for $j = 1, \ldots, p$) and $\delta^{(m)}$ are the autoregressive coefficients and innovation variances, respectively, corresponding to each set Ω_m.

3. The pth-order functional coefficient autoregressive (FAR) model of Chen and Tsay (1993):

$$Y_i - \mu_i = \sum_{j=1}^{p} f_j(\mathbf{Y}_{i-1}^*)(Y_{i-j} - \mu_{i-j}) + \epsilon_i$$

   where $\mathbf{Y}_{i-1}^*$ is the vector of mean-corrected past observations lagged no more than p times before time i or some subset of those observations, and the $f_j(\cdot)$'s are specified functions.

Note that the EXPAR(p) and TAR(p) models are special cases of the FAR(p) model. All are SAD(p) models. However, normality of the innovations in these models does not generally yield normally distributed observations. This is an important difference between these models and the linear SAD models we presented earlier in this chapter. For this reason, although the variables in these models are AD(p) and, in fact, AC(p), they are not PAC(p). Partly for this reason, partly because of their complexity, and partly because experience has not suggested that nonlinear autoregressive behavior is common for longitudinal data (for which the time series are relatively short), we believe that the usefulness of these models in this context is rather limited.

3.9 Comparisons with other models

For purposes of comparison and contrast, we devote this final section of the chapter to descriptions of two families of parametric covariance models that are *not* antedependence models: vanishing correlation (banded) models and random coefficient models. These two families have been used extensively for longitudinal data exhibiting serial correlation and may therefore be viewed as competitors to antedependence models for such data. Later in the book we will actually fit some of these models, as well as various antedependence models, to the data sets introduced in Chapter 1 and determine which models fit better than others.

3.9.1 Vanishing correlation (banded) models

Assume initially that the measurement times are equally spaced. Then, the qth-order Toeplitz, or Toeplitz(q), model for a single subject specifies that

$$\sigma_{ii} = \sigma^2 \text{ for all } i,$$

$$\rho_{ij} = \begin{cases} \rho_{i-j} & \text{if } |i-j| = 1, \ldots, q, \\ 0 & \text{if } |i-j| \geq q+1. \end{cases}$$

Here $\sigma^2, \rho_1, \ldots, \rho_q$ are parameters subject only to positive definiteness constraints. Observe that the Toeplitz(q) model is stationary, i.e., marginal variances are constant and marginal correlations are a function of only the elapsed time between measurements. Moreover, the marginal correlations vanish beyond a finite elapsed time, giving the zeros in the covariance matrix a distinct banded structure.

One way in which a Toeplitz covariance structure can arise is as a result of a moving average process. Consider, for example, the first-order moving average model

$$Y_i - \mu_i = e_i - \alpha e_{i-1}, \quad i = 1, \ldots, n, \tag{3.20}$$

where $e_0, e_1, \ldots, e_n$ are independent and identically distributed random variables with mean zero and variance $\sigma_e^2 > 0$, and α is a real-valued, unconstrained parameter. It is easily shown that the variance of Y_i is equal to

$\sigma_e^2(1 + \alpha^2)$ for $i = 1, \ldots, n$; the correlation between Y_i and Y_j is equal to $-\alpha/(1 + \alpha^2)$ whenever $|i - j| = 1$; and the correlation between Y_i and Y_j is equal to zero whenever $|i - j| > 1$. Thus the covariance matrix of $\mathbf{Y}$ is first-order Toeplitz. More generally, if $Y_1, \ldots, Y_n$ follow the qth-order moving average model

$$Y_i - \mu_i = e_i - \sum_{j=1}^{q} \alpha_{i-j} e_{i-j}, \quad i = 1, \ldots, n,$$

where the e_i's and α_i's are suitably defined, then the covariance matrix of $\mathbf{Y}$ is Toeplitz(q). However, in the finite-n situation we consider here, not all Toeplitz(q) models can be obtained from a qth-order moving average model. A case in point is provided by a Toeplitz(1) model with $n = 2$, $\sigma_{11} = \sigma_{22} = 1$ and $\rho_{12} \in (-1.0, -0.5) \cup (0.5, 1.0)$. As just noted, the correlation between Y_1 and Y_2 under the first-order moving average model (3.20) is equal to $-\alpha/(1 + \alpha^2)$, which is constrained to the interval $[-0.5, 0.5]$ even though α is unconstrained. (In the infinite-n situation, the two models *are* equivalent; see, for example, Proposition 3.2.1 of Brockwell and Davis, 1991.)

The Toeplitz(q) model can be generalized slightly, to a qth-order heterogeneous Toeplitz model, by allowing the marginal variances to be arbitrary positive numbers. In turn, this model may be generalized to a qth-order banded model by continuing to require that the correlations on off-diagonals beyond the qth are zero, but not imposing any restrictions on the remaining correlations (apart from those required for positive definiteness). The general banded model need not be stationary, nor does it require measurements to be equally spaced. Though it is more flexible than the Toeplitz(q) and heterogeneous Toeplitz(q) models, the banded(q) model has a larger number of parameters in the covariance structure [$(q + 1)(2n - q)/2$ for the banded model versus $q + 1$ for the Toeplitz(q) model] and this number increases with the number of measurement times.

The banded structure of the zero correlations distinguishes the qth-order banded model from antedependence models of any order, and the two classes of models are mutually exclusive, except when their orders are both 0 or both $n - 1$. Thus, although some marginal correlations may be zeroes under an AD(p) model, their pattern cannot be banded unless $p = 0$. On the other hand, we recall from Theorem 2.2 that the precision matrix corresponding to an AD(p) model has just such a banded pattern of zeros for any p. In fact, any positive definite banded(q) matrix has a positive definite AD(q) matrix for its inverse, and vice versa. Equivalently, the partial correlations of the banded(q) model have the same structure as the marginal correlations of the AD(q) model, and the marginal correlations of the banded(q) model have the same structure as the partial correlations of the AD(q) model.

More highly structured vanishing correlation models exist that may be used with unequally spaced data, yet are stationary and more parsimonious than the general banded model. Such models are commonly used as covariance functions for spatial data (Cressie, 1993), but they can be used for longitudinal data as well. These include the triangular model

$$\sigma_{ii} = \sigma^2, \quad \rho_{ij} = \begin{cases} 1 - |t_i - t_j|/\alpha, & \text{if } |t_i - t_j| < \alpha \\ 0, & \text{otherwise,} \end{cases}$$

and the spherical model

$$\sigma_{ii} = \sigma^2, \quad \rho_{ij} = \begin{cases} 1 - \frac{3}{2}|t_i - t_j|/\alpha + \frac{1}{2}(|t_i - t_j|/\alpha)^3, & \text{if } |t_i - t_j| < \alpha \\ 0, & \text{otherwise.} \end{cases}$$

3.9.2 Random coefficient models

A rather general random coefficient model for multiple subjects is

$$\mathbf{Y}_s = \mathbf{X}_s \boldsymbol{\beta} + \mathbf{Z}_s \mathbf{u}_s + \mathbf{e}_s, \quad s = 1, \ldots, N, \tag{3.21}$$

where $\mathbf{Z}_1, \ldots, \mathbf{Z}_N$ are specified matrices; $\mathbf{u}_1, \ldots, \mathbf{u}_N$ are vectors of random coefficients distributed independently as $N(\mathbf{0}, \mathbf{G}_s)$; $\mathbf{G}_1, \ldots, \mathbf{G}_N$ are positive definite but otherwise unstructured matrices; and $\mathbf{e}_1, \ldots, \mathbf{e}_N$ are distributed independently (of the $\mathbf{u}_s$'s and of each other) as $N(\mathbf{0}, \sigma^2 \mathbf{I}_{n_s})$. Typically the $\mathbf{G}_s$ are assumed to be equal, in which case the covariance matrix of $\mathbf{Y}_s$ is given by $\boldsymbol{\Sigma}_s = \mathbf{Z}_s \mathbf{G} \mathbf{Z}_s^T + \sigma^2 \mathbf{I}_{n_s}$. Special cases include the linear random coefficient (RCL) and quadratic random coefficient (RCQ) models. In the linear case, $\mathbf{Z}_s = [\mathbf{1}_{n_s}, \mathbf{t}_s]$ and

$$\mathbf{G} = \begin{pmatrix} \gamma_{00} & \gamma_{01} \\ \gamma_{01} & \gamma_{11} \end{pmatrix}.$$

In the quadratic case, $\mathbf{Z}_s = [\mathbf{1}_{n_s}, \mathbf{t}_s, (t_{s1}^2, t_{s2}^2, \ldots, t_{sn_s}^2)^T]$ and

$$\mathbf{G} = \begin{pmatrix} \gamma_{00} & \gamma_{01} & \gamma_{02} \\ \gamma_{01} & \gamma_{11} & \gamma_{12} \\ \gamma_{02} & \gamma_{12} & \gamma_{22} \end{pmatrix}.$$

Note also that compound symmetry, given by (1.2), is, apart from a slightly expanded parameter space, equivalent to the case $\mathbf{Z}_s = \mathbf{1}_{n_s}, \mathbf{G} = \gamma_{00}$.

Two convenient features of random coefficient models are their parsimony (note that the number of covariance structure parameters is unrelated to the number of measurement times) and their applicability to situations in which the measurement times are unequally spaced or unbalanced.

Random coefficient models have often been considered as distinct from parametric covariance models, probably because their conceptual motivation is usually a consideration of regressions that vary across subjects rather than a consideration of within-subject marginal covariance structure. Nevertheless, random coefficient models yield marginal covariance structures that generally have nonconstant variances and nonstationary correlations, a fact that does not appear to be widely appreciated. For example, it is easy to show that the marginal variances and correlations of the RCL model for a subject observed at equally-spaced time points $t_1 = 1, \ldots, t_n = n$ are given by

$$\sigma_{ii} = \sigma^2 + \gamma_{00} + 2\gamma_{01}i + \gamma_{11}i^2,$$

$$\rho_{ij} = \frac{\gamma_{00} + \gamma_{01}(i+j) + \gamma_{11}ij}{\sqrt{\sigma^2 + \gamma_{00} + 2\gamma_{01}i + \gamma_{11}i^2}\sqrt{\sigma^2 + \gamma_{00} + 2\gamma_{01}j + \gamma_{11}j^2}},$$

$$i \geq j = 1, \ldots, n.$$

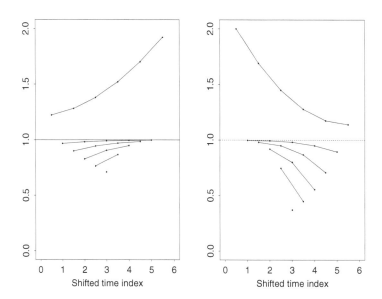

Figure 3.7 *Diagonal cross-section plots of marginal variances and correlations of the linear random coefficient model. Left panel:* $\gamma_{00} = 10$, $\gamma_{01} = 0$, $\gamma_{11} = 1$; *right panel:* $\gamma_{00} = 20$, $\gamma_{01} = -3$, $\gamma_{11} = 0.5$.

This structure is flexible enough to permit several kinds of variance and correlational behavior, including increasing or decreasing variances, and correlations of mixed sign. Figure 3.7, a diagonal cross-section plot of the marginal covariance structure, displays some possibilities. However, the model does have some notable limitations. Since γ_{11} is a variance and hence positive, we see that the model precludes the variance from being a concave-down function of time. Furthermore, the variances and correlations share parameters and thus their behaviors are inextricably intertwined, which, for example, precludes the variances from being constant if same-lag correlations are not.

CHAPTER 4

Informal Model Identification

The previous two chapters introduced antedependence models for time-ordered random variables, first in their most general, or unstructured, form, and subsequently in more parsimonious, or structured, forms. Now, we begin to consider statistical inference for such models. In this chapter especially, we describe informal methods for identifying the mean and covariance structures of normal linear antedependence models for longitudinal data. Such methods include the examination of useful summary statistics and graphical diagnostics and are conducted prior to actually fitting any antedependence models. More formal methods of inference, including likelihood-based parameter estimation, hypothesis tests, and model selection criteria, will be considered in later chapters.

It is appropriate to review and clarify the sampling framework(s) within which our inferences, be they formal or informal, are valid. For some structured antedependence models, it is possible to estimate parameters consistently from just one realization of the sequence of variables $Y_1, \ldots, Y_n$, or even from a subsequence thereof; multiple realizations are not necessary. An example is the stationary AR(p) model, for which there is a well-established literature on estimation; for an overview see, for example, Fuller (1976). For other structured antedependence models, however, and for an unstructured antedependence model of any order, the parameters of the covariance structure are too numerous to be estimable from merely one realization. Instead, multiple realizations are required, and the estimation is simplified greatly if these copies are mutually independent. The sampling framework that most naturally gives rise to independent realizations of time-ordered, correlated variables is longitudinal sampling, hence it is within the general longitudinal sampling framework introduced in Chapter 1 that we will consider inference for antedependence model parameters.

A brief review of this sampling framework and some associated notation is as follows. A response variable, Y, is measured on several occasions for each of N randomly selected subjects from a population of interest. The number of measurement occasions and the corresponding times of measurement may vary across subjects, and observations from different subjects are independent. For $s = 1, \ldots, N$ let n_s be the number of measurement occasions, let $Y_{s1}, Y_{s2}, \ldots, Y_{sn_s}$ be the measurements of the response (in chronological order) on subject s, and let $t_{s1}, t_{s2}, \ldots, t_{sn_s}$ be the corresponding measurement times. Also, let $\mathbf{x}_{si}$ be a $q \times 1$ vector of observed covariates possibly associated with Y_{si} $(i = 1, \ldots, n_s)$.

4.1 Identifying mean structure

We begin with a consideration of mean structure. To a degree, the choice of mean structure will be guided naturally by the scientific objectives and design of the study. For example, if the study's main objective is to determine how the growth of animals is affected over time by different treatments, and no covariates other than time are observed, then an initial specification of the mean structure for the response would probably consist of treatment effects, one or more time effects, and effects for time-by-treatment interaction. A separate effect could be used for each measurement time (i.e., the saturated mean structure), or the dependence on time could be modeled more parsimoniously using, say, linear and quadratic terms. The more parsimonious approach is necessary, of course, if one of the objectives is to estimate the mean response at times where no measurements were taken (interpolation or extrapolation).

Notwithstanding the importance of the study's scientific objectives for pointing to a preliminary specification of a mean structure, the analyst undoubtedly will want to use the observed data to attempt to improve upon the preliminary specification, i.e., to determine whether the specified mean structure includes covariates that do not help to explain the mean response, or to suggest additional functions of the observed covariates that do. For these purposes, means of the response variable at each measurement time should be computed and examined, as should correlations between the response and each covariate (if there are any) at each measurement time. Graphical diagnostics related to these summary statistics are generally more informative, however. The *profile plot*, for example, which was introduced in Chapter 1, can help to identify how the mean varies over time. Plots of responses against covariates other than time may also be useful. After choosing a provisional mean structure and fitting it, a *residual plot*, i.e., a plot of the fitted residuals versus time and/or other covariates can confirm, or indicate needed modifications to, the provisional mean structure. In the plot of residuals versus time, those residuals corresponding to successive measurements from the same subject may be connected with line segments to highlight how each subject's residuals vary over time. All of these plots are applicable generally, regardless of whether the data are balanced. However, they can become excessively cluttered if the number of subjects is large, in which case some pruning may be desirable. Diggle et al. (2002, pp. 33–45) present some interesting strategies for selecting subsets of individuals whose response profiles will be included for display, and those strategies could be used for displaying residual profiles as well.

Example 1: 100-km race data

Figure 1.3 is a profile plot of the split times for each 10-km section of the race, for all 80 runners. The overall mean profile (obtained by averaging the split times for each section, plotting these averages and connecting them with line segments) shows that the mean split time tends to increase rather steadily over the first 80 km of the race, but then levels off or actually decreases slightly. This last feature may be at least partly explained by the phenomenon of "kicking" (faster-than-average running for a relatively short period of time near the end of a race) that is common among well-conditioned runners. Of course, variation in topography along the course of the race may also affect a runner's performance on different sections, but unfortunately this information is not available. In any case, it appears that a mean structure with only linear and quadratic time effects will certainly not be adequate for capturing the perceptible (though small) undulations in the mean profile, and it seems unlikely that a model with a cubic time effect added will make it so.

The 100-km race data include observations not only on split time but also on a covariate, age. Figure 4.1 displays scatterplots of split time versus age, one for each of the first two and last two sections, for the 76 runners whose ages were observed. No linear effect of age is evident, but a Lowess smooth of the data indicates a possible quadratic effect in the last two sections. In those last two sections, split times appear to increase as runners' ages either increase or decrease from about 40 years. These plots thus suggest the possibility that there is a quadratic effect of age on split time in the later sections of the race, with middle-aged runners performing slightly better, on average, than either younger or older runners in those sections.

Example 2: Speech recognition data

Figure 1.4 displays profile plots of the speech recognition test scores for each of the two types of cochlear implants. Although 41 subjects were tested initially (20 for implant A and 21 for implant B), the plots clearly show that many subjects dropped out over the course of the study. The overall mean profiles for each implant increase as the study progresses, indicating improvement in speech recognition. Furthermore, the mean profile for implant type A appears to be more-or-less uniformly higher than that of implant type B for the duration of the study.

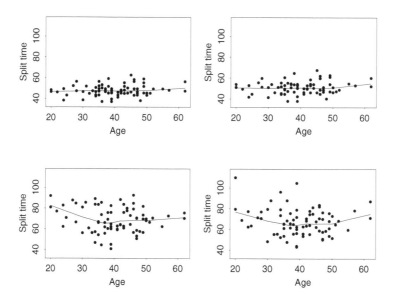

Figure 4.1 *Scatterplots of split time (minutes) versus age of runner (years) for the first (top left panel), second (top right panel), ninth (bottom left panel), and tenth (bottom right panel) 10-km sections of the 100-km race. The superimposed line is a Lowess scatterplot smooth.*

4.2 Identifying covariance structure: Summary statistics

Next we turn our attention to methods for identifying the covariance structure of normal antedependence models for longitudinal data. We assume that a particular linear mean structure has already been selected, at least provisionally, and we focus our attention solely on the covariance structure, partly for its own sake but also because an appropriate, parsimonious choice of covariance structure can substantially improve the efficiency of inferences made about the mean structure and provide better estimates of standard errors of estimated mean parameters. As with methods for identifying a mean structure, useful informal methods for identifying a covariance structure include the examination of basic summary statistics and graphical diagnostics. However, informal methods aimed at identifying the covariance structure are considerably richer in variety and complexity than those for identifying the mean, so we devote significantly more space to them here than was given to their counterparts in the previous section. Furthermore, in contrast to the situation with mean structure identification, the scientific objectives of the study are usually of no help whatsoever for choosing a covariance structure.

In this section we describe what can be learned about the data's covariance structure merely from examining certain summary statistics, and then in the following section we describe various graphical diagnostics and demonstrate what they can add to the summary statistics. Throughout both sections, we base the identification of the data's covariance structure on residuals fitted to a chosen mean structure. Initially at least, and perhaps even finally, the mean structure should be taken to be as saturated as possible (within groups, if there are any) to prevent misidentification of the covariance structure due to underfitting the mean structure. Also, we let n generally denote the number of distinct measurement times. It is assumed, unless noted otherwise, that the amount of replication across subjects at each measurement time is sufficient to support the computation and reliable interpretation of all summary statistics and graphical diagnostics. If there are groups, the entire covariance model identification enterprise should be done separately for each group first, so that it can be determined whether the covariance structure is sufficiently homogeneous across groups that the within-group covariance matrices can be pooled.

Recall from Chapter 2 that antedependence models are nested, i.e.,

$$AD(0) \subset AD(1) \subset AD(2) \subset \cdots \subset AD(n-1),$$

where the class of covariance matrices for a normal $AD(n-1)$ model is simply the class of all symmetric positive definite matrices. Consequently, every normally distributed n-variate response vector follows a normal antedependence model of sufficiently high order. So, if consideration is limited to unstructured, constant-order normal antedependence models, the problem of choosing the covariance structure is merely to choose the order, p, of the model. Recall also that $AD(p_1, \ldots, p_n) \subset AD(\max p_i)$, so if consideration is expanded to unstructured variable-order antedependence models, then the problem is to choose $p_1, \ldots p_n$. If the analyst wishes instead, or in addition, to consider structured antedependence models, then the problem involves choosing not only the order of the model but also its particular formulation (marginal, intervenor-adjusted, precision matrix, or autoregressive) plus a parsimonious model within the chosen formulation.

Which summary statistics are useful for identifying the data's covariance structure? If the data are balanced and the number of subjects is not too small, the standard sample covariance matrix and a suite of relevant matrices derived from it, such as the sample correlation matrix, the matrix of sample intervenor-adjusted partial correlations, the matrix of sample partial correlations, and the modified Cholesky decomposition of the sample precision matrix, may be computed. If the data are not balanced, the matrix of sample variances and covariances, which for simplicity we will also call the sample covariance matrix, may be computed from the nonmissing data at each time (for the variances) or pair of times (for the covariances), and from it the same suite of matrices can be derived. It is worth mentioning, however, that in the unbalanced case the sample covariance matrix, as just defined, is not guaranteed to be positive definite, no matter how large the sample size. Hence some of the derived matrices may have a few quirks, for example correlations larger than unity, which should just be ignored. In any case, an examination of the elements of the derived matrices can inform the choice of a marginal, intervenor-adjusted, precision matrix, or autoregressive formulation of an antedependence model for the data, as we now describe.

First, to identify the order of antedependence, off-diagonals of the matrix of sample intervenor-adjusted partial correlations and/or the sample partial correlation matrix should be examined for the positions of near-zero elements. To judge whether an arbitrary off-diagonal element, say r, of these matrices is too large for the corresponding population parameter, ρ, to equal zero, classical likelihood-based inference procedures for partial correlation coefficients (e.g., Section 4.3 of Anderson, 1984, and Section 6.1 of this book) may be applied. Alternatively, a simple rule of thumb may be used, which is based on the asymptotic distributional result

$$\sqrt{N^*}r \to \mathrm{N}(0,1) \quad \text{when } \rho = 0.$$

Here N^* represents the number of observations from which r is calculated, minus the number of variables the partial correlation adjusts for. The rule of thumb is to regard a sample partial correlation (of either type) as near-zero if it is smaller than $2/\sqrt{N^*}$ in absolute value. However, due to the large number of these determinations to be made, the analyst may wish to modify this rule of thumb to control the overall Type I error rate. Since the sample partial correlation coefficients (of either type) are correlated in a rather complicated way, a Bonferroni-based approach for controlling error rate may be the most practical. If, in the end, all elements on off-diagonals $p+1, \ldots, n-1$ of these matrices are judged to be near-zero, but one or more elements on the pth off-diagonals are deemed to be different from zero, then by Definition 2.3 and Theorem 2.2, this suggests that the covariance structure is that of an AD(p) model. More generally, if the first m_i elements in the ith row ($i = 1, \ldots, n$) of these matrices are deemed near-zero, but the (m_i+1)th element is not, then the covariance structure is indicated to be that of a variable-order AD($p_1, \ldots, p_n$) model, with $p_i = i - m_i$. Note that the sample correlation matrix, while important in its own right, is not as prescriptive for determining the order of antedependence as the other matrices just described for two reasons. First, in contrast to near-zero sample partial correlations or intervenor-adjusted partial correlations, near-zero marginal correlations are not indicative of any form of antedependence (unless all of them are near-zero, in which case an AD(0), or complete independence, model is indicated). Second, it is virtually impossible, except perhaps in the first-order case, to determine from a mere examination of the sample correlations whether equations (2.32) or (2.44) hold to a good approximation for some p or some $(p_1, \ldots, p_n)$. Nevertheless, the sample correlations should still be examined, and the same rule of thumb (with N^* equal to N) may be applied to judge whether any of them are near-zero.

The sample autoregressive coefficients (i.e., the subdiagonal elements of the unit lower triangular matrix of the sample precision matrix's modified Cholesky decomposition are in principle as prescriptive for antedependence as the partial correlations and intervenor-adjusted partial correlations; hence they should likewise be examined for near-zero elements. However, the autoregressive coefficients are not calibrated as easily as the two kinds of partial correlations because they are not required to lie between -1 and 1. Consequently, it is difficult to judge whether an autoregressive coefficient is near-zero merely from its magnitude or some simple sample-size-based multiple of its magnitude. Nevertheless, a rule (which is actually a size-0.05 likelihood ratio hypothesis testing approach) for making this determination can be given as follows. We judge a sample autoregressive coefficient, $\tilde{\phi}_{ij}$, to be near-zero if and only if

$$\frac{|\tilde{\phi}_{ij}|}{(\tilde{\delta}_i c_{ii,jj})^{1/2}} < t_{.025, N-i}, \tag{4.1}$$

where $\tilde{\delta}_i$ is the ith sample innovation variance, $c_{ii,jj}$ is the jth diagonal element of the inverse of the upper left $i \times i$ submatrix of the sample covariance matrix, and $t_{.025, N-i}$ is the 97.5% percentile of Student's t distribution with $N - i$ degrees of freedom. The rationale for this approach will be given in Section 6.1.

If an examination of the aforementioned matrices of summary statistics suggests a plausible order, p, of antedependence, then the same statistics in the first p subdiagonals of those matrices should be examined for trends or other patterns in order to perhaps identify plausible structured antedependence models. For example, we may observe that the sample variances, correlations, partial correlations, or autoregressive coefficients are increasing over time, which may point to the use of one or more of the structured antedependence models described in Chapter 3.

Example 1: Treatment A cattle growth data
Table 1.2 is the matrix of sample variances and correlations of the Treatment A cattle growth data. We described several interesting aspects of these statistics in Section 1.5, which we briefly review here. The variances increase, from the beginning of the study to its end, by a factor of about four. The correlations are all positive and decay more-or-less monotonically within columns; furthermore, same-lag correlations tend to increase as the study progresses.

To augment these results, we also examine the intervenor-adjusted partial correlations, the partial variances and partial correlations, and the autoregressive coefficients and innovation variances (Table 4.1). All lag-one marginal correlations exceed the rule-of-thumb value of $2/\sqrt{30} \doteq 0.365$; likewise $r_{68 \cdot 7}$ exceeds $2/\sqrt{29} \doteq 0.371$. The remaining intervenor-adjusted partial correlations, however, are not significantly different from zero. As for the partial correlations, only three of them, all lag-one, are significantly different from zero (the rule-of-thumb value for all partial correlations is $2/\sqrt{21} \doteq 0.436$). The statistically significant autoregressive coefficients correspond to exactly the same lags for which the intervenor-adjusted partial correlations were deemed significant. Overall, these statistics suggest that a first-order antedependence model may adequately characterize the data's covariance structure, except possibly for the eighth measurement occasion, for which the antedependence may need to be extended to second order.

Example 2: 100-km race data

Table 4.2 is the matrix of sample variances and correlations among the split times of the complete set of 80 competitors. The table reveals several interesting features of the marginal covariance structure: (a) the variances tend to increase over the course of the race, except for a reversal from the sixth to the seventh 10-km sections; (b) the correlations are positive and quite large, all exceeding the rule-of-thumb value of $2/\sqrt{80} \doteq 0.22$; (c) the correlations between the split time for any fixed 10-km section and split times for successive sections tend to decrease monotonically; and (d) correlations between split times of consecutive sections are not as large near the end of the race as they are earlier.

To supplement these findings, we also examine the intervenor-adjusted partial correlations, the partial variances and partial correlations, and the autoregressive coefficients and innovation variances (Table 4.3). Using the rule of thumb, all lag-one marginal correlations are judged to be significantly different from zero, as are some intervenor-adjusted partial correlations [see part (a) of the table]: $r_{35 \cdot 4}$, $r_{68 \cdot 7}$, $r_{69 \cdot 78}$, $r_{79 \cdot 8}$, $r_{5,10 \cdot 6789}$, and $r_{7,10 \cdot 89}$. Not quite as many partial correlations [part (b) of the table] are deemed to be different from zero, but those that are lie mostly in the same rows and columns of entries that are significant in part (a). The partial variances, like the marginal variances, are mostly increasing over the course of the race; so too are the innovation variances in part (c), except for the first which is larger than the next five. The autoregressive coefficients in part (c) exhibit a very interesting phenomenon. The lag-one coefficients for the first seven sections of the race are large in comparison to higher-lag coefficients on those sections; however, on sections 8, 9, and 10 the lag-one coefficients are matched or exceeded (in absolute value) by several for greater lags, including those that are lagged back to the very first section. Some of the signs of these large higher-lag coefficients are negative,

Table 4.1 *Summary statistics for the covariance structure of the Treatment A cattle growth data: (a) sample variances, along the main diagonal, and intervenor-adjusted partial correlations, below the main diagonal; (b) sample partial variances, along the main diagonal, and partial correlations, below the main diagonal; (c) sample innovation variances, along the main diagonal, and autoregressive coefficients, below the main diagonal. Off-diagonal entries in parts (a) and (b) of the table deemed to be significant using the rule of thumb of Section 4.2 are set in bold type; so also are autoregressive coefficients in part (c) whose corresponding t-ratios are significant at the 0.05 level.*

(a)

106										
.82	155									
.07	**.91**	165								
−.24	.03	**.93**	185							
.03	.02	.07	**.94**	243						
.01	−.23	−.04	.23	**.94**	284					
.16	−.17	−.12	−.18	−.04	**.93**	307				
−.06	.01	.01	−.20	.07	**.57**	**.93**	341			
.26	−.01	.09	−.22	−.23	−.30	.35	**.97**	389		
−.22	−.07	.21	.02	−.08	−.09	−.24	.15	**.96**	470	
.19	−.25	.03	.27	.16	−.24	−.18	−.28	.20	**.98**	445

(b)

26.2										
.55	16.6									
.22	.39	12.7								
−.20	.18	.41	10.5							
.00	.13	−.01	.37	16.1						
.01	−.13	.13	.30	.26	11.4					
.03	−.16	.02	.12	.04	.25	24.7				
−.17	.10	−.05	.07	.17	.22	−.05	10.8			
.24	−.02	−.17	−.15	−.01	.16	.41	**.48**	8.8		
−.28	.29	.03	−.28	−.06	.28	.01	.16	.02	9.1	
.19	−.31	.13	.25	.01	−.38	−.11	−.05	.36	**.82**	9.4

(c)

106										
1.00	51									
.06	**.89**	31								
−.22	.16	**.97**	28							
.02	.00	.05	**1.02**	32						
.01	−.25	.15	.33	**.79**	34					
.18	−.32	−.04	−.06	.16	**1.01**	47				
−.06	.05	.02	−.27	.23	**.61**	**.42**	38			
.20	−.13	.02	−.21	−.07	.00	.32	**.93**	22		
−.22	.07	.36	−.21	−.11	−.07	−.15	.32	**.99**	40	
.11	−.23	.11	.24	.01	−.34	−.07	−.05	.38	**.83**	14

Table 4.2 *Sample variances, along the main diagonal, and correlations, below the main diagonal, for the 100-km race data.*

27									
.95	36								
.84	.89	50							
.79	.83	.92	58						
.62	.64	.76	.89	90					
.62	.63	.73	.84	.94	147				
.53	.54	.61	.69	.75	.84	107			
.48	.51	.62	.70	.79	.84	.78	151		
.54	.53	.58	.67	.74	.77	.70	.76	149	
.41	.44	.47	.51	.54	.66	.72	.66	.78	168

while others are positive, and some of the coefficients that are statistically significant (as determined by the corresponding t-ratio) are considerably smaller in magnitude than others that are not. The coefficients that are statistically significant indicate that performance of a runner on the last three sections of the race is strongly associated not merely with performance on the immediately preceding section, but also with performance on several earlier sections even after conditioning on performance on the intervening sections. For the most part, these conditional associations are positive; the lone negative one occurs between the fifth and tenth split times. The positive conditional associations are not surprising, but the negative one is somewhat of a surprise. It appears to suggest that competitors who run slow on the fifth section, relative to other competitors and also to their own performance on the sixth through ninth sections, are able to run relatively faster on the last section. A possible physical explanation is that saving energy (by running relatively slower) just before the halfway point of the race enables those competitors to run relatively faster at the end.

4.3 Identifying covariance structure: Graphical methods

Graphical diagnostics for identifying the covariance structure can add substantially to what is learned from an examination of summary statistics. In addition to displaying visually the information conveyed by the sample covariance matrix and relevant matrices derived from it, graphical diagnostics may reveal nonlinear relationships among variables, anomalous observations, clustering of observations into distinct groups, and other interesting data attributes not

Table 4.3 *Summary statistics for the covariance structure of the 100-km race split times: (a) sample variances, along the main diagonal, and intervenor-adjusted partial correlations, below the main diagonal; (b) sample partial variances, along the main diagonal, and partial correlations, below the main diagonal; (c) sample innovation variances, along the main diagonal, and autoregressive coefficients, below the main diagonal. Off-diagonal entries in parts (a) and (b) of the table deemed to be significant using the rule of thumb of Section 4.2 are set in bold type; so also are autoregressive coefficients in part (c) whose corresponding t-ratios are significant at the 0.05 level.*

(a)

27									
.95	36								
−.02	**.89**	50							
.02	.03	**.92**	58						
.07	−.20	**−.34**	**.89**	90					
.06	.10	.03	.07	**.94**	147				
.02	.03	.03	.04	−.17	**.84**	107			
−.11	−.15	.10	−.05	.04	**.55**	**.78**	151		
.16	.15	.01	.02	.10	**.29**	**.25**	**.76**	149	
−.19	−.00	.06	.04	**−.35**	−.17	**.36**	.17	**.78**	168

(b)

2.4									
.81	2.2								
.00	**.33**	4.6							
−.05	.13	**.59**	3.9						
−.06	−.07	−.07	**.47**	6.6					
.09	−.02	−.06	.02	**.64**	11				
.06	.01	−.09	.05	−.02	**.31**	24			
−.13	−.03	.20	−.08	.00	**.26**	**.23**	36		
.22	−.08	−.13	.03	**.25**	−.04	−.20	**.29**	33	
−.16	.10	.07	−.04	**−.31**	.19	**.39**	−.03	**.60**	46

(c)

27									
1.09	3.5								
−.03	**1.07**	11							
.03	.01	**.97**	9.2						
.19	−.45	−.32	**1.57**	17					
.16	.02	−.05	.00	**1.16**	19				
.05	.02	.02	−.01	−.30	**.91**	33			
−.44	.02	.49	−.30	.12	**.61**	**.31**	44		
.73	−.19	−.27	−.04	.27	.17	.08	**.41**	56	
−.83	.67	.18	−.13	**−.81**	.39	**.50**	−.01	**.72**	51

discernible from summary statistics. Furthermore, some graphical diagnostics may more readily suggest particular structured antedependence models or other parsimonious models to fit to the data than a mere examination of elements in a matrix can. Subsequent to actually fitting models, these diagnostics may also be used to assess whether the model judged to fit "best" (according, typically, to some numerical criterion) actually provides a reasonable fit.

We now describe several useful graphical methods for identifying the covariance structure. In keeping with a pattern of organization used in previous chapters, we organize the presentation of these methods according to the particular formulation of antedependence with which they are most closely associated.

4.3.1 Marginal structure

A marginal covariance structure can, of course, be decomposed into two components: a variance structure and a correlation structure. Since the marginal variances in every unstructured antedependence model — even one of order zero — are arbitrary (apart from the requirement that they be positive), graphical diagnostics for variances are used to guide not the choice of order of antedependence but the selection of either a variance-stabilizing transformation (if one wants to avoid modeling variances) or an explicit parsimonious model for the dependence of the variances on time, such as that given by (3.5), i.e.,

$$\sigma_{ii} = \exp(\mathbf{v}_i^T \boldsymbol{\psi}), \quad i = 1, \ldots, n.$$

Here $\mathbf{v}_i$ is a vector of functions of time, evaluated at time t_i. Many graphical diagnostics that are useful for identifying mean structure also convey useful information about the behavior of the variances over time. The profile plot, for example, can indicate how the marginal variances behave. If subjects' profiles tend to fan out as time progresses, the response variances are revealed to be heteroscedastic, increasing with time. This type of behavior is common, especially in growth studies, and often occurs in conjunction with an increase in the mean over time. The sample variances themselves (computed from the non-missing observations at each measurement time) may be plotted against time, in order to determine whether, and how, time should enter as a covariate in a model such as (3.5), or whether a different model may be more appropriate. This plot is the sample analogue of the top portion of the diagonal cross-section plot introduced in Chapter 3. Evidence of heteroscedasticity in scatterplots of the response versus each of the other covariates (if there are any) can help to determine whether the variances depend on any covariates other than time (and functions thereof).

For identifying the marginal correlation structure, a simple plot of the same-lag correlations versus time (the sample analogue of the bottom portion of the diagonal cross-section plot) may be helpful. The most widely used graphical diagnostic, however, is the *ordinary scatterplot matrix* (OSM). This is a two-dimensional array of pairwise scatterplots, each one a plot of standardized responses at one time against standardized responses at another time, arranged in the same manner as the correlations in the sample correlation matrix. Typically, the standardization employed subtracts the sample mean and divides by the sample standard deviation of responses at the corresponding time, but variations on this, such as subtracting a fitted mean that includes covariates other than time, are possible. With the typical standardization, however, the OSM is a graphical manifestation of the sample correlation matrix, minus its main diagonal of ones. As such, the OSM is of very limited value for determining the order of antedependence; rather, its value is in revealing outliers and other features of the data that the sample correlation matrix cannot reveal, for the overall purpose of possibly identifying a suitable marginally formulated SAD or other parsimonious model for the data. Note that only the scatterplots in the lower or upper "triangle" of the OSM actually need to be included, due to the OSM's symmetry about its main diagonal of plots. Note also that if the measurement times are equally spaced and the data arise from a stationary process, then the population correlations corresponding to a given off-diagonal of the correlation matrix are equal to each other, in which case all scatterplots along the corresponding diagonal of the OSM can be superimposed to yield one plot. The results of such superimpositions are $n - 1$ scatterplots whose correlations coincide with those of the sample autocorrelation function, a diagnostic frequently used by time series analysts to guide choices of (a) the amount of differencing of the series needed to achieve stationarity, and (b) the order of a moving average model of the differenced data (see, for example, Box and Jenkins, 1976).

For additional graphical diagnostics for identifying the marginal correlation structure of longitudinal data, the reader may consult Diggle et al. (2002, pp. 46–53), Dawson et al. (1997), and Verbeke et al. (1998).

Example 1: 100-km race data
Figure 1.3, the profile plot of the complete set of split times, reveals more than just how the mean split time changes over the course of the race; it shows some interesting features of the marginal covariance structure as well. The fanning-out of profiles from left to right indicates that the variances tend to increase as the race progresses, as we noted previously from our examination of the marginal variances (Table 4.2). The plot shows that the sharp decline in the split-time variance from the sixth to the seventh sections, which we also noted previously, is largely due to a group of about a dozen slower-than-average runners who ran substantially faster on the seventh 10-km section than on the sixth. Finally, the plot reveals that the behavior of many of the runners is more erratic, in the sense that consecutive same-runner split times fluctuate more, in

the later sections of the race. This last feature comports with the decrease in same-lag correlations in the latter part of the race that we noted previously.

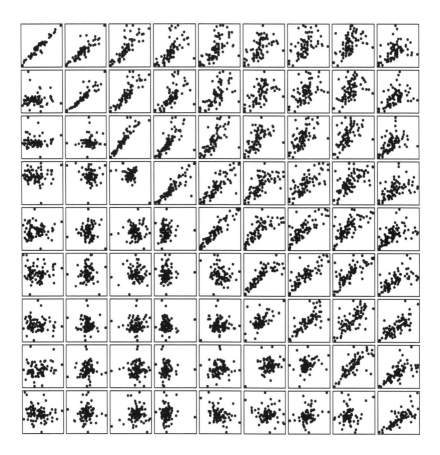

Figure 4.2 *Ordinary scatterplot matrix (upper triangle, including main diagonal) and PRISM (lower triangle, including main diagonal) of 10-km split times for the 100-km race data.*

Figure 4.2 (upper triangle) is the ordinary scatterplot matrix of these data (after standardization). It reveals the same features of the marginal correlations noted immediately above, as well as those that were observed from our previous examination of the marginal correlations. However, it also reveals an outlier, visible in the scatterplots in rows 1 through 4 of column 4, among others. The location of this outlier in each scatterplot suggests that the corresponding marginal correlations, which are already quite large, would be even larger if the outlier were not present. Furthermore, the tendency of many of the scatterplots to fan out somewhat as both variables increase indicates that the temporal correlation is perhaps somewhat stronger among faster-than-average runners than among slower-than-average runners. In other words, the faster competitors run at a more consistent speed (relative to the average speed for the section) than the slower competitors.

Example 2: Fruit fly mortality data
Figure 4.3 displays the (marginal) sample variances and lag-one correlations for the fruit fly mortality data, each computed from the available data at the corresponding measurement time(s). These quantities are also listed in Table 4.4(a). The marginal variances exhibit approximately piecewise quadratic dependence on time, increasing up to the fourth measurement time and then decreasing. The lag-one marginal correlations exhibit approximately concave-down, quadratic behavior with a point of maximum near the study's midpoint. The remaining marginal correlations, plus the intervenor-adjusted partial correlations, innovation variances, and autoregressive coefficients, are listed in Table 4.4 as well. Intervenor-adjusted partial correlations of order two and higher are close to zero and do not appear to behave in any systematic fashion; the same is true of the autoregressive coefficients beyond lag one. In light of these results, plausible models for the data would certainly include structured first-order antedependence models in which the marginal variances are modeled as quadratic or cubic functions of time and the marginal correlations are modeled as quadratic functions of time.

4.3.2 Intervenor-adjusted structure

The intervenor-adjusted formulation may be parameterized by the marginal variances, the marginal lag-one correlations, and the partial correlations between pairs of variables lagged two or more measurement times apart, adjusted for intervenors. The graphical diagnostics discussed in the previous section could be used for identifying the marginal variances and lag-one correlations, so it remains to consider the identification of the partial correlations adjusted for intervenors. One diagnostic that complements an examination of the matrix of intervenor-adjusted partial correlations is a diagonal cross-section plot of them. Another is the Partial Regression-on-Intervenors Scatterplot Matrix (PRISM) (Zimmerman, 2000), which is a rectangular array (or a subset thereof) of certain partial regression plots (also known as added variable plots). Although it is an acronym, the term "PRISM" also has semantic substance: as a prism separates visible light into its components, so a PRISM can separate the dependence structure among within-subject responses into components that are much easier to understand.

Table 4.4 *Summary statistics for the covariance structure of the fruit fly mortality data: (a) sample variances, along the main diagonal, and correlations, below the main diagonal; (b) sample variances, along the main diagonal, and intervenor-adjusted partial correlations, below the main diagonal; (c) sample innovation variances, along the main diagonal, and autoregressive coefficients, below the main diagonal. Off-diagonal entries in parts (a) and (b) of the table deemed to be significant using the rule of thumb of Section 4.2 are set in bold type; so also are autoregressive coefficients in part (c) whose corresponding t-ratios are significant at the 0.05 level.*

(a)

0.70										
.59	1.08									
.53	**.71**	1.66								
.48	**.62**	**.78**	2.61							
.34	**.52**	**.59**	**.79**	2.17						
.33	**.39**	**.44**	**.54**	**.82**	1.72					
.21	**.25**	.21	**.28**	**.57**	**.78**	1.05				
.01	.11	.08	.03	**.31**	**.51**	**.74**	0.63			
−.02	.07	−.02	.04	.05	.05	**.40**	**.59**	0.43		
−.01	.16	−.16	.17	.01	.03	**.31**	**.33**	**.46**	0.34	
.09	−.05	−.15	.28	.18	.19	.25	**.29**	**.29**	**.37**	0.42

(b)

0.70										
.59	1.08									
.13	**.71**	1.66								
.01	.19	**.78**	2.61							
.03	.09	−.08	**.79**	2.17						
.07	.06	−.04	−.22	**.82**	1.72					
−.13	.06	.04	−.20	−.16	**.78**	1.05				
−.01	−.13	.11	−.08	−.04	−.10	**.74**	0.63			
.11	.14	−.14	.00	.15	−.27	−.03	**.59**	0.43		
.08	.32	−.15	.13	.07	−.06	.10	.10	**.46**	0.34	
−.07	.01	−.34	.10	.17	.21	.08	.14	.09	**.37**	0.42

(c)

.70										
.73	.72									
.19	**.90**	.93								
.02	.31	**.81**	1.01							
.03	.14	−.03	**.71**	.72						
.06	.10	−.05	−.11	**.69**	.39					
−.11	.04	−.01	−.15	−.13	**1.08**	.39				
−.03	−.10	.13	−.05	−.02	.14	**.45**	.29			
.10	.04	−.18	−.02	.21	−.39	.09	**.78**	.33		
.08	.18	−.24	−.12	.06	.24	−.02	.07	.09	.19	
−.13	.10	−.49	.35	.06	.14	−.40	.52	.22	.16	.40

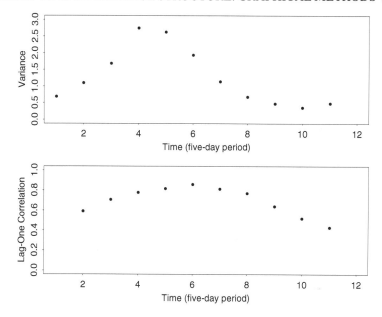

Figure 4.3 *Sample variances (top panel) and lag-one correlations (bottom panel) for the fruit fly mortality data.*

The PRISM is laid out as indicated in Table 4.5, where, as for the OSM, it suffices, due to symmetry, to display only one of the two triangles of plots. The main diagonal of plots consists of ordinary scatterplots of a standardized response against its immediate predecessor; thus this diagonal coincides with the main diagonal of the OSM. The second diagonal of plots in the PRISM comprises partial regression plots of standardized responses lagged two times apart, adjusted for the standardized response at the intervening time; that is, plots of residuals from the ordinary least squares regression (with intercept) of a standardized variable on its predecessor against residuals from the ordinary least squares regression (with intercept) of the standardized variable lagged two times back from the original standardized variable, on its successor. The third diagonal of plots comprises partial regression plots of standardized responses lagged three times apart, adjusted for the standardized responses at the two intervening times, and so on. In general, the plot in row i and column j $(i \geq j)$ is the partial regression plot of standardized response variables Y_{i+1} and Y_j adjusted for standardized responses at the intervening times $t_{j+1}, t_{j+2}, \ldots, t_i$. Thus, the PRISM is the graphical equivalent of the matrix of sample intervenor-adjusted partial correlations: the (i, j)th plot displays points whose ordinary correlation is the sample partial correlation between Y_{i+1} and

Table 4.5 *Layout of partial regression plots in a PRISM. Here* $\mathbf{Y}_{-I}$*, where* $I \subset \{1, 2, \ldots, n\}$*, is the set of all response variables except those in* I*.*

Y_2 vs. Y_1			
$Y_3\|Y_2$ vs. $Y_1\|Y_2$	Y_3 vs. Y_2		
$Y_4\|Y_2, Y_3$ vs. $Y_1\|Y_2, Y_3$	$Y_4\|Y_3$ vs. $Y_2\|Y_3$	Y_4 vs. Y_3	

$\vdots$ $\ddots$

$Y_n\|\mathbf{Y}_{-\{1,n\}}$ vs. $Y_1\|\mathbf{Y}_{-\{1,n\}}$	$Y_n\|\mathbf{Y}_{-\{1,2,n\}}$ vs. $Y_2\|\mathbf{Y}_{-\{1,2,n\}}$	$\cdots$	Y_n vs. Y_{n-1}

Y_j adjusted for all standardized responses at intervening times. It therefore augments the matrix of sample intervenor-adjusted partial correlations in the same way that the OSM augments the sample correlation matrix. Random scatter in the (i, j)th plot indicates that Y_{i+1} and Y_j are partially uncorrelated (conditionally independent under normality), adjusted for the intervening responses, whereas departures from random scatter indicate non-negligible partial correlation between Y_{i+1} and Y_j adjusted for those intervenors. Counting, from left to right in the ith row, the number of plots before one with a discernible linear association is encountered yields, upon subtraction from i, the order p_{i+1} of antedependence of Y_{i+1}.

If the measurement times are equally spaced and the data arise from a stationary process, then scatterplots along any given diagonal of the PRISM may, as for the OSM, be superimposed. The results of these superimpositions are $n-1$ scatterplots whose ordinary correlations coincide with those given by the sample partial autocorrelation function, a diagnostic commonly used by time series analysts to help determine the order of a stationary autoregressive model (see, for example, Box and Jenkins, 1976).

Examples with simulated data
To highlight the relative strength of the PRISM to the OSM for identifying the order of antedependence, we examine these scatterplot matrices for several

simulated longitudinal data sets, each having a different covariance structure. In each case, the data are a random sample of size $N = 100$ from a six-dimensional normal distribution with zero means and unit variances, and the measurement times are $t_i = i$ for $i = 1, \ldots, 6$.

Figures 4.4 and 4.5 show the OSMs (upper triangles) and PRISMs (lower triangles) of data simulated from AR(1) and AR(2) models, respectively. For the AR(1) model, we set $\phi_1 = 0.8$, which yields matrices of true correlations and true intervenor-adjusted partial correlations of

$$\begin{pmatrix} .80 & & & & \\ .64 & .80 & & & \\ .51 & .64 & .80 & & \\ .41 & .51 & .64 & .80 & \\ .33 & .41 & .51 & .64 & .80 \end{pmatrix} \quad \text{and} \quad \begin{pmatrix} .80 & & & & \\ .00 & .80 & & & \\ .00 & .00 & .80 & & \\ .00 & .00 & .00 & .80 & \\ .00 & .00 & .00 & .00 & .80 \end{pmatrix},$$

respectively. For the AR(2) model, we set $\phi_1 = 0.3$ and $\phi_2 = 0.5$, yielding the analogous matrices

$$\begin{pmatrix} .60 & & & & \\ .68 & .60 & & & \\ .50 & .68 & .60 & & \\ .49 & .50 & .68 & .60 & \\ .40 & .49 & .50 & .68 & .60 \end{pmatrix} \quad \text{and} \quad \begin{pmatrix} .60 & & & & \\ .50 & .60 & & & \\ .00 & .50 & .60 & & \\ .00 & .00 & .50 & .60 & \\ .00 & .00 & .00 & .50 & .60 \end{pmatrix}.$$

Note that the two OSMs appear very similar. Both show the persistence, over time, of the correlations, which attenuate (eventually) but do not vanish as one moves up and to the right from the main diagonal. Both also reveal a constancy of correlation strength within diagonals, which is a manifestation of the stationarity of these two models. The corresponding PRISMs, however, differ markedly with respect to their second diagonals. Plots in the second diagonal of the AR(1)'s PRISM exhibit random scatter, but as a consequence of the AR(2)'s nonzero values of $\rho_{i,i-2 \cdot i-1}$, their counterparts in the AR(2)'s PRISM do not. The third and higher diagonals of both PRISMs exhibit random scatter.

The OSMs and PRISMs corresponding to two more general antedependence structures, AD(1) and AD(0,1,1,2,1,3), are given in Figures 4.6 and 4.7. For the AD(1) model, we set $\phi_{i,i-1} = 0.3 + 0.1i$, which yields matrices of true correlations and true intervenor-adjusted partial correlations of

$$\begin{pmatrix} .50 & & & & \\ .30 & .60 & & & \\ .21 & .42 & .70 & & \\ .17 & .34 & .56 & .80 & \\ .15 & .30 & .50 & .72 & .90 \end{pmatrix} \quad \text{and} \quad \begin{pmatrix} .50 & & & & \\ .00 & .60 & & & \\ .00 & .00 & .70 & & \\ .00 & .00 & .00 & .80 & \\ .00 & .00 & .00 & .00 & .90 \end{pmatrix},$$

respectively. For the AD(0,1,1,2,1,3) model, we set $\phi_{i,i-1} = 0.5$, $\phi_{42} = \phi_{64} =$

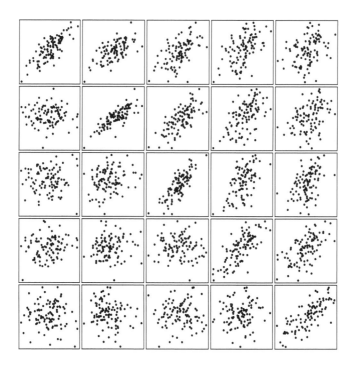

Figure 4.4 *Ordinary scatterplot matrix (upper triangle, including main diagonal) and PRISM (lower triangle, including main diagonal) of data simulated from an AR(1) process.*

0.4, and $\phi_{63} = 0.3$, yielding the analogous matrices

$$
\begin{pmatrix}
.50 & & & & \\
.25 & .50 & & & \\
.33 & .65 & .70 & & \\
.16 & .33 & .35 & .50 & \\
.29 & .57 & .76 & .86 & .81
\end{pmatrix}
\quad \text{and} \quad
\begin{pmatrix}
.50 & & & & \\
.00 & .50 & & & \\
.00 & .43 & .70 & & \\
.00 & .00 & .00 & .50 & \\
.00 & .00 & .79 & .89 & .81
\end{pmatrix}.
$$

Again, the two OSMs are rather similar: both show the persistence of correlation as one moves up and to the right from the main diagonal but, in contrast to those of the AR models, they also indicate variation in correlation strength as one moves down any particular diagonal. The PRISMs also are quite similar, except for three scatterplots: row 3, column 2; row 5, column 3; and row 5 column 4. These three scatterplots exhibit random scatter for the AD(1) but

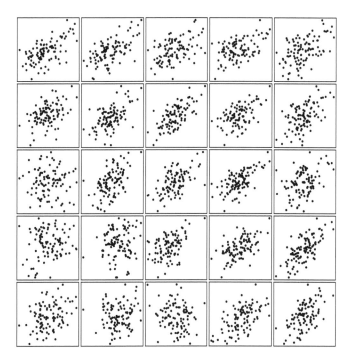

Figure 4.5 *Ordinary scatterplot matrix (upper triangle, including main diagonal) and PRISM (lower diagonal, including main diagonal) of data simulated from an AR(2) process.*

not for the variable-order model, reflecting the latter's nonzero values of $\rho_{42 \cdot 3}$, $\rho_{63 \cdot 45}$, and $\rho_{64 \cdot 5}$.

The previous examples clearly demonstrate the superior ability of the PRISM to identify antedependence structures. Lest the reader be misled, however, it is important to point out that the OSM may be much more informative than the PRISM for identifying non-antedependent structures. For example, compound symmetry and vanishing correlation models are much more easily diagnosed from the OSM than from the PRISM. This suggests that a better diagnosis of correlation structure may occurs if the OSM and PRISM are used in tandem. Doing so would conform to good statistical practice in time series analysis, where it is recommended that both the sample autocovariance function and sample partial autocovariance function be examined.

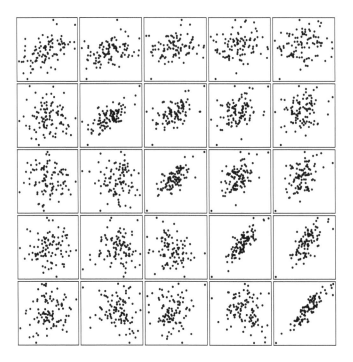

Figure 4.6 *Ordinary scatterplot matrix (upper triangle, including main diagonal) and PRISM (lower diagonal, including main diagonal) of data simulated from an AD(1) process.*

Example: 100-km race data

Figure 4.2 (lower triangle) displays the PRISM of the standardized split times for the complete data. Apart from scatterplots on the main diagonal, the plot in row 7, column 6, and possibly a few others, the plots do not indicate appreciable partial associations. This agrees with the identification of near-zero intervenor-adjusted partial correlations given in Table 4.3(a). Furthermore, the scatterplot in row 4, column 3 suggests that the significant negative estimate of $\rho_{53 \cdot 4}$ observed in the aforementioned table is due to one aberrant observation, corresponding to subject 54 listed in Table 1.8, and that it might therefore be discounted. Note that among all competitors, the 54th subject ran slowest on each of the first three sections of the race and third slowest on the fourth section, but much faster than average on section 5, and it is his dramatic change of pace on the fifth section that appears to be responsible for the corresponding

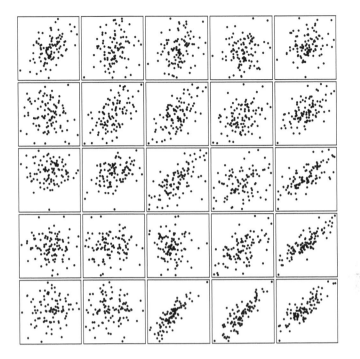

Figure 4.7 *Ordinary scatterplot matrix (upper triangle, including main diagonal) and PRISM (lower triangle, including main diagonal) of data simulated from an AD(0,1,1,2,1,3) process.*

point being so isolated in all of the column-four scatterplots of the PRISM. As expected, removal of this subject from the data changes the estimate of $\rho_{53 \cdot 4}$ substantially (from –0.34 to –0.07). Whether removal of this subject appreciably affects how well various models actually fit is a matter to be taken up by further analyses of these data presented later.

Together, the matrix of sample intervenor-adjusted partial correlations and the PRISM suggest that antedependence models up to order three should certainly be fit to the data, and perhaps that antedependence models of orders four and five should be fit as well. Variable-order antedependence models up to order five should also be considered. Later in this book we will fit and compare several of these models, and the fits will be seen to comport well with the graphical analysis presented here. The PRISM also points to some possible further investigations, such as identifying and deciding how to deal with the outlier visible in the fourth row and fourth column of scatterplots, and seeking a plausible explanation for the positive, significantly-different-from-zero partial correlations adjusted for one or more intervenors in the last few sections of the race.

4.3.3 Precision matrix structure

The precision matrix formulation may be parameterized by the partial variances, i.e., the residual variances from the regressions of Y_i on all $n-1$ of the other responses, and the partial correlations, i.e., the correlations between residuals from regressions of Y_i and Y_j on all $n-2$ other responses. The former are merely the reciprocals of the main diagonal elements of the precision matrix, and the latter can be viewed as negatives of ordinary correlations between variables having Σ^{-1} as their covariance matrix. Thus, potentially useful graphical diagnostics for this formulation include a plot of the sample partial variances against time and a plot of the sample partial correlations against time or lag. The relevant scatterplot matrix is the aptly named *partial scatterplot matrix* (Davison and Sardy, 2000), which is laid out as indicated in Table 4.6. The (i, j)th scatterplot in this matrix plots the residuals from the regression of Y_{i+1} on all other variables save Y_j, on the residuals from the regression of Y_j on all other variables save Y_{i+1}. As such, this scatterplot matrix is the graphical equivalent of the sample partial correlation matrix. However, if the OSM and PRISM have already been constructed it is generally not worth the trouble to construct the partial scatterplot matrix, for it usually does not reveal anything important about the data not already evident in the OSM and PRISM.

4.3.4 Autoregressive structure

A graphical diagnostic that complements an examination of the elements of the modified Cholesky decomposition of the sample precision matrix was proposed by Pourahmadi (1999, 2002), who called it the *regressogram*. Pourahmadi's regressogram is really two plots: a plot of the autoregressive coefficients versus lag, and a plot of the log innovation variances versus time. Here, we use "regressogram" to mean only the first of these plots, and we call the second plot the *innovariogram*. Furthermore, we distinguish two types of regressograms, corresponding to choosing either lag (Pourahmadi's original prescription) or time as the variable on the horizontal axis.

The relevant scatterplot matrix for identifying autoregressive structure is the partial regression-on-predecessors scatterplot matrix, which is depicted in Table 4.7. The (i, j)th scatterplot in this matrix is a plot of the residuals from the

Table 4.6 *Layout of partial regression plots in a partial scatterplot matrix. Here* $\mathbf{Y}_{-I}$, *where* $I \subset \{1, 2, \ldots, n\}$, *is the set of all response variables except those in* I.

$Y_2\|\mathbf{Y}_{-\{1,2\}}$ vs. $Y_1\|\mathbf{Y}_{-\{1,2\}}$				
$Y_3\|\mathbf{Y}_{-\{1,3\}}$ vs. $Y_1\|\mathbf{Y}_{-\{1,3\}}$	$Y_3\|\mathbf{Y}_{-\{2,3\}}$ vs. $Y_2\|\mathbf{Y}_{-\{2,3\}}$			
$Y_4\|\mathbf{Y}_{-\{1,4\}}$ vs. $Y_1\|\mathbf{Y}_{-\{1,4\}}$	$Y_4\|\mathbf{Y}_{-\{2,4\}}$ vs. $Y_2\|\mathbf{Y}_{-\{2,4\}}$	$Y_4\|\mathbf{Y}_{-\{3,4\}}$ vs. $Y_3\|\mathbf{Y}_{-\{3,4\}}$		
$\vdots$			$\ddots$	
$Y_n\|\mathbf{Y}_{-\{1,n\}}$ vs. $Y_1\|\mathbf{Y}_{-\{1,n\}}$	$Y_n\|\mathbf{Y}_{-\{2,n\}}$ vs. $Y_2\|\mathbf{Y}_{-\{2,n\}}$	$\cdots$		$Y_n\|\mathbf{Y}_{-\{n-1,n\}}$ vs. $Y_{n-1}\|\mathbf{Y}_{-\{n-1,n\}}$

regression of Y_{i+1} on all its predecessors save Y_j, versus the residuals from the regression of Y_j on all those same predecessors of Y_{i+1}. Like the partial scatterplot matrix, however, this scatterplot matrix generally does not add anything to what can be learned from the PRISM.

Example 1: Treatment A cattle growth data
Figure 4.8 displays the innovariogram and regressogram (with lag along the horizontal axis) for the cattle weights. These plots suggest that the cattle weights might be modeled reasonably well by an unconstrained linear SAD model of order between 2 and 10. Their smoothly undulating character also suggests that we take the model's log innovation variances to be a cubic function of time and, if an antedependent model of order higher than eight is adopted, that it also take the autoregressive coefficients to be a cubic function of lag. We shall fit such a model, among many others, to these data in the next chapter.

Example 2: 100-km race data
Figure 4.9 displays the innovariogram and regressogram for lags one to three (with section number along the horizontal axis) for the complete set of split times. Apart from the initial section, the innovation variances are increasing over the course of the race, leveling off near its conclusion. It would thus appear that the possibility exists for modeling the innovation variances parsimoniously. The lag-one autoregressive coefficients fluctuate somewhat but

Table 4.7 *Layout of partial regression plots in a partial regression-on-predecessors scatterplot matrix. Here* $\mathbf{Y}_{-I}$, *where* $I \subset \{1, 2, \ldots, n\}$, *is the set of all response variables except those in* I.

Y_2 vs. Y_1		
$Y_3\|Y_2$ vs. $Y_1\|Y_2$	$Y_3\|Y_1$ vs. $Y_2\|Y_1$	
$Y_4\|Y_2,Y_3$ vs. $Y_1\|Y_2,Y_3$	$Y_4\|Y_1,Y_3$ vs. $Y_2\|Y_1,Y_3$	$Y_4\|Y_1,Y_2$ vs. $Y_3\|Y_1,Y_2$

$\vdots$ $\ddots$

| $Y_n\|\mathbf{Y}_{-\{1,n\}}$ vs. $Y_1\|\mathbf{Y}_{-\{1,n\}}$ | $Y_n\|\mathbf{Y}_{-\{2,n\}}$ vs. $Y_2\|\mathbf{Y}_{-\{2,n\}}$ | $\cdots$ | $Y_n\|\mathbf{Y}_{-\{n-1,n\}}$ vs. $Y_{n-1}\|\mathbf{Y}_{-\{n-1,n\}}$ |

do appear to be smaller over the last three sections than earlier in the race. Same-lag autoregressive coefficients for lags two and three, however, do not display any systematic behavior, but instead seem to have one or two relatively extreme values haphazardly intermingled with near-zero values. The same is true for same-lag autoregressive coefficients of higher order, for which we do not show regressogram plots. Consequently, parametric modeling of autoregressive coefficients beyond those for the first lag is unlikely to be successful.

4.4 Concluding remarks

We have presented several summary statistics and graphical methods that can help identify the mean and covariance structures of longitudinal data. Some of these, especially those that explore the data's mean structure and marginal covariance structure (e.g., profile plot, sample covariance matrix, ordinary scatterplot matrix) are used routinely by analysts of longitudinal data, as indeed they should be. Unfortunately, however, those statistics and graphs that shed light on the data's antedependence or partial covariance structure are used much less frequently. We believe that the examples presented in this chapter

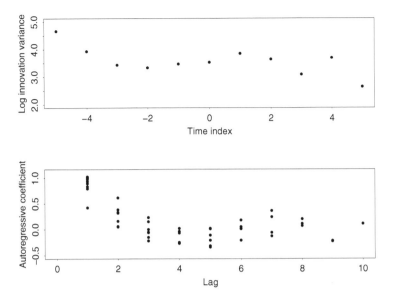

Figure 4.8 *Sample log innovariogram and regressogram for the Treatment A cattle growth data.*

make a strong case that such summary statistics as sample intervenor-adjusted partial correlations and autoregressive coefficients, and graphical tools like the PRISM, should be used just as routinely. A suite of R functions for this purpose, written by the first author, are available for download from his Web page, www.stat.uiowa.edu/~dzimmer.

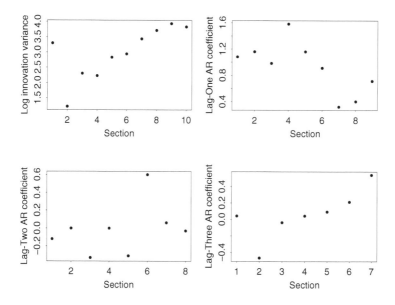

Figure 4.9 *Sample innovariogram and regressogram for lags one to three for the 100-km race split times.*

CHAPTER 5

Likelihood-Based Estimation

Suppose that the mean structure and an antedependent covariance structure for a set of longitudinal data have been tentatively identified, possibly using the informal methods described in the previous chapter. The next step of the data analysis will naturally be to estimate the tentative model's parameters. In this chapter, we consider the estimation of the parameters of this model by likelihood-based methods, under the assumption that the joint distribution of responses on each subject is multivariate normal. Under appropriate regularity conditions, maximum likelihood estimates have several good properties, including consistency, asymptotic efficiency, and asymptotic normality. Some alternative estimation methods for antedependence models are described briefly in Chapter 9.

In the first section of this chapter, we formulate a general normal linear model with antedependent covariance structure for the longitudinal responses, $\{Y_{s1}, \ldots, Y_{sn_s} : s = 1, \ldots, N\}$, within the sampling framework described at the beginning of Chapter 4. This model serves as the basis for estimation in this chapter, as well as for other inference procedures in subsequent chapters. Following that, we describe in detail two likelihood-based estimation procedures, maximum likelihood and residual maximum likelihood (REML), for the parameters of this general model. We then specialize these procedures to several practically important special cases of unstructured antedependence models, for which it is possible to either express the estimators in closed form or otherwise obtain them more efficiently than in the general case. Finally, we specialize the procedures to structured antedependence models.

5.1 Normal linear AD(p) model

5.1.1 Model

The *normal linear AD(p) model* is a special case of the general linear model with parametric covariance structure described in Chapter 1. Like that model, this model allows the mean of Y_{si} to depend on the covariates in $\mathbf{x}_{si}$, but only as a linear function of q parameters. Thus

$$E(Y_{si}) = \mathbf{x}_{si}^T \boldsymbol{\beta}$$

for some vector $\boldsymbol{\beta} \in R^q$. The covariance structure of $Y_{s1}, \ldots, Y_{sn_s}$ in this model is taken to be pth-order antedependent, where p is common across subjects and is assumed, in this chapter, to be known; formal methods for choosing p in practice will be considered in Chapter 6.

Recall that we have considered four distinct parameterizations of an AD(p) covariance structure. Each of these has its own particular advantages for modeling, so we describe further aspects of the model in terms of each. For this purpose let

$$\mathbf{Y}_s = \begin{pmatrix} Y_{s1} \\ Y_{s2} \\ \vdots \\ Y_{sn_s} \end{pmatrix} \quad \text{and} \quad \mathbf{X}_s = \begin{pmatrix} \mathbf{x}_{s1}^T \\ \mathbf{x}_{s2}^T \\ \vdots \\ \mathbf{x}_{sn_s}^T \end{pmatrix}.$$

The marginal parameterization of the normal linear AD(p) model asserts that for $s = 1, \ldots, N$,

$$\mathbf{Y}_s \sim \text{ independent } \mathrm{N}_{n_s}\left(\mathbf{X}_s\boldsymbol{\beta}, \boldsymbol{\Sigma}_s(\boldsymbol{\theta}_\sigma)\right), \tag{5.1}$$

where $\boldsymbol{\Sigma}_s(\boldsymbol{\theta}_\sigma)$ is AD(p); that is, the elements of $\boldsymbol{\Sigma}_s(\boldsymbol{\theta}_\sigma)$ on off-diagonals $p + 1, \ldots, n_s - 1$ can be expressed as functions of elements on the main diagonal and first p off-diagonals by applying equation (2.32). Here we have written the marginal covariance matrices as functions of a parameter vector $\boldsymbol{\theta}_\sigma$, for it is assumed that elements $\{\sigma_{sij}\}$ on the main diagonals and first p off-diagonals of these matrices are given by a parametric function of the measurement times, i.e.,

$$\sigma_{sij} = \sigma(t_{si}, t_{sj}; \boldsymbol{\theta}_\sigma) \quad \text{for } |i - j| \leq p.$$

Alternatively, the marginal covariance matrices may be specified by functions for the variances and correlations, quantities which are easier to interpret than the covariances. In any case, to complete the model formulation, the parameter vectors $\boldsymbol{\beta}$ and $\boldsymbol{\theta}_\sigma$ are assumed to be elements of a specified joint parameter space $\{(\boldsymbol{\beta}, \boldsymbol{\theta}_\sigma): \boldsymbol{\beta} \in R^q, \boldsymbol{\theta}_\sigma \in \Theta_\sigma\}$ where Θ_σ is the set of $\boldsymbol{\theta}_\sigma$-values for which $\boldsymbol{\Sigma}_s(\boldsymbol{\theta}_\sigma)$ is positive definite for all s.

The normal linear AD(p) model can also be parameterized in terms of its intervenor-adjusted formulation. Let $\boldsymbol{\Xi}_s$ denote the matrix of marginal variances, lag-one marginal correlations, and intervenor-adjusted partial correlations in their natural positions, for the sth subject. Let H denote the mapping (which is one-to-one) of the marginal covariance matrix to this matrix, i.e.,

$$H(\boldsymbol{\Sigma}_s) = \boldsymbol{\Xi}_s. \tag{5.2}$$

Then the model may be specified as

$$\mathbf{Y}_s \sim \text{ independent } \mathrm{N}_{n_s}\left(\mathbf{X}_s\boldsymbol{\beta}, H^{-1}[\boldsymbol{\Xi}_s(\boldsymbol{\theta}_\xi)]\right), \quad s = 1, \dots, N, \quad (5.3)$$

where the elements $\{\xi_{sij}\}$ of $\boldsymbol{\Xi}_s$ satisfy $\xi_{sij} = 0$ if $|i-j| > p$ and are otherwise given by the following parametric function of the measurement times:

$$\xi_{sij} = \xi(t_{si}, t_{sj}; \boldsymbol{\theta}_\xi) \quad \text{for } |i - j| \leq p.$$

Here $\boldsymbol{\beta}$ and $\boldsymbol{\theta}_\xi$ are taken to be elements of a specified joint parameter space $\{(\boldsymbol{\beta}, \boldsymbol{\theta}_\xi): \boldsymbol{\beta} \in R^q, \boldsymbol{\theta}_\xi \in \Theta_\xi\}$ where Θ_ξ is the set of $\boldsymbol{\theta}_\xi$-values for which $H^{-1}[\boldsymbol{\Xi}_s(\boldsymbol{\theta}_\xi)]$ is positive definite for all s.

Yet another alternative is to parameterize the model in terms of precision matrices, as follows:

$$\mathbf{Y}_s \sim \text{ independent } \mathrm{N}_{n_s}\left(\mathbf{X}_s\boldsymbol{\beta}, [\boldsymbol{\Sigma}_s^{-1}(\boldsymbol{\theta}_\gamma)]^{-1}\right), \quad s = 1, \dots, N, \quad (5.4)$$

where $\boldsymbol{\Sigma}_s^{-1}$ is the precision matrix of $\mathbf{Y}_s$, with elements $\{\sigma^{sij}\}$ satisfying $\sigma^{sij} = 0$ if $|i - j| > p$. All remaining elements of the precision matrix are given by a parametric function of the measurement times, i.e.,

$$\sigma^{sij} = \gamma(t_{si}, t_{sj}; \boldsymbol{\theta}_\gamma) \quad \text{for } |i - j| \leq p;$$

or, functions may be given for the more interpretable partial variances and partial correlations. To complete the formulation, $\boldsymbol{\beta}$ and $\boldsymbol{\theta}_\gamma$ are taken to be elements of a specified joint parameter space $\{(\boldsymbol{\beta}, \boldsymbol{\theta}_\gamma): \boldsymbol{\beta} \in R^q, \boldsymbol{\theta}_\gamma \in \Theta_\gamma\}$, where Θ_γ is the set of $\boldsymbol{\theta}_\gamma$-values for which $\boldsymbol{\Sigma}_s^{-1}(\boldsymbol{\theta}_\gamma)$ is positive definite for all s.

Finally, the model can be written in terms of an autoregressive parameterization, which itself can be formulated in two distinct ways. First, in a manner similar to (5.1), (5.3), and (5.4), the joint distribution of each $\mathbf{Y}_s$ may be specified:

$$\mathbf{Y}_s \sim \text{ independent } \mathrm{N}_{n_s}\left(\mathbf{X}_s\boldsymbol{\beta}, [\mathbf{T}_s(\boldsymbol{\theta}_\phi)]^{-1}\mathbf{D}_s(\boldsymbol{\theta}_\delta)[\mathbf{T}_s^T(\boldsymbol{\theta}_\phi)]^{-1}\right),$$
$$s = 1, \dots, N. \quad (5.5)$$

Here $\mathbf{T}_s(\boldsymbol{\theta}_\phi)$ and $\mathbf{D}_s(\boldsymbol{\theta}_\delta)$ are the unit lower triangular and diagonal matrices, respectively, of the modified Cholesky decomposition of the precision matrix of $\mathbf{Y}_s$. The subdiagonal elements $\{-\phi_{sij}\}$ of $\mathbf{T}_s(\boldsymbol{\theta}_\phi)$ satisfy $\phi_{sij} = 0$ if $i - j > p$, owing to the assumed pth-order antedependence of the variables in $\mathbf{Y}_s$. Furthermore, the innovation variances in $\mathbf{D}_s(\boldsymbol{\theta}_\delta)$ and the nonzero autoregressive coefficients in $\mathbf{T}_s(\boldsymbol{\theta}_\phi)$ are expressible as parametric functions of the measurement times, i.e.,

$$\delta_{si} = \delta(t_{si}; \boldsymbol{\theta}_\delta), \quad (5.6)$$
$$\phi_{si,i-k} = \phi(t_{si}, t_{s,i-k}; \boldsymbol{\theta}_\phi) \quad \text{for } k = 1, \dots, p_i \quad (5.7)$$

where we recall that $p_i = \min(i - 1, p)$. Here, $\boldsymbol{\theta}_\delta$ and $\boldsymbol{\theta}_\phi$ are vectors of unknown parameters belonging to specified parameter spaces Θ_δ and Θ_ϕ, the former being a set within which δ_{si} is positive for all s and i. Thus, the joint parameter space is $\{\boldsymbol{\beta} \in R^q, \boldsymbol{\theta}_\delta \in \Theta_\delta, \boldsymbol{\theta}_\phi \in \Theta_\phi\}$. A second formulation expressed in terms of an autoregressive parameterization uses (2.21) to specify the distribution of each element of $\mathbf{Y}_s$ conditionally on its predecessors, as follows:

$$Y_{si}|Y_{s1}, \ldots, Y_{s,i-1} \quad \sim \quad \mathrm{N}\left(\mathbf{x}_{si}^T\boldsymbol{\beta} + \sum_{k=1}^{p_i} \phi_{si,i-k}(Y_{s,i-k} - \mathbf{x}_{s,i-k}^T\boldsymbol{\beta}), \delta_{si}\right),$$
$$i = 1, \ldots, n_s; \, s = 1, \ldots, N. \tag{5.8}$$

To complete this formulation, we also require $\mathbf{Y}_1, \ldots, \mathbf{Y}_N$ to be independent, and we specify the same parametric functions, (5.6) and (5.7), for the innovation variances and autoregressive coefficients, with the same parameter space as in the first formulation.

5.1.2 Estimability

Because the general longitudinal sampling framework imposes no restrictions on subjects' measurement times, the potential exists for the model's covariance structure to be overparameterized. In the worst case, no subjects have any measurement times in common and the pth-order antedependence is unstructured. In this case the number of parameters in the covariance structure is, using (2.7),

$$\sum_{s=1}^{N} \frac{(2n_s - p)(p+1)}{2}.$$

This is far too many parameters to estimate consistently from the observed data. Two factors can improve this situation: replication of measurement times across subjects and a more structured type of antedependence.

In the special case of balanced data, for example, measurement times are common across subjects, with the consequence that the number of covariance structure parameters for multiple subjects is the same as for one subject. Therefore, it is possible in this case to estimate covariance parameters consistently. In fact, the covariance structure for even the least parsimonious pth-order antedependence model, the unstructured AD(p) model, is estimable in this situation, provided that the mean structure is sufficiently simple and N is sufficiently large. Further details for this important case are given in Section 5.3.

Although balancedness (rectangularity) of the data is sufficient for ensuring estimability of an unstructured antedependence model, it is not strictly necessary. That is, an unstructured antedependence model can sometimes be used effectively with unbalanced data, provided that the degree of imbalance is not too extreme. To illustrate, suppose that the response variable could potentially be

measured at four times t_1, t_2, t_3, t_4 for each subject and that the corresponding four variables are AD(1). Further suppose, however, that the response is
actually only measured at times t_1, t_2, t_3 for half of the subjects and at times
t_1, t_2, t_4 for the other half. Then, using (2.38), the covariance matrix among
observed responses for those subjects observed at times t_1, t_2, t_3 is given by

$$\Sigma_{123} = \begin{pmatrix} \sigma_{11} & & \\ (\sigma_{11}\sigma_{22})^{1/2}\rho_1 & \sigma_{22} & \\ (\sigma_{11}\sigma_{33})^{1/2}\rho_1\rho_2 & (\sigma_{22}\sigma_{33})^{1/2}\rho_2 & \sigma_{33} \end{pmatrix},$$

while that for the other half of subjects is given by

$$\Sigma_{124} = \begin{pmatrix} \sigma_{11} & & \\ (\sigma_{11}\sigma_{22})^{1/2}\rho_1 & \sigma_{22} & \\ (\sigma_{11}\sigma_{44})^{1/2}\rho_1\rho_2\rho_3 & (\sigma_{22}\sigma_{44})^{1/2}\rho_2\rho_3 & \sigma_{44} \end{pmatrix}.$$

Inspection of these matrices indicates that all parameters of the covariance
structure are consistently estimable from the data actually observed. Note that
this would not be so if the covariance structure among the four variables was
completely arbitrary, for then σ_{34} would be nonestimable. As another important side note, observe that each of Σ_{123} and Σ_{124} retains an AD(1) covariance
structure (i.e., the correlation on the second off-diagonal is equal to the product
of the two correlations on the first off-diagonal), despite the "missing" observations. This is a manifestation of the preservation of first-order antedependence
for any subsequence of normal AD(1) variables, which we noted in Section
2.5.

Unfortunately, however, some departures from balance result in the non-es-
timability of at least some parameters of an unstructured antedependent covariance matrix. The previous illustration can be modified to demonstrate this.
Suppose that the actual measurement times were t_1, t_3 for half of the subjects
and t_2, t_4 for the other half. Then the two marginal covariance matrices are

$$\Sigma_{13} = \begin{pmatrix} \sigma_{11} & \\ (\sigma_{11}\sigma_{33})^{1/2}\rho_1\rho_2 & \sigma_{33} \end{pmatrix}$$

and

$$\Sigma_{24} = \begin{pmatrix} \sigma_{22} & \\ (\sigma_{22}\sigma_{44})^{1/2}\rho_2\rho_3 & \sigma_{44} \end{pmatrix},$$

and we can see that none of ρ_1, ρ_2, ρ_3 are estimable. Nevertheless, in cases such
as this one a structured AD model may be used successfully. In fact, some SAD
models are sufficiently parsimonious that their covariance parameters are estimable regardless of the degree of imbalance. Consider, for example, adapting
the continuous-time AR(1) model given by (3.4) to the general longitudinal
sampling framework, as follows:

$$\text{var}(Y_{si}) = \sigma^2, \quad \text{corr}(Y_{si}, Y_{sj}) = \phi^{|t_{si}-t_{sj}|},$$

The two covariance parameters of this model, σ^2 and ϕ, are estimable even if no two subjects have any measurement times in common.

5.2 Estimation in the general case

Now we consider likelihood-based estimation of the parameters of the normal linear AD(p) model. In this section we denote the parameters of the covariance structure generically by $\boldsymbol{\theta} \in \Theta$, which can represent either $\boldsymbol{\theta}_\sigma$, $\boldsymbol{\theta}_\xi$, $\boldsymbol{\theta}_\gamma$, or $(\boldsymbol{\theta}_\delta^T, \boldsymbol{\theta}_\phi^T)^T$, according to how the covariance structure is parameterized, and we assume that $\boldsymbol{\theta}$ is consistently estimable.

The likelihood function associated with any of the joint formulations (5.1), (5.3), (5.4), or (5.5) of the model is given by

$$
L(\boldsymbol{\beta}, \boldsymbol{\theta}) = (2\pi)^{-n_+/2} \prod_{s=1}^{N} |\boldsymbol{\Sigma}_s(\boldsymbol{\theta})|^{-1/2}
$$

$$
\times \exp\left\{ -\frac{1}{2} \sum_{s=1}^{N} (\mathbf{Y}_s - \mathbf{X}_s\boldsymbol{\beta})^T [\boldsymbol{\Sigma}_s(\boldsymbol{\theta})]^{-1} (\mathbf{Y}_s - \mathbf{X}_s\boldsymbol{\beta}) \right\},
$$

$$(5.9)$$

where $n_+ = \sum_{s=1}^{N} n_s$. Therefore, the log-likelihood function is

$$
\log L(\boldsymbol{\beta}, \boldsymbol{\theta}) = -\frac{n_+}{2} \log 2\pi - \frac{1}{2} \sum_{s=1}^{N} \log |\boldsymbol{\Sigma}_s(\boldsymbol{\theta})|
$$

$$
-\frac{1}{2} \sum_{s=1}^{N} (\mathbf{Y}_s - \mathbf{X}_s\boldsymbol{\beta})^T [\boldsymbol{\Sigma}_s(\boldsymbol{\theta})]^{-1} (\mathbf{Y}_s - \mathbf{X}_s\boldsymbol{\beta}).
$$

$$(5.10)$$

Maximum likelihood estimates of $\boldsymbol{\beta}$ and $\boldsymbol{\theta}$ are any values of these parameters (in R^q and Θ, respectively) that maximize $L(\boldsymbol{\beta}, \boldsymbol{\theta})$ or (equivalently) $\log L(\boldsymbol{\beta}, \boldsymbol{\theta})$.

Now assume, without loss of generality, that $\mathbf{X}_s$ is of full column rank q for at least one s. For any $\boldsymbol{\theta}_0 \in \Theta$ define

$$
\hat{\boldsymbol{\beta}}(\boldsymbol{\theta}_0) = \left(\sum_{s=1}^{N} \mathbf{X}_s^T [\boldsymbol{\Sigma}_s(\boldsymbol{\theta}_0)]^{-1} \mathbf{X}_s \right)^{-1} \sum_{s=1}^{N} \mathbf{X}_s^T [\boldsymbol{\Sigma}_s(\boldsymbol{\theta}_0)]^{-1} \mathbf{Y}_s, \quad (5.11)
$$

which would be the generalized least squares estimator of $\boldsymbol{\beta}$ if var($\mathbf{Y}_s$) was equal to $\boldsymbol{\Sigma}_s(\boldsymbol{\theta}_0)$ for each s. By employing the standard device of adding and subtracting $\mathbf{X}_s\hat{\boldsymbol{\beta}}(\boldsymbol{\theta})$ to and from each occurrence of $(\mathbf{Y}_s - \mathbf{X}_s\boldsymbol{\beta})$ in (5.10) and expanding the resulting quadratic form, the log-likelihood may be reexpressed as follows:

$$
\begin{aligned}
\log L(\boldsymbol{\beta}, \boldsymbol{\theta}) \;=\; & -\frac{n_+}{2} \log 2\pi - \frac{1}{2} \sum_{s=1}^{N} \log |\boldsymbol{\Sigma}_s(\boldsymbol{\theta})| \\
& -\frac{1}{2} \sum_{s=1}^{N} (\mathbf{Y}_s - \mathbf{X}_s \hat{\boldsymbol{\beta}}(\boldsymbol{\theta}))^T [\boldsymbol{\Sigma}_s(\boldsymbol{\theta})]^{-1} (\mathbf{Y}_s - \mathbf{X}_s \hat{\boldsymbol{\beta}}(\boldsymbol{\theta})) \\
& -\frac{1}{2} (\hat{\boldsymbol{\beta}}(\boldsymbol{\theta}) - \boldsymbol{\beta})^T \left\{ \sum_{s=1}^{N} \mathbf{X}_s^T [\boldsymbol{\Sigma}_s(\boldsymbol{\theta})]^{-1} \mathbf{X}_s \right\} (\hat{\boldsymbol{\beta}}(\boldsymbol{\theta}) - \boldsymbol{\beta}).
\end{aligned}
$$

$$(5.12)$$

Clearly, for each $\boldsymbol{\theta} \in \Theta$ this is maximized with respect to $\boldsymbol{\beta}$ by $\hat{\boldsymbol{\beta}}(\boldsymbol{\theta})$. Substituting $\hat{\boldsymbol{\beta}}(\boldsymbol{\theta})$ for $\boldsymbol{\beta}$ in (5.12) yields the profile log-likelihood function,

$$
\begin{aligned}
\log L^*(\boldsymbol{\theta}) \;=\; & -\frac{n_+}{2} \log 2\pi - \frac{1}{2} \sum_{s=1}^{N} \log |\boldsymbol{\Sigma}_s(\boldsymbol{\theta})| \\
& -\frac{1}{2} \sum_{s=1}^{N} (\mathbf{Y}_s - \mathbf{X}_s \hat{\boldsymbol{\beta}}(\boldsymbol{\theta}))^T [\boldsymbol{\Sigma}_s(\boldsymbol{\theta})]^{-1} (\mathbf{Y}_s - \mathbf{X}_s \hat{\boldsymbol{\beta}}(\boldsymbol{\theta})).
\end{aligned}
$$

$$(5.13)$$

Thus, a maximum likelihood estimate of $\boldsymbol{\theta}$ is any $\hat{\boldsymbol{\theta}} \in \Theta$ at which $\log L^*(\boldsymbol{\theta})$ attains its maximum, and the corresponding maximum likelihood estimate of $\boldsymbol{\beta}$ is $\hat{\boldsymbol{\beta}}(\hat{\boldsymbol{\theta}})$.

The problem of maximizing $\log L^*(\boldsymbol{\theta})$ is a constrained (over the set Θ) nonlinear optimization problem for which a closed-form solution exists only in special cases. Some practically important special cases will be considered in subsequent sections. In general, however, maximum likelihood estimates of $\boldsymbol{\theta}$ must be obtained numerically. One possible numerical approach is a grid search, but this is only effective when the parameter space for $\boldsymbol{\theta}$ is low-dimensional, or in other words, when the antedependence is highly structured. Alternatively, iterative algorithms can be used. Two important classes of iterative algorithms are gradient algorithms and the Nelder-Mead simplex algorithm (Nelder and Mead, 1965).

In a gradient algorithm, the $(l+1)$st iterate $\boldsymbol{\theta}^{(l+1)}$ is computed by updating the lth iterate $\boldsymbol{\theta}^{(l)}$ according to the equation

$$\boldsymbol{\theta}^{(l+1)} = \boldsymbol{\theta}^{(l)} + \eta^{(l)} \mathbf{M}^{(l)} \mathbf{g}^{(l)},$$

where $\eta^{(l)}$ is a scalar, $\mathbf{M}^{(l)}$ is an $m \times m$ matrix, m is the dimensionality of $\boldsymbol{\theta}$, and $\mathbf{g}^{(l)}$ is the gradient vector of $\log L^*(\boldsymbol{\theta})$ evaluated at $\boldsymbol{\theta} = \boldsymbol{\theta}^{(l)}$, i.e., $\mathbf{g}^{(l)} = \partial \log L^*(\boldsymbol{\theta})/\partial\boldsymbol{\theta}\big|_{\boldsymbol{\theta}=\boldsymbol{\theta}^{(l)}}$. The matrix product of $\mathbf{M}^{(l)}$ and $\mathbf{g}^{(l)}$ can be thought of as defining the search direction (relative to the lth iterate $\boldsymbol{\theta}^{(l)}$), while $\eta^{(l)}$

defines the size of the step to be taken in that direction. Two gradient algorithms commonly used in conjunction with maximizing a log-likelihood function are the Newton-Raphson and Fisher scoring procedures. In the Newton-Raphson procedure, $\mathbf{M}^{(l)}$ is the inverse of the $m \times m$ matrix whose (i,j)th element is $-\partial^2 \log L^*(\boldsymbol{\theta})/\partial\theta_i\partial\theta_j|_{\boldsymbol{\theta}=\boldsymbol{\theta}^{(l)}}$. In the Fisher scoring algorithm, $\mathbf{M}^{(l)} = (\mathbf{B}^{(l)})^{-1}$ where $\mathbf{B}^{(l)}$ is the Fisher information matrix associated with $\log L^*(\boldsymbol{\theta})$ evaluated at $\boldsymbol{\theta}^{(l)}$, i.e., $\mathbf{B}^{(l)}$ is the $m \times m$ matrix whose (i,j)th element is $E\{-\partial^2 \log L^*(\boldsymbol{\theta})/\partial\theta_i\partial\theta_j|_{\boldsymbol{\theta}=\boldsymbol{\theta}^{(l)}}\}$. For both algorithms, one may set $\eta^{(l)}$ equal to 1.0. Thus, Fisher scoring is identical to Newton-Raphson except that the second-order partial derivatives are replaced by their expectations. General expressions for the gradient vector and the second-order partial derivatives of $\log L^*(\boldsymbol{\theta})$ and $\log L(\boldsymbol{\beta},\boldsymbol{\theta})$ and their expectations may be found in Harville (1975).

The Nelder-Mead simplex algorithm (Nelder and Mead, 1965) minimizes a given function of g variables, in our case the negative of the profile log-likelihood function (5.13), by comparing the function's values at the $(g+1)$ vertices of a general simplex, followed by the replacement of the vertex with the highest value by another point. The algorithm adapts itself to the local landscape, elongating down long inclined planes (i.e., extension), changing direction upon encountering a valley at an angle (i.e., reflection), and contracting in the neighborhood of a minimum (i.e., contraction).

There are a number of practical issues to consider when implementing an iterative algorithm for obtaining maximum likelihood estimates of $\boldsymbol{\theta}$. These include choices of parameterization, starting value for $\boldsymbol{\theta}$, convergence criterion, and methods for accommodating constraints on $\boldsymbol{\theta}$. Some guidance on these and other implementation issues for mixed linear models in general is provided by Harville (1977). For antedependence models in particular, the autoregressive parameterization usually is more convenient computationally than the marginal, intervenor-adjusted, or precision matrix parameterizations. This is due partly to the simple positivity constraints on the innovation variances (which, for that matter, may be dispensed with by reparameterizing the innovation variances to their logs) and the complete absence of constraints on the autoregressive coefficients, and partly to particularly simple expressions for the determinants and inverses of covariance matrices in terms of their modified Cholesky decompositions. More details on this point will be provided in Section 5.5.4.

In general, there is no guarantee of uniqueness of the maximum likelihood estimator of $\boldsymbol{\theta}$, nor is it assured that all local maxima of the profile likelihood function are global maxima. In any case, a reasonable practical strategy for determining whether a local maximum obtained by an iterative algorithm is likely to be the unique global maximum is to repeat the algorithm from several widely dispersed starting values.

Although the maximum likelihood estimator of θ has several desirable properties, it has a well-known shortcoming: it is biased as a consequence of the "loss
in degrees of freedom" from estimating β (Harville, 1977). This bias may be
substantial even for samples of moderate size if either the correlation among
temporally proximate observations is strong or q (the dimensionality of β) is
appreciable relative to n_+ (the overall sample size). However, the bias can be
reduced substantially, and in some cases eliminated completely, by employing
the variant of maximum likelihood estimation known as residual maximum
likelihood (REML) estimation. In REML estimation, the likelihood function
(or equivalently the log-likelihood function) associated with $n_+ - q$ linearly
independent linear combinations of the observations known as error contrasts,
rather than the likelihood of the observations, is maximized. An error contrast
is a linear combination of the observations that has expectation zero for all β
and all $\theta \in \Theta$; furthermore, two error contrasts

$$\sum_{s=1}^{N} \mathbf{c}_s^T \mathbf{Y}_s \quad \text{and} \quad \sum_{s=1}^{N} \mathbf{d}_s^T \mathbf{Y}_s$$

are said to be linearly independent if $(\mathbf{c}_1^T, \ldots, \mathbf{c}_N^T)^T$ and $(\mathbf{d}_1^T, \ldots, \mathbf{d}_N^T)^T$ are
linearly independent vectors. It turns out that these contrasts need not be ob-
tained explicitly because the log-likelihood function associated with any set of
$n_+ - q$ linearly independent contrasts differs by at most an additive constant
(which does not depend on β or θ) from the function

$$\log L_R(\boldsymbol{\theta}) = -\frac{n_+ - q}{2} \log 2\pi - \frac{1}{2} \sum_{s=1}^{N} \log |\boldsymbol{\Sigma}_s(\boldsymbol{\theta})|$$

$$-\frac{1}{2} \sum_{s=1}^{N} (\mathbf{Y}_s - \mathbf{X}_s \hat{\boldsymbol{\beta}}(\boldsymbol{\theta}))^T [\boldsymbol{\Sigma}_s(\boldsymbol{\theta})]^{-1} (\mathbf{Y}_s - \mathbf{X}_s \hat{\boldsymbol{\beta}}(\boldsymbol{\theta}))$$

$$-\frac{1}{2} \log \left| \sum_{s=1}^{N} \mathbf{X}_s^T [\boldsymbol{\Sigma}_s(\boldsymbol{\theta})]^{-1} \mathbf{X}_s \right| \tag{5.14}$$

(Harville, 1977). A REML estimate of θ is any value $\tilde{\theta} \in \Theta$ at which $\log L_R(\boldsymbol{\theta})$
attains its maximum. This estimate generally must be obtained via the same
kinds of numerical procedures used to obtain a maximum likelihood estimate.
Expressions for the gradient vector and the second-order partial derivatives of
$\log L_R(\boldsymbol{\theta})$ and their expectations may be found in Harville (1977). The uniqueness of a REML estimate, like that of a maximum likelihood estimate, is not
generally guaranteed. Once a REML estimate of θ is obtained, the corresponding estimate of β is obtained as its generalized least squares estimator evaluated at $\theta = \tilde{\theta}$, i.e., $\tilde{\beta} = \beta(\tilde{\theta})$.

5.3 Unstructured antedependence: Balanced data

In this section, we specialize the results of the previous section to the case in which the antedependence is unstructured and the data are balanced. It turns out that closed-form expressions for the maximum likelihood and REML estimators of the parameters may exist in this case, depending on the model's mean structure.

Because the unstructured $AD(p)$ covariance structure may be parameterized in four distinct ways, we may write it as either $\Sigma(\boldsymbol{\theta}_\sigma)$, $\Sigma(\boldsymbol{\theta}_\xi)$, $\Sigma(\boldsymbol{\theta}_\gamma)$, or $\Sigma(\boldsymbol{\theta}_\delta, \boldsymbol{\theta}_\phi)$, where

$$
\begin{aligned}
\boldsymbol{\theta}_\sigma &= (\sigma_{11}, \ldots, \sigma_{nn}, \sigma_{21}, \ldots, \sigma_{n,n-p})^T, \\
\boldsymbol{\theta}_\xi &= (\sigma_{11}, \ldots, \sigma_{nn}, \rho_{21}, \ldots, \rho_{n,n-1}, \rho_{31\cdot2}, \ldots, \\
&\qquad \rho_{n,n-p\cdot\{n-p+1:n-1\}})^T, \\
\boldsymbol{\theta}_\gamma &= (\sigma^{11}, \ldots, \sigma^{nn}, \sigma^{21}, \ldots, \sigma^{n,n-p})^T, \\
\begin{pmatrix} \boldsymbol{\theta}_\delta \\ \boldsymbol{\theta}_\phi \end{pmatrix} &= (\delta_1, \ldots, \delta_n, \phi_{21}, \ldots, \phi_{n,n-p})^T.
\end{aligned}
$$

For each parameterization, the parameter space for the covariance structure is the subset Θ of $[(2n - p)(p + 1)/2]$-dimensional Euclidean space for which Σ is positive definite. As was noted in Chapter 2, the parameter spaces for $\boldsymbol{\theta}_\sigma$, $\boldsymbol{\theta}_\xi$, and $\boldsymbol{\theta}_\gamma$ cannot be expressed as a Cartesian product of intervals on individual parameters, while the parameter space for $(\boldsymbol{\theta}_\delta^T, \boldsymbol{\theta}_\gamma^T)^T$ is merely $\{\delta_i : \delta_i > 0$ for $i = 1, \ldots, n\}$. The simplicity of the latter parameter space may be advantageous computationally, but certain theoretical results are conveniently expressed or derived in terms of two of the other parameterizations, so we do not limit consideration to merely the autoregressive parameterization.

5.3.1 Saturated mean

Suppose that the model's mean is saturated. In this case, the general linear mean structure for subject s, which appeared in previous expressions as $\mathbf{X}_s\boldsymbol{\beta}$, specializes (since $\mathbf{X}_s = \mathbf{I}_n$ — the $n \times n$ identity matrix — for all s) to simply $\boldsymbol{\beta}$, which we write, using more conventional notation for this case, as $\boldsymbol{\mu}$. Furthermore, the balancedness ensures that the covariance matrix is common across subjects. Thus the general model (5.1) specializes to

$$
\mathbf{Y}_s \sim \text{ iid } N_n(\boldsymbol{\mu}, \Sigma(\boldsymbol{\theta})), \quad s = 1, \ldots, N. \tag{5.15}
$$

The mean vector $\boldsymbol{\mu}$ is unrestricted in R^n, but the positive definite covariance matrix $\Sigma(\boldsymbol{\theta})$ is assumed to be unstructured $AD(p)$, where p is known. Henceforth we refer to model (5.15) as the normal saturated-mean, unstructured $AD(p)$ model.

We first consider the estimation of μ. Substituting $\mathbf{I}_n$ for $\mathbf{X}_s$ in expression (5.11), we find that the maximum likelihood estimator of μ is simply the sample mean vector,

$$\overline{\mathbf{Y}} = (\overline{Y}_i) = \frac{1}{N} \sum_{s=1}^{N} \mathbf{Y}_s,$$

regardless of the value of θ.

Next consider the estimation of the covariance structure. Define

$$\mathbf{A} = (a_{ij}) = \frac{1}{N} \sum_{s=1}^{N} (\mathbf{Y}_s - \overline{\mathbf{Y}})(\mathbf{Y}_s - \overline{\mathbf{Y}})^T \tag{5.16}$$

and

$$\mathbf{S} = (s_{ij}) = \left(\frac{N}{N-1}\right) \mathbf{A}. \tag{5.17}$$

It is well known (e.g., Theorem 3.2.1 of Anderson, 1984) that $\mathbf{A}$ and $\mathbf{S}$ are the maximum likelihood and REML estimators, respectively, of the covariance matrix when it is positive definite but otherwise arbitrary, $N > n$, and the mean is saturated. This is no longer true when the covariance matrix is unstructured AD(p); however, it turns out that $\mathbf{A}$ and $\mathbf{S}$ are still highly relevant, as maximum likelihood and REML estimators of the elements of the unstructured AD(p) covariance matrix are functions of the elements of $\mathbf{A}$ and $\mathbf{S}$.

Specializing equation (5.13) for use with model (5.15), we find that the profile log-likelihood function is given by

$$\log L^*(\boldsymbol{\theta}) \quad = \quad -\frac{nN}{2} \log 2\pi - \frac{N}{2} \log |\boldsymbol{\Sigma}(\boldsymbol{\theta})|$$

$$-\frac{1}{2} \sum_{s=1}^{N} (\mathbf{Y}_s - \overline{\mathbf{Y}})^T [\boldsymbol{\Sigma}(\boldsymbol{\theta})]^{-1} (\mathbf{Y}_s - \overline{\mathbf{Y}}).$$

The sum of quadratic forms in this expression, being scalar-valued, is equal to its trace. Hence, using the invariance of the trace of the product of two matrices of dimensions $a \times b$ and $b \times a$ with respect to order of multiplication, we may rewrite the profile log-likelihood function as

$$\log L^*(\boldsymbol{\theta}) \quad = \quad -\frac{nN}{2} \log 2\pi - \frac{N}{2} \log |\boldsymbol{\Sigma}(\boldsymbol{\theta})|$$

$$-\frac{N}{2} \text{tr}(\mathbf{A}[\boldsymbol{\Sigma}(\boldsymbol{\theta})]^{-1}). \tag{5.18}$$

We are now ready to state and prove a theorem giving the maximum likelihood estimators of the parameters of this model. We provide two proofs of the theorem. The first, which is based on a precision matrix parameterization, is essentially the same as the classical proof of Gabriel (1962) and we include it here in homage to him. The second, which is similar to that of Macchiavelli (1992), is based on a product-of-conditionals, autoregressively parameterized form of the likelihood function. The second proof offers some theoretically advantageous representations and is considerably easier to extend to more complex situations.

Theorem 5.1. *If $\mathbf{Y}_1, \ldots, \mathbf{Y}_N$ are balanced and follow the normal saturated-mean, unstructured AD(p) model, and $N - 1 > p$, then:*

(a) *The maximum likelihood estimators of $\boldsymbol{\mu}$ and $\boldsymbol{\theta}_\sigma$ are, respectively, $\overline{\mathbf{Y}}$ and*

$$\hat{\boldsymbol{\theta}}_\sigma = (a_{11}, a_{22}, \ldots, a_{nn}, a_{21}, \ldots, a_{n,n-p})^T,$$

where $\mathbf{A} = (a_{ij})$ is given by (5.16);

(b) *The maximum likelihood estimator of $\boldsymbol{\Sigma}$ is $\hat{\boldsymbol{\Sigma}} = \boldsymbol{\Sigma}(\hat{\boldsymbol{\theta}}_\sigma)$, and elements of this matrix on off-diagonals $p + 1, \ldots, n$ may be obtained recursively by applying equation (2.32) of Theorem 2.4 to $\hat{\boldsymbol{\Sigma}}$;*

(c) *The maximum likelihood estimators of $\boldsymbol{\Xi}$ and $\boldsymbol{\Sigma}^{-1}$ are $\hat{\boldsymbol{\Xi}} = H(\hat{\boldsymbol{\Sigma}})$ and $\hat{\boldsymbol{\Sigma}}^{-1}$, respectively, and elements on off-diagonals $p + 1, \ldots, n - 1$ of these matrices are equal to zero;*

(d) *The maximum likelihood estimators of the non-trivial autoregressive coefficients $\{\phi_{ij}\}$ and innovation variances $\{\delta_i\}$ are given by*

$$(\hat{\phi}_{i,i-p_i}, \hat{\phi}_{i,i-p_i+1}, \ldots, \hat{\phi}_{i,i-1}) = \mathbf{a}_{i-p_i:i-1,i}^T \mathbf{A}_{i-p_i:i-1}^{-1}, \quad i = 2, \ldots, n,$$

and

$$\hat{\delta}_i = \begin{cases} a_{11} & \text{for } i = 1 \\ a_{ii} - \mathbf{a}_{i-p_i:i-1,i}^T \mathbf{A}_{i-p_i:i-1}^{-1} \mathbf{a}_{i-p_i:i-1,i} & \text{for } i = 2, \ldots, n. \end{cases}$$

Proof. That $\overline{\mathbf{Y}}$ is the maximum likelihood estimator of $\boldsymbol{\mu}$ was established earlier in this section. To show that $\hat{\boldsymbol{\theta}}_\sigma$ is the maximum likelihood estimator of $\boldsymbol{\theta}_\sigma$, let us parameterize the profile log-likelihood function in terms of the nonzero elements of $\boldsymbol{\Sigma}^{-1}$; that is, in terms of

$$\boldsymbol{\theta}_\gamma = (\sigma^{11}, \ldots, \sigma^{nn}, \sigma^{21}, \ldots, \sigma^{n,n-p})^T.$$

By (5.18), the profile log-likelihood function in these terms is given by

$$\log L^*(\boldsymbol{\theta}_\gamma) = -\frac{N}{2} \left(n \log 2\pi + \log |\boldsymbol{\Sigma}(\boldsymbol{\theta}_\gamma)| + \text{tr}(\mathbf{A}[\boldsymbol{\Sigma}(\boldsymbol{\theta}_\gamma)]^{-1}) \right).$$

Now

$$\begin{aligned} \log |\boldsymbol{\Sigma}(\boldsymbol{\theta}_\gamma)| + \text{tr}(\mathbf{A}[\boldsymbol{\Sigma}(\boldsymbol{\theta}_\gamma)]^{-1}) &= -\log |\boldsymbol{\Sigma}^{-1}(\boldsymbol{\theta}_\gamma)| + \sum_{i=1}^n \sum_{j=1}^n a_{ij}\sigma^{ij} \\ &= -\log |\boldsymbol{\Sigma}^{-1}(\boldsymbol{\theta}_\gamma)| + \sum_{i=1}^n a_{ii}\sigma^{ii} \end{aligned}$$

$$+2 \sum \sum_{0 < i-j \leq p} a_{ij} \sigma^{ij}.$$

Let $(\mathbf{\Sigma}^{-1}(\boldsymbol{\theta}_\gamma))_{ij}$ denote the cofactor of σ^{ij}, i.e., $(-1)^{i+j}$ times the determinant of the submatrix of $\mathbf{\Sigma}^{-1}(\boldsymbol{\theta}_\gamma)$ obtained by deleting its ith row and jth column. Using Theorems A.1.4 and A.1.5, we obtain, for $|i - j| \leq p$,

$$\frac{\partial}{\partial \sigma^{ij}} \log L^*(\boldsymbol{\theta}_\gamma) = \begin{cases} (N/2) \left(\dfrac{(\mathbf{\Sigma}^{-1}(\boldsymbol{\theta}_\gamma))_{ii}}{|\mathbf{\Sigma}^{-1}(\boldsymbol{\theta}_\gamma)|} - a_{ii} \right) & \text{if } i = j \\[3mm] N \left(\dfrac{(\mathbf{\Sigma}^{-1}(\boldsymbol{\theta}_\gamma))_{ij}}{|\mathbf{\Sigma}^{-1}(\boldsymbol{\theta}_\gamma)|} - a_{ij} \right) & \text{if } i \neq j \end{cases}$$

$$= \begin{cases} (N/2)(\sigma_{ii} - a_{ii}) & \text{if } i = j \\ N(\sigma_{ij} - a_{ij}) & \text{if } i \neq j. \end{cases}$$

Setting these equations equal to zero, we find that $\boldsymbol{\theta}_\sigma$ is equal to

$$(a_{11}, a_{22}, \ldots, a_{nn}, a_{21}, \ldots, a_{n,n-p})^T$$

at the unique stationary point of $\log L^*(\boldsymbol{\theta}_\gamma)$. Using Theorems A.1.4 and A.1.5 again, it can be verified that the matrix of second-order partial derivatives of $\log L^*(\boldsymbol{\theta}_\gamma)$ with respect to $\boldsymbol{\theta}_\gamma$, at the stationary point, is given by

$$\text{diag}(-\sigma_{11}^2, \ldots, -\sigma_{nn}^2, -4\sigma_{21}^2, \ldots, -4\sigma_{n,n-p}^2),$$

which is clearly negative definite. Hence the stationary point is, in fact, the unique point of maximum. This proves part (a) of the theorem. Part (b) then follows immediately from Theorem 2.4; part (c) follows from Definition 2.3, expression (5.2), and Theorem 2.2; and part (d) follows from expressions (2.22) and (2.23) relating the autoregressive parameterization to the marginal parameterization, the almost sure positiveness of $a_{ii} - \mathbf{a}_{i-p_i:i-1,i}^T \mathbf{A}_{i-p_i:i-1}^{-1} \mathbf{a}_{i-p_i:i-1,i}$ when $N - 1 > p$, and the invariance property of maximum likelihood estimators. □

Alternative proof. The joint density of $\mathbf{Y}_s$, say $f(\mathbf{Y}_s)$, may be written as

$$f(\mathbf{Y}_s) = \prod_{i=1}^{n} f_i(Y_{si}|Y_{s,i-1}, \ldots, Y_{s1})$$

$$= \prod_{i=1}^{n} f_i(Y_{si}|Y_{s,i-1}, \ldots, Y_{s,i-p_i}) \qquad (5.19)$$

where the second equality follows from the definition of pth-order antedependence. Since $\mathbf{Y}_s$ is multivariate normal, each conditional density in expression (5.19) is normal as well; more specifically, it follows from (5.8) that $f_i(Y_{si}|Y_{s,i-1}, \ldots, Y_{s,i-p_i})$ is the density of a

$$N\left(\mu_i + \sum_{k=1}^{p_i} \phi_{i,i-k}(Y_{s,i-k} - \mu_{i-k}), \delta_i \right)$$

random variable. Therefore the log-likelihood function may be written in terms of the autoregressive parameterization as follows:

$$
\log L(\boldsymbol{\mu}, \boldsymbol{\theta}_\delta, \boldsymbol{\theta}_\phi) = \log \prod_{s=1}^{N} \prod_{i=1}^{n} f_i(Y_{si}|Y_{s,i-1}, \ldots, Y_{s,i-p_i}) \tag{5.20}
$$

$$
= \sum_{i=1}^{n} \sum_{s=1}^{N} \log \left\{ (2\pi\delta_i)^{-1/2} \exp \left\{ - \left[Y_{si} - \mu_i \right.\right.\right.
$$
$$
\left.\left.\left. - \sum_{k=1}^{p_i} \phi_{i,i-k}(Y_{s,i-k} - \mu_{i-k}) \right]^2 \middle/ 2\delta_i \right\} \right\} \tag{5.21}
$$

$$
= -\frac{nN}{2} \log 2\pi - \frac{1}{2} \sum_{i=1}^{n} \sum_{s=1}^{N} \left\{ \log \delta_i + \left[Y_{si} - \mu_i \right.\right.
$$
$$
\left.\left. - \sum_{k=1}^{p_i} \phi_{i,i-k}(Y_{s,i-k} - \mu_{i-k}) \right]^2 \middle/ \delta_i \right\}. \tag{5.22}
$$

Now define

$$
\mu_i^* = \mu_i - \sum_{k=1}^{p_i} \phi_{i,i-k}\mu_{i-k}
$$

and put $\boldsymbol{\mu}^* = (\mu_i^*)$. Observe that $\boldsymbol{\mu}^* = \mathbf{T}\boldsymbol{\mu}$, where $\mathbf{T}$ is the unit lower triangular matrix of the modified Cholesky decomposition of $\boldsymbol{\Sigma}^{-1}$. The transformation from $\boldsymbol{\mu}$ to $\boldsymbol{\mu}^*$ is one-to-one, as $\mathbf{T}$ is nonsingular. Therefore, we may maximize a reparameterized version of (5.22), i.e.,

$$
\log L(\boldsymbol{\mu}^*, \boldsymbol{\theta}_\delta, \boldsymbol{\theta}_\phi) = -\frac{nN}{2} \log 2\pi - \frac{1}{2} \sum_{i=1}^{n} \sum_{s=1}^{N} \left\{ \log \delta_i + \left(Y_{si} - \mu_i^* \right.\right.
$$
$$
\left.\left. - \sum_{k=1}^{p_i} \phi_{i,i-k}Y_{s,i-k} \right)^2 \middle/ \delta_i \right\}, \tag{5.23}
$$

to obtain maximum likelihood estimates $\hat{\boldsymbol{\mu}}^* = (\hat{\mu}_i^*)$, $\hat{\boldsymbol{\theta}}_\delta$, and $\hat{\boldsymbol{\theta}}_\phi$ and then, using the invariance property of maximum likelihood estimation, obtain the maximum likelihood estimator of $\boldsymbol{\mu}$ as

$$
\hat{\boldsymbol{\mu}} = [\mathbf{T}(\hat{\boldsymbol{\theta}}_\phi)]^{-1}\hat{\boldsymbol{\mu}}^*. \tag{5.24}
$$

Now, the ith term of the first sum in (5.23) is, apart from an additive constant, simply the log-likelihood corresponding to an ordinary least squares regression, with intercept, of the ith response variable on the p_i response variables immediately preceding it. There are n of these regressions, each based on N observations of the variables. The $\{\phi_{i,i-k}\}$, $\{\delta_i\}$, and $\{\mu_i^*\}$ are the slope coefficients, error variances, and intercepts, respectively, in these regressions. Note that each error variance is estimable since $N > p + 1$. Therefore, it follows from standard regression theory that the maximum likelihood estimators of the $\{\phi_{i,i-k}\}$ and $\{\delta_i\}$ are given by the expressions in part (d) of the theorem. Also, standard regression theory tells us that the maximum likelihood estimators of the intercepts are given by their least squares estimators, i.e.,

$$\hat{\mu}_i^* = \overline{Y}_i - \sum_{k=1}^{p_i} \hat{\phi}_{i,i-k} \overline{Y}_{i-k},$$

or equivalently in matrix form, $\hat{\boldsymbol{\mu}}^* = \mathbf{T}(\hat{\boldsymbol{\theta}}_\phi)\overline{\mathbf{Y}}$. By (5.24), we obtain $\hat{\boldsymbol{\mu}} = \overline{\mathbf{Y}}$. The remaining parts of the theorem follow by expressions (2.16) and (2.17) relating the marginal and precision matrix parameterizations to the autoregressive parameterization, and the invariance property of maximum likelihood estimators. □

To illustrate Theorem 5.1, consider the first-order case. Recalling from equation (2.39) that the elements of $\boldsymbol{\Sigma}$ on off-diagonals $2, \ldots, n$ may be expressed as ratios of products of elements on the first off-diagonal and main diagonal in the first-order case, part (b) of the theorem yields

$$\hat{\sigma}_{ii} = a_{ii}, \quad i = 1, \ldots, n,$$

$$\hat{\sigma}_{i,i-1} = a_{i,i-1}, \quad i = 2, \ldots, n,$$

$$\hat{\sigma}_{ij} = \frac{\prod_{m=j}^{i-1} a_{m+1,m}}{\prod_{m=j+1}^{i-1} a_{mm}}, \quad i = 3, \ldots, n; \ j = 1, \ldots, i - 2.$$

Thus, in the first-order case, the maximum likelihood estimates of the elements of the covariance matrix may be expressed as explicit functions of the maximum likelihood estimates of certain elements (namely those on the main diagonal and first off-diagonal) of the covariance matrix under general multivariate dependence.

As a further illustration of part (b) of Theorem 5.1, let us revisit the case of an unstructured AD(2) model with $n = 5$ that was used in Section 2.3.3 to illustrate the recursive computation of elements on off-diagonals $p+1, \ldots, n$ of the covariance matrix from those on previous diagonals. The maximum likelihood estimator of $\boldsymbol{\Sigma}$ in this case is

$$\hat{\boldsymbol{\Sigma}} = \begin{pmatrix} a_{11} & & & & symm \\ a_{21} & a_{22} & & & \\ a_{31} & a_{32} & a_{33} & & \\ \hat{\sigma}_{41} & a_{42} & a_{43} & a_{44} & \\ \hat{\sigma}_{51} & \hat{\sigma}_{52} & a_{53} & a_{54} & a_{55} \end{pmatrix}$$

where, using expressions (2.29) and (2.31),

$$\hat{\sigma}_{41} = (a_{21}, a_{31}) \begin{pmatrix} a_{22} & a_{32} \\ a_{23} & a_{33} \end{pmatrix}^{-1} \begin{pmatrix} a_{42} \\ a_{43} \end{pmatrix}$$

and

$$
\begin{pmatrix} \hat{\sigma}_{51} \\ \hat{\sigma}_{52} \end{pmatrix} = \begin{pmatrix} a_{31} & \hat{a}_{41} \\ a_{32} & a_{42} \end{pmatrix} \begin{pmatrix} a_{33} & a_{43} \\ a_{34} & a_{44} \end{pmatrix}^{-1} \begin{pmatrix} a_{53} \\ a_{54} \end{pmatrix}.
$$

As this example illustrates, maximum likelihood estimates of the elements of antedependent covariance matrices of order two and higher are also functions of maximum likelihood estimates of the elements of the covariance matrix under general multivariate dependence, but the functional dependence is not as explicit as it is in the first-order case.

Next we consider REML estimation.

Theorem 5.2. *If* $\mathbf{Y}_1, \ldots, \mathbf{Y}_N$ *are balanced and follow the normal saturated-mean, unstructured AD(p) model, and* $N - 1 > p$, *then:*

(a) *The residual maximum likelihood estimator of* $\boldsymbol{\theta}_\sigma$ *is*

$$
\tilde{\boldsymbol{\theta}}_\sigma = (s_{11}, s_{22}, \ldots, s_{nn}, s_{21}, \ldots, s_{n,n-p})^T,
$$

   where $\mathbf{S} = (s_{ij})$ *is given by (5.17), and the corresponding generalized least squares estimator of* $\boldsymbol{\mu}$ *is* $\overline{\mathbf{Y}}$;

(b) *The residual maximum likelihood estimator of* $\boldsymbol{\Sigma}$ *is* $\tilde{\boldsymbol{\Sigma}} = \boldsymbol{\Sigma}(\tilde{\boldsymbol{\theta}}_\sigma)$, *and elements of this matrix on off-diagonals* $p + 1, \ldots, n$ *may be obtained recursively by applying equation (2.32) of Theorem 2.4 to* $\tilde{\boldsymbol{\Sigma}}$;

(c) *The residual maximum likelihood estimators of* $\boldsymbol{\Xi}$ *and* $\boldsymbol{\Sigma}^{-1}$ *are* $H(\tilde{\boldsymbol{\Sigma}})$ *and* $\tilde{\boldsymbol{\Sigma}}^{-1}$, *respectively, and elements on off-diagonals* $p + 1, \ldots, n - 1$ *of these matrices are equal to zero;*

(d) *The residual maximum likelihood estimators of the non-trivial autoregressive coefficients* $\{\phi_{ij}\}$ *and innovation variances* $\{\delta_i\}$ *are given by*

$$
(\tilde{\phi}_{i,i-p_i}, \tilde{\phi}_{i,i-p_i+1}, \ldots, \tilde{\phi}_{i,i-1}) = \mathbf{s}^T_{i-p_i:i-1,i} \mathbf{S}^{-1}_{i-p_i:i-1}, \quad i = 2, \ldots, n,
$$

   and

$$
\tilde{\delta}_i = \begin{cases} s_{11} & \text{for } i = 1 \\ s_{ii} - \mathbf{s}^T_{i-p_i:i-1,i} \mathbf{S}^{-1}_{i-p_i:i-1} \mathbf{s}_{i-p_i:i-1,i} & \text{for } i = 2, \ldots, n. \end{cases}
$$

Proof. By (5.14) and (5.18), the residual log-likelihood function is given by

$$
\begin{aligned} \log L_R(\boldsymbol{\theta}) &= -\frac{n(N-1)}{2} \log 2\pi - \frac{n}{2} \log N - \frac{N-1}{2} \log |\boldsymbol{\Sigma}(\boldsymbol{\theta})| \\ &\quad - \frac{N-1}{2} \text{tr}(\mathbf{S}[\boldsymbol{\Sigma}(\boldsymbol{\theta})]^{-1}). \end{aligned}
$$

A very similar approach to that used to maximize the profile log-likelihood function in the first proof of Theorem 5.1 may be used to show that the unique point of maximum of $\log L_R(\boldsymbol{\theta})$ is $\tilde{\boldsymbol{\theta}}_\sigma = (s_{11}, s_{22}, \ldots, s_{nn}, s_{21}, \ldots, s_{n,n-p})^T$. This, plus our earlier observation that $\overline{\mathbf{Y}}$ is the generalized least squares estimator of $\boldsymbol{\mu}$ regardless of the value of $\boldsymbol{\theta}$, proves part (a) of the theorem. The remaining parts follow easily from the invariance property of REML estimators. $\square$

By comparing Theorems 5.1 and 5.2, it is evident that the maximum likelihood estimators and REML estimators are proportional to each other. That is,

$$\tilde{\boldsymbol{\theta}}_\sigma = [N/(N-1)]\hat{\boldsymbol{\theta}}_\sigma, \quad \tilde{\boldsymbol{\Sigma}} = [N/(N-1)]\hat{\boldsymbol{\Sigma}}, \quad \tilde{\boldsymbol{\Sigma}}^{-1} = [(N-1)/N]\hat{\boldsymbol{\Sigma}}^{-1},$$

and for all i and j,

$$\tilde{\phi}_{ij} = \hat{\phi}_{ij}, \quad \tilde{\delta}_i = [N/(N-1)]\hat{\delta}_i.$$

Moreover, as a consequence of this proportionality, the maximum likelihood and REML estimators of a correlation — be it marginal, partial, or intervenor-adjusted — are equal. Proportionality among maximum likelihood and REML estimators of the elements and/or parameters of a covariance matrix does not generally occur, so it is an unusual and fortuitous circumstance that it does occur for unstructured AD(p) models.

The next two theorems and their corollaries establish some statistical properties of $\overline{\mathbf{Y}}$, $\tilde{\boldsymbol{\theta}}_\sigma$, and $\tilde{\boldsymbol{\Sigma}}$ under the model considered in this section.

Theorem 5.3. *If $\mathbf{Y}_1, \ldots, \mathbf{Y}_N$ are balanced and follow the normal saturated-mean, unstructured AD(p) model, and $N - 1 > p$, then $\overline{\mathbf{Y}}$ and the REML estimator $\tilde{\boldsymbol{\Sigma}}$ are unbiased.*

Proof. Showing that $\overline{\mathbf{Y}}$ is unbiased is trivial. To show that $\tilde{\boldsymbol{\Sigma}}$ is unbiased, first note from Theorem 5.2(b) that $\tilde{\sigma}_{i,i-j} = s_{i,i-j}$ for $j = 0, \ldots, p$. It follows from the well-known unbiasedness of $\mathbf{S}$ that

$$E(\tilde{\sigma}_{i,i-j}) = \sigma_{i,i-j} \quad \text{for } j = 0, \ldots, p. \tag{5.25}$$

To establish the unbiasedness of the remaining elements of $\tilde{\boldsymbol{\Sigma}}$, we exploit the product-of-conditionals form (5.22) of the log-likelihood function and equation (2.27) for recursively obtaining elements on off-diagonals $p + 1, \ldots, n$ of $\boldsymbol{\Sigma}$ from those on previous diagonals. As noted in the alternative proof of Theorem 5.1, the maximum likelihood estimators $(\hat{\phi}_{i,i-p_i}, \ldots, \hat{\phi}_{i,i-1})$ of the non-trivial autoregressive coefficients are merely the estimated regression coefficients from the ordinary least squares regression of the response variable at time i on the responses at the previous p_i times (with an intercept). As such, we know from standard regression theory that, conditional on all responses prior to time i, $(\hat{\phi}_{i,i-p_i}, \ldots, \hat{\phi}_{i,i-1})$ is unbiased and hence, since the maximum likelihood estimators and REML estimators of the autoregressive coefficients coincide, that

$$E(\tilde{\phi}_{i,i-p_i}, \ldots, \tilde{\phi}_{i,i-1} | \mathbf{Y}_1^*, \ldots, \mathbf{Y}_{i-1}^*) = (\phi_{i,i-p_i}, \ldots, \phi_{i,i-1}). \tag{5.26}$$

Here $\mathbf{Y}_j^* = (Y_{1j}, Y_{2j}, \ldots, Y_{Nj})^T$, i.e., the set of observations from all subjects at time j. It follows from equation (2.27) that for $i = 1, \ldots, n$ and

$j = p_i + 1, \ldots, i - 1$,

$$\tilde{\sigma}_{i,i-j} = \sum_{k=1}^{p_i} \tilde{\phi}_{i,i-k} \tilde{\sigma}_{i-k,i-j}. \qquad (5.27)$$

Consider the case $j = p_i + 1$ of (5.27):

$$\tilde{\sigma}_{i,i-p_i-1} = \sum_{k=1}^{p_i} \tilde{\phi}_{i,i-k} \tilde{\sigma}_{i-k,i-p_i-1}.$$

Now using (5.26), (5.25), and (2.27) (in this order), we obtain

$$
\begin{aligned}
E(\tilde{\sigma}_{i,i-p_i-1}) &= E\{E(\tilde{\sigma}_{i,i-p_i-1} | \mathbf{Y}_1^*, \ldots, \mathbf{Y}_{i-1}^*)\} \\
&= E\left\{ \sum_{k=1}^{p_i} \tilde{\sigma}_{i-k,i-p_i-1} E(\tilde{\phi}_{i,i-k} | \mathbf{Y}_1^*, \ldots, \mathbf{Y}_{i-1}^*) \right\} \\
&= E\left\{ \sum_{k=1}^{p_i} \tilde{\sigma}_{i-k,i-p_i-1} \phi_{i,i-k} \right\} \\
&= \sum_{k=1}^{p_i} \sigma_{i-k,i-p_i-1} \phi_{i,i-k} \\
&= \sigma_{i,i-p_i-1};
\end{aligned}
$$

hence $\tilde{\sigma}_{i,i-p_i+1}$ is unbiased. Unbiasedness for arbitrary j may be proved by induction. Because we have just shown that unbiasedness holds for the case $j = p_i + 1$, it suffices to show that if it holds for $j = p_i + m$, then it holds for $j = p_i + m + 1$. But this can be shown by essentially the same conditioning argument as used above for the case $j = p_i + 1$. Thus $\tilde{\boldsymbol{\Sigma}}$ is unbiased. $\square$

Theorem 5.4. *If* $\mathbf{Y}_1, \ldots, \mathbf{Y}_N$ *are balanced and follow the normal saturated-mean, unstructured AD(p) model, and* $N - 1 > p$, *then* $\overline{\mathbf{Y}}$ *and the REML estimator* $\tilde{\boldsymbol{\theta}}_\sigma = (s_{11}, s_{22}, \ldots, s_{nn}, s_{21}, \ldots, s_{n,n-p})^T$ *are a set of complete sufficient statistics for* $\boldsymbol{\mu}$ *and* $\boldsymbol{\theta}_\sigma$, *and so are* $\overline{\mathbf{Y}}$ *and* $(\tilde{\delta}_1, \tilde{\delta}_2, \ldots, \tilde{\delta}_n, \tilde{\phi}_{21}, \ldots, \tilde{\phi}_{n,n-p})^T$.

Proof. Using a precision matrix parameterization, the likelihood function (5.9) for the general model may be simplified and reexpressed in this special case as

$$
\begin{aligned}
L(\boldsymbol{\mu}, \boldsymbol{\theta}_\gamma) &= (2\pi)^{-nN/2} |\boldsymbol{\Sigma}(\boldsymbol{\theta}_\gamma)|^{-N/2} \exp\left\{ -\frac{N}{2} \boldsymbol{\mu}^T [\boldsymbol{\Sigma}(\boldsymbol{\theta}_\gamma)]^{-1} \boldsymbol{\mu} \right\} \\
&\quad \times \exp\left\{ -\frac{1}{2} \text{tr}(\mathbf{B}[\boldsymbol{\Sigma}(\boldsymbol{\theta}_\gamma)]^{-1}) + N \overline{\mathbf{Y}}^T [\boldsymbol{\Sigma}(\boldsymbol{\theta}_\gamma)]^{-1} \boldsymbol{\mu} \right\}
\end{aligned}
$$

where $\mathbf{B} = (b_{ij}) = \sum_{s=1}^{N} \mathbf{Y}_s \mathbf{Y}_s^T$. Now, by the same argument used to reexpress the trace of $\mathbf{A}[\boldsymbol{\Sigma}(\boldsymbol{\theta}_\gamma)]^{-1}$ in the first proof of Theorem 5.1, we obtain

$$\text{tr}(\mathbf{B}[\boldsymbol{\Sigma}(\boldsymbol{\theta}_\gamma)]^{-1}) = \sum_{i=1}^{n} b_{ii}\sigma^{ii} + 2\sum\sum_{0<i-j\leq p} b_{ij}\sigma^{ij}.$$

Thus, the joint density of the observations may be expressed as follows:

$$\begin{aligned}
L(\boldsymbol{\mu},\boldsymbol{\theta}_\gamma) &= (2\pi)^{-nN/2}|\boldsymbol{\Sigma}(\boldsymbol{\theta}_\gamma)|^{-N/2}\exp\left\{-\frac{N}{2}\boldsymbol{\mu}^T[\boldsymbol{\Sigma}(\boldsymbol{\theta}_\gamma)]^{-1}\boldsymbol{\mu}\right\} \\
&\quad \times \exp\left\{-\frac{1}{2}\left(\sum_{i=1}^{n} b_{ii}\sigma^{ii} + 2\sum\sum_{0<i-j\leq p} b_{ij}\sigma^{ij}\right)\right. \\
&\quad \left. +N\overline{\mathbf{Y}}^T[\boldsymbol{\Sigma}(\boldsymbol{\theta}_\gamma)]^{-1}\boldsymbol{\mu}\right\}.
\end{aligned}$$

Written in this form, the family of joint densities of the observations is seen to be an exponential family. Furthermore, the joint parameter space for $\boldsymbol{\mu}$ and $\boldsymbol{\theta}_\gamma$, while not expressible as the Cartesian product of intervals for individual parameters, nevertheless is easily seen to contain an open rectangle; therefore, by a well-known result (e.g., Theorem 6.2.25 of Casella and Berger, 2002), the set of statistics $(\overline{\mathbf{Y}}^T, b_{11}, b_{22}, \ldots, b_{nn}, b_{21}, \ldots, b_{n,n-p})^T$ is complete. Clearly, this set of statistics is sufficent as well by the Factorization Theorem (Theorem 6.2.6 of Casella and Berger, 2002). Finally, since

$$s_{ij} = \frac{1}{N-1}\left(b_{ij} - N\overline{Y}_i\overline{Y}_j\right)$$

for all i and j, the mapping of $(\overline{\mathbf{Y}}^T, b_{11}, b_{22}, \ldots, b_{nn}, b_{21}, \ldots, b_{n,n-p})^T$ to $(\overline{\mathbf{Y}}^T, s_{11}, s_{22}, \ldots, s_{nn}, s_{21}, \ldots, s_{n,n-p})^T$ and the mapping of $(\overline{\mathbf{Y}}^T, s_{11}, s_{22}, \ldots, s_{nn}, s_{21}, \ldots, s_{n,n-p})^T$ to $(\overline{\mathbf{Y}}^T, \tilde{\delta}_1, \tilde{\delta}_2, \ldots, \tilde{\delta}_n, \tilde{\phi}_{21}, \ldots, \tilde{\phi}_{n,n-p})^T$ are one-to-one, yielding the result. $\square$

Since $\overline{\mathbf{Y}}$ and $\tilde{\boldsymbol{\Sigma}}$ are unbiased (Theorem 5.3) and these estimators are functions of the set of complete sufficient statistics (Theorem 5.4), we have (by, e.g., Theorem 7.3.23 of Casella and Berger, 2002) the following corollary. The first-order case of part (a) of the corollary was established by Byrne and Arnold (1983), the general case by Johnson (1989). The proofs of parts (b) and (c) of the corollary are trivial.

Corollary 5.4.1. *If $\mathbf{Y}_1, \ldots, \mathbf{Y}_N$ are balanced and follow the normal saturated-mean, unstructured AD(p) model, and $N-1 > p$, then:*

(a) $\overline{\mathbf{Y}}$ *and* $\tilde{\boldsymbol{\Sigma}}$ *are the uniformly minimum variance unbiased estimators of* $\boldsymbol{\mu}$ *and* $\boldsymbol{\Sigma}$;

(b) $\overline{\mathbf{Y}}$ *is distributed as* $N(\boldsymbol{\mu}, \frac{1}{N}\boldsymbol{\Sigma})$;

(c) *The covariance matrix of $\overline{\mathbf{Y}}$ may be estimated unbiasedly by $\frac{1}{N}\tilde{\Sigma}$.*

It follows immediately from Corollary 5.4.1(c) that a reasonable estimate of the standard error of $\overline{Y}_i$ is given by

$$\widehat{se}(\overline{Y}_i) = \left(\frac{1}{N}\tilde{\sigma}_{ii}\right)^{1/2},$$

and that this quantity is invariant to the model's order of antedependence. Moreover, from Corollary 5.4.1(b) and well-known results on the joint distribution of $\overline{Y}_i$ and $\tilde{\sigma}_{ii}$, we obtain the following $100(1-\alpha)\%$ confidence interval for μ_i:

$$\overline{Y}_i \pm t_{\alpha/2,N-1}\widehat{se}(\overline{Y}_i).$$

(Here $t_{\alpha/2,N-1}$ is defined, similarly as in (4.1), as the $100(1-\alpha/2)$th percentile of Student's t distribution with $N-1$ degrees of freedom.) In fact, for any i, inference for $\mu_{i:i+p}$ (and any subvector or linear combination thereof) is identical under the normal saturated-mean, unstructured AD(p) model and the general multivariate dependence model.

It should be clear to the reader that Theorems 5.1 through 5.4 and Corollary 5.4.1 extend, with appropriate modifications, to variable-order AD models. Recall that an unstructured AD($p_1, p_2, \ldots, p_n$) model can be parameterized by either the non-redundant marginal covariance matrix elements

$$\{\sigma_{ij} : i = 1, \ldots, n; \ j = i - p_i, \ldots, i\},$$

the variances and non-trivial intervenor-adjusted partial correlations

$$\{\sigma_{ii} : i = 1, \ldots, n\} \text{ and } \{\rho_{ij \cdot \{j+1:i-1\}} : i = 1, \ldots, n; \ j = i - p_i, \ldots, i-1\},$$

or the innovation variances and non-trivial autoregressive coefficients

$$\{\delta_i : i = 1, \ldots, n\} \text{ and } \{\phi_{ij} : i = 1, \ldots, n; \ j = i - p_i, \ldots, i-1\}.$$

Let us extend our previous notation for $\boldsymbol{\theta}_\sigma$, $\boldsymbol{\theta}_\xi$, and $(\boldsymbol{\theta}_\delta^T, \boldsymbol{\theta}_\phi^T)^T$, respectively, to represent these three sets of parameters. We give the following theorem without proof. Note that there is an important difference between part (d) of this theorem and part (d) of Theorem 5.1.

Theorem 5.5. *If $\mathbf{Y}_1, \ldots, \mathbf{Y}_N$ are balanced and follow the normal saturated-mean, unstructured variable-order AD($p_1, \ldots, p_n$) model, and $N-1 > \max_i p_i$, then:*

(a) *The maximum likelihood estimators of $\boldsymbol{\mu}$ and $\boldsymbol{\theta}_\sigma$ are, respectively, $\overline{\mathbf{Y}}$ and*

$$\hat{\boldsymbol{\theta}}_\sigma = (a_{ij} : i = 1, \ldots, n; \ j = i - p_i, \ldots, i)$$

where $\mathbf{A} = (a_{ij})$ is given by (5.16);

(b) *The maximum likelihood estimator of Σ is $\hat{\Sigma} = \Sigma(\hat{\theta}_\sigma)$, and the redundant elements of this matrix may be obtained recursively by applying equation (2.44) of Theorem 2.8 to $\hat{\Sigma}$;*

(c) *The maximum likelihood estimator of Ξ is $\hat{\Xi} = H(\hat{\Sigma})$ where H is defined by (2.6), and those elements of $\hat{\Xi}$ for which $i = 1, \ldots, n$ and $j = 1, \ldots, i - p_i - 1$ are equal to zero;*

(d) *The maximum likelihood estimators of the non-trivial autoregressive coefficients $\{\phi_{ij}\}$ and innovation variances $\{\delta_i\}$ are given by*

$$\hat{\phi}_i^T \equiv (\hat{\phi}_{i,i-p_i}, \hat{\phi}_{i,i-p_i+1}, \ldots, \hat{\phi}_{i,i-1}) \;=\; \mathbf{a}_{i-p_i:i-1,i}^T \hat{\mathbf{A}}_{i-p_i:i-1}^{-1},$$
$$i = 2, \ldots, n,$$

and

$$\hat{\delta}_i = \begin{cases} a_{11} & \text{for } i = 1 \\ a_{ii} - \mathbf{a}_{i-p_i:i-1,i}^T \hat{\mathbf{A}}_{i-p_i:i-1}^{-1} \mathbf{a}_{i-p_i:i-1,i} & \text{for } i = 2, \ldots, n, \end{cases}$$

where $\hat{\mathbf{A}}_{i-p_i:i-1}$ is the matrix consisting of rows $i - p_i$ through $i - 1$, and columns $i - p_i$ through $i - 1$, of

$$\hat{\mathbf{T}}_{i-1}^{-1} \hat{\mathbf{D}}_{i-1} (\hat{\mathbf{T}}_{i-1}^T)^{-1};$$

$\hat{\mathbf{T}}_{i-1}$ is the lower triangular matrix with ones on its main diagonal and $-\hat{\phi}_j^T$ for the elements in its jth row immediately preceding that row's main diagonal element, and its remaining elements all equal to zero; and $\hat{\mathbf{D}}_{i-1}$ is the diagonal matrix with elements $\hat{\delta}_1, \ldots, \hat{\delta}_{i-1}$ on its main diagonal;

(e) *The residual maximum likelihood estimators of the parameters listed in parts (a)-(d) above are given by expressions exactly the same as those of the maximum likelihood estimators except that elements of $\mathbf{A}$ are replaced by the corresponding elements of the matrix $\mathbf{S}$ defined by (5.17);*

(f) *$\overline{\mathbf{Y}}$ and $\tilde{\Sigma}$ (the REML estimator of Σ) are the uniformly minimum variance unbiased estimators of μ and Σ, $\overline{\mathbf{Y}}$ is distributed as $N(\mu, \frac{1}{N}\Sigma)$, and the covariance matrix of $\overline{\mathbf{Y}}$ may be estimated unbiasedly by $\frac{1}{N}\tilde{\Sigma}$.*

Example: Treatment A cattle growth data

Table 5.1 presents REML estimates of marginal variances and correlations for normal saturated-mean, unstructured antedependence models of orders 10, 1, and 2 for the Treatment A cattle growth data. Note that: (a) because there are 11 measurement times, REML estimates for the tenth-order model are merely the ordinary sample variances and correlations; (b) because there are 30 cattle, maximum likelihood estimates of the variances (not shown) under each model are equal to merely 29/30 times their REML counterparts, while the maximum likelihood and REML estimates of the correlations coincide; (c) REML estimates for the AD(1) model coincide with those for the AD(10) model on the main diagonal and first subdiagonal, while estimates for the AD(2) model coincide with those for the AD(10) model on the main diagonal and first two subdiagonals. Comparison of estimates from the models of orders 1 and 10 indicates that the correlations of the first-order model match the sample correlations quite well, the largest discrepancies being due to smaller estimates, under the AD(1) model, of correlations between the response at time 8 and several of its predecessors. The estimated correlations of the second-order model match the sample correlations between the response at time 8 and its predecessors better, but at the expense of producing many overly large estimated correlations at large lags. So it is not clear which model, AD(1) or AD(2), fits best. In Chapter 6 we will revisit this issue using formal hypothesis testing.

5.3.2 Multivariate regression mean

The results of the previous section can be extended to a model having a somewhat more general mean structure, namely that of the classical multivariate regression model. In this model, the same covariate values are used as explanatory variables for all responses from the same subject, or equivalently

$$\mathbf{X}_s = \mathbf{z}_s^T \otimes \mathbf{I}_n, \quad s = 1, \dots, N, \tag{5.28}$$

for some $m \times 1$ covariate vectors $\mathbf{z}_1, \dots, \mathbf{z}_N$, where $\otimes$ denotes the Kronecker product. Note, importantly, that this does not require that the regression coefficients be identical across time, but only that the covariates are so. This, in turn, requires the measurement times to be common to all subjects (balanced data), and the covariates to be time-independent (which precludes using the time of measurement as a covariate). This may seem overly restrictive, but in fact these restrictions are satisfied sufficiently often for this case to be of some importance. The saturated-mean case already considered corresponds to putting $\mathbf{z}_s \equiv 1$ into expression (5.28).

Thus the assumed model in this section is

$$\mathbf{Y}_s \sim \text{independent } N_n \left((\mathbf{z}_s^T \otimes \mathbf{I}_n)\boldsymbol{\beta}, \boldsymbol{\Sigma}(\boldsymbol{\theta}) \right), \quad s = 1, \dots, N, \tag{5.29}$$

where $\mathbf{z}_s$ is $m \times 1$ and $\boldsymbol{\Sigma}(\boldsymbol{\theta})$ is an unstructured AD(p) covariance matrix. We refer to this model as the normal multivariate regression, unstructured AD(p) model.

Let

$$\mathbf{Z} = \begin{pmatrix} \mathbf{z}_1^T \\ \vdots \\ \mathbf{z}_N^T \end{pmatrix}$$

and assume that $\mathbf{Z}$ is of full column rank m. Under model (5.29), we find

Table 5.1 *REML estimates of marginal variances (along the main diagonal) and correlations (off the main diagonal) for the Treatment A cattle growth data, under the normal saturated-mean unstructured AD(p) model: (a) p = 10; (b) p = 1; (c) p = 2. Estimates in (b) and (c) are differences between the REML estimate under the lower-order model and the AD(10) model.*

(a)

106										
.82	155									
.76	.91	165								
.66	.84	.93	185							
.64	.80	.88	.94	243						
.59	.74	.85	.91	.94	284					
.52	.63	.75	.83	.87	.93	307				
.53	.67	.77	.84	.89	.94	.93	341			
.52	.60	.71	.77	.84	.90	.93	.97	389		
.48	.58	.70	.73	.80	.87	.88	.94	.96	470	
.48	.55	.68	.71	.77	.83	.86	.92	.96	.98	445

(b)

0										
.00	0									
-.01	.00	0								
.03	.00	.00	0							
.01	-.01	-.01	.00	0						
.02	.00	.03	-.02	.00	0					
.05	.06	.01	.00	.01	.00	0				
.00	-.02	-.06	-.07	-.07	-.07	.00	0			
.00	.03	-.02	-.02	-.05	-.06	-.03	.00	0		
.02	.02	-.04	-.01	-.04	-.06	-.01	-.01	.00	0	
.01	.04	-.03	.00	-.02	-.03	.00	.00	-.01	.00	0

(c)

0										
.00	0									
.00	.00	0								
.05	.00	.00	0							
.04	.00	.00	.00	0						
.06	.04	.00	.00	.00	0					
.09	.09	.04	.02	.00	.00	0				
.08	.06	.03	.02	.00	.00	.00	0			
.08	.12	.08	.07	.03	.03	.00	.00	0		
.11	.12	.06	.09	.04	.03	.02	.00	.00	0	
.10	.14	.08	.10	.07	.06	.04	.02	.00	.00	0

that for any $\boldsymbol{\theta}_0 \in \Theta$, the generalized least squares estimator of β given by expression (5.11) specializes as follows:

$$
\begin{aligned}
\hat{\beta}(\boldsymbol{\theta}_0) &= \left\{ (\mathbf{Z} \otimes \mathbf{I}_n)^T [\mathbf{I}_N \otimes \boldsymbol{\Sigma}(\boldsymbol{\theta}_0)]^{-1} (\mathbf{Z} \otimes \mathbf{I}_n) \right\}^{-1} \\
&\quad \times (\mathbf{Z} \otimes \mathbf{I}_n)^T [\mathbf{I}_N \otimes \boldsymbol{\Sigma}(\boldsymbol{\theta}_0)]^{-1} \begin{pmatrix} \mathbf{Y}_1 \\ \vdots \\ \mathbf{Y}_N \end{pmatrix} \\
&= [(\mathbf{Z}^T \mathbf{Z})^{-1} \mathbf{Z}^T \otimes \mathbf{I}_n] \begin{pmatrix} \mathbf{Y}_1 \\ \vdots \\ \mathbf{Y}_N \end{pmatrix} \quad\quad (5.30) \\
&\equiv \hat{\beta}.
\end{aligned}
$$

We have written this estimator simply as $\hat{\beta}$ since it does not depend on $\boldsymbol{\theta}_0$, and for the same reason it is the maximum likelihood estimator of β regardless of the maximum likelihood estimate of $\boldsymbol{\theta}$. Observe that it coincides with the ordinary least squares estimator.

Now we extend our previous definitions (5.16) and (5.17) of $\mathbf{A}$ and $\mathbf{S}$ as follows:

$$
\mathbf{A} = (a_{ij}) = \frac{1}{N} \sum_{s=1}^{N} [\mathbf{Y}_s - (\mathbf{z}_s^T \otimes \mathbf{I}_n)\hat{\beta}][\mathbf{Y}_s - (\mathbf{z}_s^T \otimes \mathbf{I}_n)\hat{\beta}]^T \quad\quad (5.31)
$$

and

$$
\mathbf{S} = (s_{ij}) = \left(\frac{N}{N - m} \right) \mathbf{A}. \quad\quad (5.32)
$$

It is easily verified that these extended definitions reduce to those of the previous section when $\mathbf{z}_s \equiv 1$. It is well known (e.g., Theorem 8.2.1 of Anderson, 1984) that $\mathbf{A}$ and $\mathbf{S}$ are the maximum likelihood and REML estimators, respectively, of the covariance matrix when this matrix is positive definite but otherwise arbitrary and the mean structure is that of multivariate linear regression. Furthermore, upon specializing the profile log-likelihood function (5.13) for use with model (5.29), we find that for this model

$$
\log L^*(\boldsymbol{\theta}) = -\frac{nN}{2} \log 2\pi - \frac{N}{2} \log |\boldsymbol{\Sigma}(\boldsymbol{\theta})| - \frac{N}{2} \mathrm{tr}(\mathbf{A}[\boldsymbol{\Sigma}(\boldsymbol{\theta})]^{-1}), \quad (5.33)
$$

which is identical to expression (5.18) except, of course, that $\mathbf{A}$ is defined more generally here. Similarly, upon specializing the residual log-likelihood function (5.14) to this model and using the result

$$
\begin{aligned}
\log \left| \sum_{s=1}^{N} (\mathbf{z}_s^T \otimes \mathbf{I}_n)^T [\boldsymbol{\Sigma}(\boldsymbol{\theta})]^{-1} (\mathbf{z}_s^T \otimes \mathbf{I}_n) \right| &= \log |\mathbf{Z}^T \mathbf{Z} \otimes [\boldsymbol{\Sigma}(\boldsymbol{\theta})]^{-1}| \\
&= \log \{ |\mathbf{Z}^T \mathbf{Z}|^n |[\boldsymbol{\Sigma}(\boldsymbol{\theta})]^{-1}|^m \}
\end{aligned}
$$

(where the last equality follows by Theorem A.1.6), we find that

$$
\begin{aligned}
\log L_R(\boldsymbol{\theta}) \quad = \quad & -\frac{n(N-m)}{2}\log 2\pi - \frac{n}{2}\log|\mathbf{Z}^T\mathbf{Z}| \\
& -\frac{N-m}{2}\log|\boldsymbol{\Sigma}(\boldsymbol{\theta})| - \frac{N-m}{2}\mathrm{tr}(\mathbf{S}[\boldsymbol{\Sigma}(\boldsymbol{\theta})]^{-1}).
\end{aligned}
$$

It is also worth noting that the conditional form of the log-likelihood function for this model may be written in terms of the autoregressive parameterization as

$$
\begin{aligned}
\log L(\boldsymbol{\beta}^*, \boldsymbol{\theta}_\delta, \boldsymbol{\theta}_\phi) \quad = \quad & -\frac{nN}{2}\log 2\pi - \frac{1}{2}\sum_{i=1}^{n}\sum_{s=1}^{N}\left\{\log\delta_i + \left(Y_{si} - \mathbf{z}_s^T\boldsymbol{\beta}_i^*\right.\right. \\
& \left.\left. -\sum_{k=1}^{p_i}\phi_{i,i-k}Y_{s,i-k}\right)^2 \middle/ \delta_i \right\},
\end{aligned}
$$

where

$$
\boldsymbol{\beta}_i^* = \boldsymbol{\beta}_i - \sum_{k=1}^{p_i}\phi_{i,i-k}\boldsymbol{\beta}_{i-k}.
$$

As a consequence of these results, Theorems 5.1 through 5.4 and Corollary 5.4.1 may be extended to balanced data following the normal multivariate regression, unstructured AD(p) model of this section as follows. Proofs of these extensions completely mimic those of the original results, hence they are omitted.

Theorem 5.6. *If* $\mathbf{Y}_1, \ldots, \mathbf{Y}_N$ *are balanced and follow the normal multivariate regression, unstructured AD(p) model with m covariates, and* $N - m > p$, *then:*

(a) *The maximum likelihood estimators of* $\boldsymbol{\beta}$ *and* $\boldsymbol{\theta}_\sigma$ *are, respectively,* $\hat{\boldsymbol{\beta}}$ *and*

$$
\hat{\boldsymbol{\theta}}_\sigma = (a_{11}, a_{22}, \ldots, a_{nn}, a_{21}, \ldots, a_{n,n-p})^T
$$

where $\hat{\boldsymbol{\beta}}$ *is given by expression (5.30) and* $\mathbf{A} = (a_{ij})$ *is given by expression (5.31);*

(b) *The maximum likelihood estimator of* $\boldsymbol{\Sigma}$ *is* $\hat{\boldsymbol{\Sigma}} = \boldsymbol{\Sigma}(\hat{\boldsymbol{\theta}}_\sigma)$, *and the redundant elements of this matrix may be obtained recursively by applying equation (2.32) of Theorem 2.4 to* $\hat{\boldsymbol{\Sigma}}$;

(c) *The maximum likelihood estimators of* $\boldsymbol{\Xi}$ *and* $\boldsymbol{\Sigma}^{-1}$ *are* $H(\hat{\boldsymbol{\Sigma}})$ *and* $\hat{\boldsymbol{\Sigma}}^{-1}$, *respectively, and elements on off-diagonals* $p+1, \ldots, n-1$ *of these matrices are equal to zero;*

(d) *The maximum likelihood estimators of the non-trivial autoregressive coefficients* $\{\phi_{ij}\}$ *and innovation variances* $\{\delta_i\}$ *are given by*

$$
(\hat{\phi}_{i,i-p_i}, \hat{\phi}_{i,i-p_i+1}, \ldots, \hat{\phi}_{i,i-1}) = \mathbf{a}_{i-p_i:i-1,i}^T \mathbf{A}_{i-p_i:i-1}^{-1}, \quad i = 2, \ldots, n,
$$

and

$$\hat{\delta}_i = \begin{cases} a_{11} & \text{for } i = 1 \\ a_{ii} - \mathbf{a}_{i-p_i:i-1,i}^T \mathbf{A}_{i-p_i:i-1}^{-1} \mathbf{a}_{i-p_i:i-1,i} & \text{for } i = 2, \ldots, n; \end{cases}$$

(e) *The residual maximum likelihood estimators of the parameters listed in parts (a)-(d) above are given by expressions exactly the same as those of the maximum likelihood estimators except that elements of* $\mathbf{A}$ *are replaced by the corresponding elements of the matrix* $\mathbf{S}$ *defined by (5.32);*

(f) $\hat{\boldsymbol{\beta}}$ *and* $\tilde{\boldsymbol{\Sigma}}$ *(the REML estimator of* $\boldsymbol{\Sigma}$*) are the uniformly minimum variance unbiased estimators of* $\boldsymbol{\beta}$ *and* $\boldsymbol{\Sigma}$*,* $\hat{\boldsymbol{\beta}}$ *is distributed as* $N(\boldsymbol{\beta}, (\mathbf{Z}^T\mathbf{Z})^{-1} \otimes \tilde{\boldsymbol{\Sigma}})$*, and*

$$(\mathbf{Z}^T\mathbf{Z})^{-1} \otimes \tilde{\boldsymbol{\Sigma}}$$

is an unbiased estimator of $var(\hat{\boldsymbol{\beta}})$*.*

Let $\hat{\beta}_{(k-1)n+i}$ denote the element of $\hat{\boldsymbol{\beta}}$ corresponding to the kth covariate and ith measurement time ($k = 1, \ldots, m$; $i = 1, \ldots, n$). It follows immediately from Theorem 5.6(f) that a reasonable estimate of the standard error of $\hat{\beta}_{(k-1)n+i}$ is given by

$$\widehat{se}(\hat{\beta}_{(k-1)n+i}) = (b_{kk}\tilde{\sigma}_{ii})^{1/2},$$

where b_{kk} is the kth diagonal element of $(\mathbf{Z}^T\mathbf{Z})^{-1}$, and that this estimated standard error is invariant to the model's order of antedependence. It also follows that

$$\hat{\beta}_{(k-1)n+i} \pm t_{\alpha/2, N-m}\widehat{se}(\hat{\beta}_{(k-1)n+i})$$

is a $100(1 - \alpha)\%$ confidence interval for $\beta_{(k-1)n+i}$. In fact, for any i, inference for elements of $\boldsymbol{\beta}$ corresponding to time indices lagged no more than p units apart (and any subset or linear combination of these elements) is identical under the normal multivariate regression, unstructured AD(p) model and the general multivariate dependence model.

Theorem 5.5 of the previous section may also be extended to balanced data following the normal multivariate regression, unstructured AD($p_1, \ldots, p_n$) model. One merely replaces expressions (5.16) and (5.17) for the matrices $\mathbf{A}$ and $\mathbf{S}$ in Theorem 5.5 with their counterparts given by (5.31) and (5.32), replaces $\overline{\mathbf{Y}}$ in part (f) of Theorem 5.5 with $\hat{\boldsymbol{\beta}}$, and replaces the condition $N - 1 > \max_i p_i$ with $N - m > \max_i p_i$.

Example: 100-km race data
Consider the split times of competitors on each 10-km section of the 100-km race. In Chapter 4 we presented an exploratory analysis of these data that suggested that competitors' ages may affect their split times and that this effect may be approximately quadratic. The results of that exploratory analysis also suggested that among constant-order unstructured antedependence models, a third-order model might fit the data reasonably well. Consequently, here we

fit a model with a multivariate regression mean structure, consisting of an intercept and terms for the linear and quadratic effects of age for each section, and a third-order unstructured antedependence covariance structure, to the split times of the 76 competitors whose ages were recorded. The age variable was centered by subtracting its mean, so as to render the linear and quadratic effects nearly orthogonal. Thus our model for the mean structure was

$$E(Y_{si}) = \beta_{0i} + \beta_{1i}[age(s)] + \beta_{2i}[age(s)]^2 \qquad (5.34)$$

where i indexes the 10-km sections and $[age(s)]$ represents the centered age of subject s; the latter quantity, more specifically, is given by $age(s)-39.8815789$. Table 5.2 gives REML estimates (which in this case coincide with ordinary least squares estimates) of the parameters of the mean structure. Interestingly, the magnitudes of estimates of all three types of coefficients (intercept, linear, quadratic) in the mean structure tend to vary in rather systematic ways as the race progresses. The estimated intercepts increase over most of the race, decreasing slightly in the last two sections; this tracks the behavior that was evident from the profile plot given previously (Figure 1.3) and indicates, of course, that on average the competitors decelerate over most of the race, but maintain their speed or accelerate slightly over the final two sections. The estimated linear coefficients are slightly positive at the beginning of the race and then trend negative, indicating that early in the race the slightly younger-than-average competitors run relatively faster, but by the halfway point begin to run relatively slower, than their slightly older-than-average counterparts. Although none of the estimated linear coefficients are statistically significant individually, their reasonably consistent negative trend is compelling. The behavior of the estimated quadratic coefficients is even more interesting. Slightly positive but statistically insignificant early in the race, they become ever larger as the race progresses (apart from the seventh section) and are statistically significant on sections 5, 6, 8, 9, and 10. The positive coefficient for each section indicates that on average, middle-aged competitors run faster than relatively much younger and relatively much older competitors, and that this better relative performance is magnified as the race progresses.

Table 5.3 gives REML estimates of the marginal variances and correlations under AD(9) (general multivariate dependence), AD(3), and AD(0,1,1,1,2,1,1, 2,3,5) models, all with mean structure given by (5.34). The rationale for including this particular variable-order AD model will become evident in Example 2 of Section 6.5. Comparison of the estimated correlations beyond lag three for the two constant-order models [part (b) of the table] indicates that the AD(3) model fits very well until the last two sections of the race, where it underestimates the two highest-lag correlations with the penultimate split time and overestimates some moderate-lag correlations with the last split time. Comparison of the variable-order AD model's estimated correlations with those of the AD(9) model [part (c) of the table] shows that this relatively more

Table 5.2 *REML estimates of age effects on 100-km race split times under an unstructured AD(3) model with multivariate regression mean structure. Estimates of linear and quadratic effects more than two estimated standard errors from zero are listed in boldface.*

Section	Intercept	Linear ($\times 10^{-2}$)	Quadratic ($\times 10^{-3}$)
1	47.7	9.1	1.9
2	50.5	8.9	4.8
3	48.9	1.2	9.9
4	52.2	4.4	13.6
5	53.0	−1.3	**21.7**
6	58.2	−5.4	**25.2**
7	61.2	−4.6	15.3
8	67.2	−28.8	**27.4**
9	66.5	−16.8	**28.2**
10	64.6	−16.2	**34.8**

parsimonious covariance structure fits nearly as well as the AD(3) structure in the first nine sections of the race and rectifies the latter's overestimation of moderate-lag correlations with the last split time; however, it fits the highest-lag correlations with the last split time relatively less well.

Formal testing for the importance of the linear and quadratic effects of age will be taken up in later chapters, as will formal comparisons of the unstructured AD models with each other and with other antedependence models.

5.3.3 Arbitrary linear mean

Now let us relax the assumptions on the model's mean structure even further, by allowing it to be of arbitrary linear form. That is, we take the model now to be

$$\mathbf{Y}_s \sim \text{independent } \mathrm{N}_n\left(\mathbf{X}_s\boldsymbol{\beta}, \boldsymbol{\Sigma}(\boldsymbol{\theta})\right), \quad s = 1, \ldots, N, \qquad (5.35)$$

where $\mathbf{X}_s$ has full column rank q for at least one s, $\boldsymbol{\Sigma}(\boldsymbol{\theta})$ is an unstructured AD(p) covariance matrix, and $N - q > p$. Within the more general mean structure of this model, time of measurement and functions thereof may be included as covariates.

For any $\boldsymbol{\theta}_0 \in \Theta$, expression (5.11) for the generalized least squares estimator of $\boldsymbol{\beta}$ specializes only slightly in this case, to

$$\hat{\boldsymbol{\beta}}(\boldsymbol{\theta}_0) = \left(\sum_{s=1}^{N} \mathbf{X}_s^T[\boldsymbol{\Sigma}(\boldsymbol{\theta}_0)]^{-1}\mathbf{X}_s\right)^{-1} \sum_{s=1}^{N} \mathbf{X}_s^T[\boldsymbol{\Sigma}(\boldsymbol{\theta}_0)]^{-1}\mathbf{Y}_s.$$

Table 5.3 *REML estimates of marginal variances (along the main diagonal) and correlations (off the main diagonal) for the 100-km race split times, under normal multivariate regression-mean, unstructured antedependence models which include an overall intercept and linear and quadratic effects of age for each section:* (a) $AD(9)$; (b) $AD(3)$; (c) $AD(0, 1, 1, 1, 2, 1, 1, 2, 3, 5)$. *Estimates in (b) and (c) are differences between the REML estimates for the AD(3) and AD(9) models, and for the AD(0,1,1,1,2,1,1,2,3,5) and AD(9) models, respectively.*

(a)

27									
.95	35								
.85	.90	49							
.79	.83	.92	58						
.62	.64	.74	.88	88					
.63	.63	.71	.84	.93	145				
.53	.55	.59	.69	.74	.83	108			
.51	.52	.61	.70	.78	.84	.78	140		
.55	.53	.55	.65	.71	.75	.68	.73	136	
.41	.42	.42	.47	.48	.62	.71	.61	.74	154

(b)

0									
.00	0								
.00	.00	0							
.00	.00	.00	0						
−.01	.00	.00	.00	0					
−.04	−.02	.00	.00	.00	0				
−.04	−.03	.00	.00	.00	.00	0			
−.01	.00	−.01	.00	.00	.00	.00	0		
−.11	−.07	−.01	−.02	−.02	.00	.00	.00	0	
−.02	−.01	.05	.09	.12	.05	.00	.00	.00	0

(c)

0									
.00	0								
.00	.00	0							
−.01	.00	.00	0						
.00	.02	.00	.00	0					
−.04	−.02	−.02	−.02	.00	0				
−.05	−.03	−.02	.00	.03	.00	0			
−.02	−.01	−.03	−.02	.00	.00	.00	0		
−.11	−.07	−.03	−.03	−.01	.00	.00	.00	0	
−.11	−.10	−.06	−.04	.00	.00	.00	.00	.00	0

Thus, in contrast to what transpires when the mean structure is that of a multi-variate regression model, here the generalized least squares estimator depends on the covariance parameters. As a consequence, closed-form expressions for the maximum likelihood estimators, $\hat{\beta}$ and $\hat{\theta}$, and the REML estimators, $\tilde{\beta}$ and $\tilde{\theta}$ do not exist. Nevertheless, these estimators may be obtained by iterative maximization algorithms, such as the Nelder-Mean simplex or Newton-Raphson algorithms mentioned earlier in this chapter. Alternatively, the maximum likelihood estimators may be obtained by a simpler algorithm, which consists of iterating between

$$\hat{\beta}^{(l)} = \hat{\beta}\left(\hat{\theta}_\sigma^{(l-1)}\right)$$

and

$$\hat{\theta}_\sigma^{(l)} = \left(a_{11}^{(l)}, \ldots, a_{nn}^{(l)}, a_{21}^{(l)}, \ldots, a_{n,n-p}^{(l)}\right)^T,$$

for $l = 1, 2, \ldots$, where the $a_{ij}^{(l)}$'s are the elements of

$$\mathbf{A}^{(l)} = \frac{1}{N}\sum_{s=1}^{N}\left(\mathbf{Y}_s - \mathbf{X}_s\hat{\beta}^{(l)}\right)\left(\mathbf{Y}_s - \mathbf{X}_s\hat{\beta}^{(l)}\right)^T.$$

To initiate this iterative process, we may set

$$\hat{\theta}_\sigma^{(0)} = (\mathbf{1}_n^T, \mathbf{0}^T)^T,$$

in which case

$$\hat{\beta}^{(1)} = \left(\sum_{s=1}^{N}\mathbf{X}_s^T\mathbf{X}_s\right)^{-1}\sum_{s=1}^{N}\mathbf{X}_s^T\mathbf{Y}_s,$$

the ordinary least squares estimator of β. Upon convergence, the algorithm's final iterates are taken as maximum likelihood estimates.

Estimated standard errors of the elements of $\hat{\beta}$ are given by the square roots of the diagonal elements of the matrix

$$\left(\sum_{s=1}^{N}\mathbf{X}_s^T[\mathbf{\Sigma}(\hat{\theta})]^{-1}\mathbf{X}_s\right)^{-1}. \tag{5.36}$$

Denoting the square root of the kth of those elements by $\widehat{se}(\hat{\beta}_k)$, an approximate $100(1-\alpha)\%$ confidence interval for β_k is given by $\hat{\beta}_k \pm t_{\alpha/2,N-q}\widehat{se}(\hat{\beta}_k)$. Estimated standard errors of the elements of $\tilde{\beta}$ are given by square roots of the diagonal elements of a matrix identical to (5.36) but with $\hat{\theta}$ replaced by $\tilde{\theta}$, and a REML-based approximate $100(1-\alpha)\%$ confidence interval for β_k is given by $\tilde{\beta}_k \pm t_{\alpha/2,N-q}\widehat{se}(\tilde{\beta}_k)$.

Similar inferential procedures to those just described may also be used when the unstructured antedependence is of variable order.

Example: 100-km race data

The multivariate regression-mean, unstructured AD(3) model fitted to the split times in the previous section had 30 parameters in its mean structure. Might a model with a much more parsimonious mean structure fit nearly as well? On the basis of the profile plot of split times (Figure 1.3) and the effects of age on split time noted in the analysis of the previous section, one relatively parsimonious mean structure that may be of interest is a cubic function of (centered) section number, plus linear and quadratic effects of (centered) age. That is,

$$E(Y_{si}) = \beta_0 + \beta_1(i - 5.5) + \beta_2(i - 5.5)^2$$
$$+ \beta_3(i - 5.5)^3 + \beta_4[age(s)] + \beta_5[age(s)]^2 \qquad (5.37)$$

where again i indexes the 10-km sections and $[age(s)]$ represents the centered age of subject s. This mean is not of multivariate regression form, as it includes section-dependent covariates, namely linear, quadratic, and cubic functions of section itself. Nevertheless, REML estimates of its parameters, corresponding to an unstructured AD model of any order, may be obtained using the method described in this section. The fitted mean for the unstructured AD(3) model with this mean structure is

$$\hat{Y}_{si} = \underset{(0.933)}{58.702} + \underset{(0.234)}{3.422} \ (i - 5.5) - \underset{(0.0325)}{0.0529} \ (i - 5.5)^2$$
$$- \underset{(0.0095)}{0.0399} \ (i - 5.5)^3 + \underset{(0.0614)}{0.0529} \ [age(s)] + \underset{(0.0048)}{0.0009} \ [age(s)]^2,$$

$$(5.38)$$

where we have given the estimated standard error of each estimated coefficient directly under it in parentheses. We see that the linear and cubic effects of section are statistically significant, but in contrast to the results for the model fitted in the previous section, neither the linear nor quadratic effect of age is significant. One possible explanation for this is that smoothing the trend over sections, rather than using a mean structure saturated with respect to section, results in fitted residuals that are sufficiently large to obscure the quadratic effect of age that was evident in the previous model. That many residuals from the present model are substantially larger than those from the previous model is clear from a plot of the fitted mean model and from an examination of the REML estimates of marginal variances. Figure 5.1 plots the fitted mean as a function of section for a person of near-average age (40 years) and, for comparison, it also plots the saturated mean's profile. The modeled cubic trend reproduces the gross features of the saturated mean's profile, but some detail is lost and the agreement between the two is poor on the second and eighth sections especially. Table 5.4 gives REML estimates of marginal variances and correlations for the AD(3) residual covariance structure. The variances in this table are consistently larger (for some sections even 20% larger) than the corresponding variances in Table 5.3(a). Partly as a result, the correlations in this

Table 5.4 *REML estimates of marginal variances (along the main diagonal) and correlations (off the main diagonal) for the 100-km race split times under an unstructured AD(3) model with mean structure cubic in section and quadratic in age.*

30									
.95	44								
.74	.73	50							
.71	.71	.92	59						
.46	.43	.75	.87	97					
.47	.45	.72	.84	.91	150				
.39	.37	.59	.70	.76	.84	110			
.37	.36	.55	.65	.68	.81	.73	163		
.36	.34	.54	.63	.68	.77	.68	.74	146	
.32	.31	.48	.48	.62	.68	.72	.59	.76	172

table tend to be somewhat smaller than their counterparts computed by summing the entries in Table 5.3(a,b).

However, the additional noise induced by smoothing over sections is not the only factor responsible for the disappearance of the significant quadratic age effect. This follows from the fact that REML estimates (not shown) of linear and quadratic age effects for an AD(3) model with a mean structure that is saturated, rather than cubic, for sections actually differ very little from those in (5.38). A much more important factor is that the REML estimator of β no longer coincides with the ordinary least squares estimator, and so the within-subject correlation across sections affects the estimation of the model's mean structure. Apparently, the within-subject correlation structure accounts for some of the variability that was ascribed to age effects in the previous fitted model. The question of whether the more parsimonious AD(3) model fitted in this section is as satisfactory as the previous one cannot be resolved without a more formal approach, which we will take up in Chapter 7.

It is worth mentioning that the maximum likelihood estimates (not shown) of parameters in these models, while not equal to the REML estimates, differ very little from them. The differences are at most 1%, and usually less than 0.1%, of the estimates' magnitudes.

5.4 Unstructured antedependence: Unbalanced data

It can happen, either by design or by chance, that the data are not balanced. For example, clinical patients may return for repeated examinations or medical tests at different times than other patients, or some may drop out prior to the

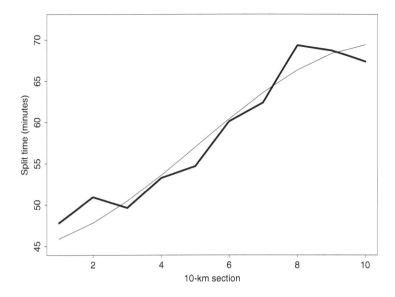

Figure 5.1 *Fitted mean split times on each 10-km section of the 100-km race under an unstructured AD(3) model with mean structure cubic in section and quadratic in age. Solid thin line: fitted mean for a competitor 40 years old; Thick line: saturated mean profile.*

end of the study and never return. An example of a study in which such "drop-outs" occur is provided by the speech recognition data of Section 1.7.3. In that study, there were 41 subjects and four intended measurement times. On the first two measurement occasions all subjects were observed, but on the third and fourth occasions only 33 and 21 subjects, respectively, were observed.

A fruitful way to deal with unbalanced data is to act as though they are an incomplete, or missing-data, version of a balanced data set. That is, we act as though the observed data are a portion of the complete balanced data set that would have resulted if we had observed all subjects at each time at which at least one subject was observed. If the complete data follow an unstructured AD(p) model, and the missing data pattern is "nice," then estimation of the model's parameters from the incomplete data can proceed using the estimation procedures described in Section 5.3 with minor modifications, as we now show.

We assume throughout this section that the missing data mechanism is independent of the observations and has no parameters in common with those of the joint distribution of the observations. Under these assumptions, the missing data mechanism is *ignorable* for the purposes of likelihood-based inference (Little and Rubin, 2002). Furthermore, we modify our notation slightly for this section by letting n be the number of distinct measurement times in the complete data set and letting $N_1, \ldots, N_n$ be the time-specific sample sizes, i.e., the number of subjects actually observed at each measurement time.

One "nice" missing data pattern in longitudinal studies is *monotone* missing data, which occurs, as for the speech recognition data, when some subjects present at the beginning of the study drop out before its conclusion. Formally, a longitudinal sample is monotone if $Y_{s,j+1}$ is missing whenever Y_{sj} is missing ($s = 1, \ldots, N$; $j = 2, \ldots, n-1$). Here n is the number of measurement times for subjects that do not drop out. The time-specific sample sizes are ordered as follows:

$$N = N_1 \geq N_2 \geq \cdots \geq N_n.$$

For simplicity of notation we assume, without loss of generality, that indices $s = 1, \ldots, N$ are assigned to subjects in such a way that the N_n subjects with complete data are listed first, then the $N_{n-1} - N_n$ subjects missing only the last measurement, then the $N_{n-2} - N_{n-1}$ subjects missing only the last two measurements, and so on.

We will consider only constant-order antedependence models in this section. However, all of the results presented can be extended easily to variable-order unstructured antedependence models.

5.4.1 Monotone missing data, saturated mean

First we consider the estimation of model parameters when the model's mean structure is saturated, i.e., when the complete data follow model (5.15).

Theorem 5.7. *If* $Y_1, \ldots, Y_N$ *have a monotone missing data pattern, the corresponding complete data follow the normal saturated-mean, unstructured AD(p) model, and* $N_i - 1 > p$ *for all* i, *then:*

(a) *The maximum likelihood estimators of the non-trivial autoregressive coefficients* $\{\phi_{ij}\}$ *and innovation variances* $\{\delta_i\}$ *are given by*

$$\hat{\boldsymbol{\phi}}_i^T \equiv (\hat{\phi}_{i,i-p_i}, \hat{\phi}_{i,i-p_i+1}, \ldots, \hat{\phi}_{i,i-1}) = \mathbf{a}_{i-p_i:i-1,i}^T \mathbf{A}_{i-p_i:i-1,i}^{-1},$$
$$i = 2, \ldots, n,$$

and

$$\hat{\delta}_i = \begin{cases} a_{11} & \text{for } i = 1 \\ a_{ii} - \mathbf{a}_{i-p_i:i-1,i}^T \mathbf{A}_{i-p_i:i-1,i}^{-1} \mathbf{a}_{i-p_i:i-1,i} & \text{for } i = 2, \ldots, n \end{cases}$$

where $\mathbf{a}_{i-p_i:i-1,i}$ and $\mathbf{A}_{i-p_i:i-1,i}$ are the indicated subvector and indicated submatrix, respectively, and a_{ij} is the ijth element, of the $i \times i$ matrix

$$\mathbf{A}_i = \frac{1}{N_i} \sum_{s=1}^{N_i} (\mathbf{Y}_{s,1:i} - \overline{\mathbf{Y}}_i)(\mathbf{Y}_{s,1:i} - \overline{\mathbf{Y}}_i)^T, \quad i = 1, \ldots, n; \, j = 1, \ldots, i,$$

$\mathbf{Y}_{s,1:i}$ is the subvector of $\mathbf{Y}_s$ consisting of its first i elements, and

$$\overline{\mathbf{Y}}_i = \frac{1}{N_i} \sum_{s=1}^{N_i} \mathbf{Y}_{s,1:i}.$$

(b) *The maximum likelihood estimator of Σ is $\hat{\Sigma} = \hat{\mathbf{T}}^{-1}\hat{\mathbf{D}}(\hat{\mathbf{T}}^T)^{-1}$, where $\hat{\mathbf{T}}$ is the lower triangular matrix with ones on its main diagonal and $-\hat{\boldsymbol{\phi}}_i^T$ for the elements in its ith row immediately preceding that row's main diagonal element, and with its remaining elements all equal to zero; and $\hat{\mathbf{D}}$ is the diagonal matrix with elements $\hat{\delta}_1, \ldots, \hat{\delta}_n$ on its main diagonal;*

(c) *The maximum likelihood estimators of Ξ and Σ^{-1} are $H(\hat{\Sigma})$ and $\hat{\Sigma}^{-1}$, respectively, and elements on off-diagonals $p+1, \ldots, n-1$ of these matrices are equal to zero;*

(d) *The maximum likelihood estimator of $\boldsymbol{\mu}$ is given by*

$$\hat{\boldsymbol{\mu}} = \hat{\mathbf{T}}^{-1}\hat{\boldsymbol{\mu}}^*$$

where $\hat{\boldsymbol{\mu}}^* = (\hat{\mu}_i^*)$ and

$$\hat{\mu}_i^* = \overline{Y}_i - \sum_{k=1}^{p_i} \hat{\phi}_{i,i-k}\overline{Y}_{i-k,i}, \quad i = 1, \ldots, n,$$

and $\overline{Y}_i$ and $\overline{Y}_{i-k,i}$ are the ith and $(i-k)$th elements, respectively, of $\overline{\mathbf{Y}}_i$;

(e) *The residual maximum likelihood estimators of the parameters listed in parts (a)-(d) above are given by expressions exactly the same as those of the maximum likelihood estimators except that elements of $\mathbf{A}_i$ are replaced by the corresponding elements of $\mathbf{S}_i = [N_i/(N_i - 1)]\mathbf{A}_i$.*

Proof. The log-likelihood function may be written in product-of-conditionals form, similarly to (5.20) through (5.22), as

$$\log L(\boldsymbol{\mu}, \boldsymbol{\theta}_\delta, \boldsymbol{\theta}_\phi) = \log \prod_{s=1}^{N} \prod_{i=1}^{n_s} f_i(Y_{si}|Y_{s,i-1}, \ldots, Y_{s,i-p_i})$$

$$= \log \prod_{i=1}^{n} \prod_{s=1}^{N_i} f_i(Y_{si}|Y_{s,i-1}, \ldots, Y_{s,i-p_i})$$

$$= \sum_{i=1}^{n} \sum_{s=1}^{N_i} \log \left\{ (2\pi\delta_i)^{-1/2} \exp \left\{ - \left[Y_{si} - \mu_i \right. \right. \right.$$

$$-\sum_{k=1}^{p_i} \phi_{i,i-k}\left(Y_{s,i-k} - \mu_{i-k}\right)\Big]^2 \Big/ 2\delta_i \bigg\}\bigg\}$$

$$= -\frac{n_+}{2} \log 2\pi - \frac{1}{2} \sum_{i=1}^{n} \sum_{s=1}^{N_i} \bigg\{ \log \delta_i + \Big[Y_{si} - \mu_i$$

$$-\sum_{k=1}^{p_i} \phi_{i,i-k}\left(Y_{s,i-k} - \mu_{i-k}\right)\Big]^2 \Big/ \delta_i \bigg\}.$$

After reparameterizing the mean structure as in the alternative proof of Theorem 5.1, we have

$$\log L(\boldsymbol{\mu}^*, \boldsymbol{\theta}_\delta, \boldsymbol{\theta}_\phi) = -\frac{n_+}{2} \log 2\pi - \frac{1}{2} \sum_{i=1}^{n} \sum_{s=1}^{N_i} \bigg\{ \log \delta_i + \Big(Y_{si} - \mu_i^*$$

$$-\sum_{k=1}^{p_i} \phi_{i,i-k} Y_{s,i-k} \Big)^2 \Big/ \delta_i \bigg\}. \tag{5.39}$$

Now, as in (5.23), the ith term of the first sum in (5.39) is, apart from an additive constant, simply the log-likelihood corresponding to an ordinary least squares regression, with intercept, of the ith normally distributed response variable on the p_i response variables immediately preceding it; here, however, the number of observations in the ith regression is N_i rather than N. Hence the maximum likelihood estimators of the $\{\phi_{ij}\}$ coincide with their least squares estimators from these regressions, while those of the $\{\delta_i\}$ are the residual sums of squares from these regressions divided by N_i; these estimators are given by the expressions in part (a) of the theorem. Furthermore, maximum likelihood estimators of the intercepts are given by the least squares estimators

$$\hat{\mu}_i^* = \overline{Y}_i - \sum_{k=1}^{p_i} \hat{\phi}_{i,i-k} \overline{Y}_{i-k,i},$$

where $\overline{Y}_i$ and $\overline{Y}_{i-k,i}$ are the ith and $(i-k)$th elements, respectively, of $\overline{\mathbf{Y}}_i$. Part (d) then follows from (5.24) and the invariance of maximum likelihood estimators. The remaining parts of the theorem follow by the same arguments used to prove the analogous results of Theorem 5.1. $\square$

The upshot of Theorem 5.7 is that when the missing data are monotone, maximum likelihood estimates of the parameters of the normal saturated-mean unstructured AD(p) model may be obtained using the same regression-on-predecessors approach used when the data are complete. The only difference is that the ith of these regressions uses only the observations that are complete up to time i.

5.4.2 *Monotone missing data, multivariate regression mean*

Now we extend Theorem 5.7 to a model with a multivariate regression mean structure; that is, we assume that the complete data follow model (5.29). Again we will find that the approach of obtaining maximum likelihood estimators by regressing on predecessors and covariates, used when the data are complete, extends to this situation as well, the only difference being that the ith of these regressions uses only the observations that are complete up to time i. Let

$$
\mathbf{Z}_i = \begin{pmatrix} \mathbf{z}_1^T \\ \vdots \\ \mathbf{z}_{N_i}^T \end{pmatrix}
$$

and assume that $\mathbf{Z}_i$ has full column rank m for all i.

Theorem 5.8. *If $\mathbf{Y}_1, \ldots, \mathbf{Y}_N$ have a monotone missing data pattern, the corresponding complete data follow the normal multivariate regression, unstructured AD(p) model with m covariates, and $N_i - m > p$ for all i, then:*

(a) *The maximum likelihood estimators of the non-trivial autoregressive coefficients $\{\phi_{ij}\}$ and innovation variances $\{\delta_i\}$ are given by*

$$
\hat{\boldsymbol{\phi}}_i^T \equiv (\hat{\phi}_{i,i-p_i}, \hat{\phi}_{i,i-p_i+1}, \ldots, \hat{\phi}_{i,i-1}) = \mathbf{a}_{i-p_i:i-1,i}^T \mathbf{A}_{i-p_i:i-1,i}^{-1},
$$
$$
i = 2, \ldots, n,
$$

and

$$
\hat{\delta}_i = \begin{cases} a_{11} & \text{for } i = 1 \\ a_{ii} - \mathbf{a}_{i-p_i:i-1,i}^T \mathbf{A}_{i-p_i:i-1,i}^{-1} \mathbf{a}_{i-p_i:i-1,i} & \text{for } i = 2, \ldots, n \end{cases}
$$

where $\mathbf{a}_{i-p_i:i-1,i}$ and $\mathbf{A}_{i-p_i:i-1,i}$ are the indicated subvector and indicated submatrix, respectively, and a_{ij} is the ijth element, of the $i \times i$ matrix

$$
\mathbf{A}_i = \frac{1}{N_i} \sum_{s=1}^{N_i} [\mathbf{Y}_{s,1:i} - (\mathbf{z}_s^T \otimes \mathbf{I}_i)\overline{\boldsymbol{\beta}}_i][\mathbf{Y}_{s,1:i} - (\mathbf{z}_s^T \otimes \mathbf{I}_i)\overline{\boldsymbol{\beta}}_i]^T,
$$
$$
i = 1, \ldots, n; \ j = 1, \ldots, i,
$$

and

$$
\overline{\boldsymbol{\beta}}_i = [(\mathbf{Z}_i^T \mathbf{Z}_i)^{-1} \mathbf{Z}_i^T \otimes \mathbf{I}_i] \begin{pmatrix} \mathbf{Y}_{1,1:i} \\ \mathbf{Y}_{2,1:i} \\ \vdots \\ \mathbf{Y}_{N_i,1:i} \end{pmatrix};
$$

(b) *The maximum likelihood estimator of $\boldsymbol{\Sigma}$ is $\hat{\boldsymbol{\Sigma}} = \hat{\mathbf{T}}^{-1}\hat{\mathbf{D}}(\hat{\mathbf{T}}^T)^{-1}$, where $\hat{\mathbf{T}}$ is the lower triangular matrix with ones on its main diagonal and $-\hat{\boldsymbol{\phi}}_i^T$ for the elements in its ith row immediately preceding that row's main diagonal* *element, and with its remaining elements all equal to zero; and $\hat{\mathbf{D}}$ is the diagonal matrix with elements $\hat{\delta}_1, \ldots, \hat{\delta}_n$ on its main diagonal;*

(c) *The maximum likelihood estimators of $\boldsymbol{\Xi}$ and $\boldsymbol{\Sigma}^{-1}$ are $H(\hat{\boldsymbol{\Sigma}})$ and $\hat{\boldsymbol{\Sigma}}^{-1}$, respectively, and elements on off-diagonals $p+1, \ldots, n-1$ of these matrices are equal to zero;*

(d) *The maximum likelihood estimator of $\boldsymbol{\beta}$ is given by*

$$\hat{\boldsymbol{\beta}} = (\hat{\mathbf{T}}^{-1} \otimes \mathbf{I}_m)\hat{\boldsymbol{\beta}}^*$$

where $\hat{\boldsymbol{\beta}}^* = (\hat{\boldsymbol{\beta}}_1^{*T}, \ldots, \hat{\boldsymbol{\beta}}_n^{*T})^T$,

$$\hat{\boldsymbol{\beta}}_i^* = (\mathbf{Z}_i^T \mathbf{Z}_i)^{-1} \mathbf{Z}_i^T \begin{pmatrix} Y_{1i} \\ \vdots \\ Y_{N_i i} \end{pmatrix} \quad \text{if } p_i = 0,$$

$$\begin{aligned}
\hat{\boldsymbol{\beta}}_i^* &= \{\mathbf{Z}_i^T[\mathbf{I}_{N_i} - \mathbf{Q}_i(\mathbf{Q}_i^T \mathbf{Q}_i)^{-1}\mathbf{Q}_i^T]\mathbf{Z}_i\}^{-1} \\
&\times \mathbf{Z}_i^T[\mathbf{I}_{N_i} - \mathbf{Q}_i(\mathbf{Q}_i^T \mathbf{Q}_i)^{-1}\mathbf{Q}_i^T] \begin{pmatrix} Y_{1i} \\ \vdots \\ Y_{N_i i} \end{pmatrix}, \quad \text{otherwise,}
\end{aligned}$$

and

$$\mathbf{Q}_i = \begin{pmatrix} Y_{1,i-1} & \cdots & Y_{1,i-p_i} \\ \vdots & & \vdots \\ Y_{N_i,i-1} & \cdots & Y_{N_i,i-p_i} \end{pmatrix}.$$

(e) *The residual maximum likelihood estimators of the parameters listed in parts (a) through (d) above are given by expressions exactly the same as those of the maximum likelihood estimators except that elements of $\mathbf{A}_i$ are replaced by the corresponding elements of $\mathbf{S}_i = [N_i/(N_i - m)]\mathbf{A}_i$.*

Proof. The log-likelihood function (5.39) may be extended to the present situation as follows:

$$\begin{aligned}
\log L(\boldsymbol{\beta}^*, \boldsymbol{\theta}_\delta, \boldsymbol{\theta}_\phi) &= -\frac{n_+}{2}\log 2\pi - \frac{1}{2}\sum_{i=1}^{n}\sum_{s=1}^{N_i}\Bigg\{\log \delta_i + \bigg(Y_{si} - \mathbf{z}_s^T\boldsymbol{\beta}_i^* \\
&\qquad - \sum_{k=1}^{p_i}\phi_{i,i-k}Y_{s,i-k}\bigg)^2 \bigg/ \delta_i\Bigg\}
\end{aligned} \tag{5.40}$$

where

$$\boldsymbol{\beta}_i^* = \boldsymbol{\beta}_i - \sum_{k=1}^{p_i}\phi_{i,i-k}\boldsymbol{\beta}_{i-k}. \tag{5.41}$$

Now, the ith term of the first sum in (5.40) is, apart from an additive constant, simply the log-likelihood corresponding to an ordinary least squares regression of the ith response variable on the p_i response variables immediately preceding it and on the m covariates, based on N_i observations. Hence the maximum likelihood estimators of the $\{\phi_{ij}\}$ and $\{\beta_i^*\}$ coincide with their least squares estimators from these regressions, and the maximum likelihood estimators of the $\{\delta_i\}$ are the residual sums of squares from these regressions divided by N_i, all of which are given by the expressions in parts (a) and (d) of the theorem. Part (d) then follows from (5.41) and the invariance of maximum likelihood estimators. The remaining parts of the theorem follow by the same arguments used to prove the analogous results of Theorem 5.1. $\square$

Example: Speech recognition data

We now illustrate the REML estimation methodology of Theorem 5.8(e) using the speech recognition data. As noted earlier in this section, this data set has dropouts, which in this case are subjects that were initially fitted with a cochlear implant and tested on one or more occasions for audiologic performance, but after the second or third such occasion did not return for further testing. Recall from Section 1.7.3 that there are two types of implants, labeled generically as types A and B, and four intended measurement occasions. Here $N_1 = N_2 = 41$, $N_3 = 33$, and $N_4 = 21$. We adopt an eight-parameter mean structure for these data that is saturated over time within each implant group, i.e.,

$$E(Y_{si}) = \begin{cases} \mu_{Ai} & \text{if subject } s \text{ has implant A} \\ \mu_{Bi} & \text{if subject } s \text{ has implant B} \end{cases}$$
$$i = 1, 2, 3, 4. \tag{5.42}$$

For the covariance structure, we assume homogeneity across the two implant groups. This assumption will be checked in Chapter 6 and shown to be supported by the data.

Table 5.5 gives REML estimates of the pooled marginal variances and correlations under unstructured AD(3) (general multivariate dependence) and AD(1) models. We observe that correlations lagged the same number of measurement occasions apart increase slightly over time, despite being further apart in actual time. This confirms a prior belief of the researchers who conducted this study, which is that a typical subject's audiologic performance becomes more consistent over time. We also observe that the estimated marginal variances corresponding to the last two measurement times are different under the two models. This contrasts with the situation with balanced data, for which estimates of all marginal variances coincide across orders of antedependence. The estimated lag-one correlations agree well (differences are less than 0.005), but only the first ones match perfectly. Comparison of the estimated correlations beyond lag one across the two models indicates that the fitted first-order model overestimates the lag-three correlation somewhat, but does not fit too badly. In Chapter 6 we will formally test for the order of antedependence of these data

Table 5.5 *REML estimates of pooled marginal variances (along the main diagonal) and correlations (below the main diagonal) for the speech recognition sentence scores data, under normal unstructured antedependence models with saturated means within each implant group: (a) AD(3); (b) AD(1).*

(a)

404			
.85	593		
.72	.90	549	
.64	.87	.95	575

(b)

404			
.85	593		
.77	.90	547	
.73	.85	.95	544

and will find that a first-order model is not rejected at traditional levels of significance. REML estimates of the mean structure parameters for the first-order model are as follows:

$$
\begin{pmatrix} \tilde{\mu}_{A1} \\ \tilde{\mu}_{A2} \\ \tilde{\mu}_{A3} \\ \tilde{\mu}_{A4} \end{pmatrix} = \begin{pmatrix} 28.52 \\ 49.14 \\ 55.87 \\ 63.57 \end{pmatrix}, \qquad
\begin{pmatrix} \tilde{\mu}_{B1} \\ \tilde{\mu}_{B2} \\ \tilde{\mu}_{B3} \\ \tilde{\mu}_{B4} \end{pmatrix} = \begin{pmatrix} 19.40 \\ 38.24 \\ 44.49 \\ 46.90 \end{pmatrix}.
$$

In Chapter 7 we will formally compare the two mean vectors.

5.4.3 Monotone missing data, arbitrary linear mean

Finally, we relax the assumptions on the mean structure to allow it to be of arbitrary linear form. That is, we suppose that the complete data follow model (5.35).

As in the case of balanced data with this mean structure, closed-form expressions for the maximum likelihood and REML estimators of β and θ do not exist, but we can obtain them by iterative maximization procedures. A Nelder-Mead simplex or Newton-Raphson algorithm can be effective. Alternatively, for the case of maximum likelihood estimation one can iterate between

$$
\hat{\beta}^{(l)} = \left(\sum_{s=1}^{N} \mathbf{X}_s^T \left[\mathbf{\Sigma}_s \left(\hat{\boldsymbol{\theta}}_\sigma^{(l-1)} \right) \right]^{-1} \mathbf{X}_s \right)^{-1} \sum_{s=1}^{N} \mathbf{X}_s^T \left[\mathbf{\Sigma}_s \left(\hat{\boldsymbol{\theta}}_\sigma^{(l-1)} \right) \right]^{-1} \mathbf{Y}_s
$$

and

$$\hat{\boldsymbol{\theta}}_{\sigma}^{(l)} = \left(a_{11}^{(l)}, \ldots, a_{nn}^{(l)}, a_{21}^{(l)}, \ldots, a_{n,n-p}^{(l)}\right)^{T},$$

for $l = 1, 2, \ldots$, where $a_{ij}^{(l)}$ is the ijth element of

$$\mathbf{A}_{i}^{(l)} = \frac{1}{N_i} \sum_{s=1}^{N_i} \left(\mathbf{Y}_{s,1:i} - \mathbf{X}_{s,1:i}\hat{\boldsymbol{\beta}}^{(l)}\right) \left(\mathbf{Y}_{s,1:i} - \mathbf{X}_{s,1:i}\hat{\boldsymbol{\beta}}^{(l)}\right)^{T},$$

$$i = 1, \ldots, n; \ j = 1, \ldots, i.$$

Here $\mathbf{X}_{s,1:i}$ is the submatrix of $\mathbf{X}_s$ consisting of its first i rows and we assume that $N_i - q > p$. Note that iterates of the sth subject's covariance matrix, $\boldsymbol{\Sigma}_s\left(\hat{\boldsymbol{\theta}}_{\sigma}^{(l)}\right)$, utilize only those elements $\{a_{ij}^{(l)}\}$ of $\hat{\boldsymbol{\theta}}_{\sigma}^{(l)}$ whose first index is less than or equal to n_s. The iterative process may be initiated with the same value, $\hat{\boldsymbol{\theta}}_{\sigma}^{(0)} = (\mathbf{1}_n^T, \mathbf{0}^T)^T$, that is used in the balanced case.

5.4.4 Other missing data patterns

To this point the only missing data pattern we have considered is that of monotone dropouts. Another relatively simple and commonly occurring missing data pattern is that of monotone drop-ins (also known as delayed entry or staggered entry), in which subjects enter the study at different times, but once entered remain until its conclusion. For such a pattern, the time-specific sample sizes are monotone increasing, i.e.,

$$N_1 \leq N_2 \leq \cdots \leq N_n = N$$

where in this case n is the number of measurement times for subjects that were present at the beginning of the study. If we define the $n_s \times n_s$ "exchange" matrix

$$\mathbf{E}_s = \begin{pmatrix} 0 & \cdots & 0 & 1 \\ 0 & \cdots & 1 & 0 \\ \vdots & & \vdots & \vdots \\ 1 & \cdots & 0 & 0 \end{pmatrix},$$

then likelihood-based estimators may be obtained by applying exactly the same procedures used for the case of dropouts, but with $\mathbf{E}_s\mathbf{Y}_s$, the vector of observations in reverse time order, in place of $\mathbf{Y}_s$, and $\mathbf{E}_s\mathbf{X}_s$ in place of $\mathbf{X}_s$. This follows from the fact that pth-order antedependent variables are also pth-order antedependent when arranged in reverse time order.

For data with an arbitrary pattern of missingness, the EM algorithm (Dempster, Laird, and Rubin, 1977) may be used to obtain maximum likelihood estimates. As applied to normal unstructured antedependence models, the EM algorithm is quite similar to that used to estimate an arbitrary positive definite covariance matrix of a normal distribution. For the sake of brevity we consider only the case where the complete data follow the normal, saturated-mean pth-order unstructured antedependence model. It is this case, with small p, for which the EM algorithm is particularly useful, for in this case we may take advantage of the relatively simple-to-compute expressions for the maximum likelihood estimates given in Theorems 5.1 and 5.6. Note that assuming pth-order antedependence for the complete data is different than assuming it for the incomplete data as we did in previous sections, though complete-data antedependence implies incomplete-data antedependence when $p = 1$. Furthermore, we present the EM algorithm for maximum likelihood estimation only; modifications for REML estimation are straightforward.

For each observational vector $\mathbf{Y}_s$ for which one or more observations of the complete vector are missing, let $\mathbf{Y}_s^+$ be the complete $n \times 1$ data vector and let $\mathbf{P}_s$ be the $n \times n$ permutation matrix which puts all the observed data first. That is,

$$\mathbf{P}_s \mathbf{Y}_s^+ = \left(\begin{array}{c} \mathbf{Y}_s^{(1)} \\ \mathbf{Y}_s^{(2)} \end{array} \right)$$

where $\mathbf{Y}_s^{(1)}$ is the $n_s \times 1$ vector of the observed measurements on subject s and $\mathbf{Y}_s^{(2)}$ is unobserved. (Note that $\mathbf{P}_s$ is a generalization of the exchange matrix $\mathbf{E}_s$.) Write

$$\left(\begin{array}{c} \mu_s^{(1)} \\ \mu_s^{(2)} \end{array} \right) \quad \text{and} \quad \left(\begin{array}{cc} \mathbf{\Sigma}_s^{(11)} & \mathbf{\Sigma}_s^{(12)} \\ \mathbf{\Sigma}_s^{(21)} & \mathbf{\Sigma}_s^{(22)} \end{array} \right)$$

for the mean and covariance matrix, respectively of $\mathbf{P}_s \mathbf{Y}_s^+$. Then the EM algorithm proceeds by iterating between a "prediction step" and an "estimation step," as follows:

Prediction step. Given estimates $\hat{\mu}$ and $\hat{\mathbf{\Sigma}}$ from the estimation step, predict $\mathbf{Y}_s^{(2)}$, $\mathbf{Y}_s^{(2)} \mathbf{Y}_s^{(2)T}$, and $\mathbf{Y}_s^{(2)} \mathbf{Y}_s^{(1)T}$ by their conditional means:

$$\widehat{\mathbf{Y}_s^{(2)}} = \hat{\mu}_s^{(2)} + \hat{\mathbf{\Sigma}}_s^{(21)} \left(\hat{\mathbf{\Sigma}}_s^{11} \right)^{-1} \left(\mathbf{Y}_s^{(1)} - \hat{\mu}_s^{(1)} \right), \qquad (5.43)$$

$$\widehat{\mathbf{Y}_s^{(2)} \mathbf{Y}_s^{(2)T}} = \hat{\mathbf{\Sigma}}_s^{(22)} - \hat{\mathbf{\Sigma}}_s^{(21)} \left(\hat{\mathbf{\Sigma}}_s^{(11)} \right)^{-1} \hat{\mathbf{\Sigma}}_s^{(12)} + \widehat{\mathbf{Y}_s^{(2)}} \widehat{\mathbf{Y}_s^{(2)T}}, (5.44)$$

$$\widehat{\mathbf{Y}_s^{(2)} \mathbf{Y}_s^{(1)T}} = \widehat{\mathbf{Y}_s^{(2)}} \mathbf{Y}_s^{(1)T}. \qquad (5.45)$$

Estimation step. Given the pseudo-complete data produced by the prediction step, update the maximum likelihood estimates of μ and $\mathbf{\Sigma}$ in accordance with Theorem 5.1:

$$\hat{\mu} = \frac{1}{N} \sum_{s=1}^{N} \mathbf{P}_s^{-1} \left(\begin{array}{c} \mathbf{Y}_s^{(1)} \\ \widehat{\mathbf{Y}_s^{(2)}} \end{array} \right),$$

$$\hat{\mathbf{A}} = (\hat{a}_{ij}) = \frac{1}{N} \sum_{s=1}^{N} \mathbf{P}_s^{-1} \begin{pmatrix} \mathbf{Y}_s^{(1)} \mathbf{Y}_s^{(1)T} & (\widehat{\mathbf{Y}_s^{(2)} \mathbf{Y}_s^{(1)T}})^T \\ \widehat{\mathbf{Y}_s^{(2)} \mathbf{Y}_s^{(1)T}} & \widehat{\mathbf{Y}_s^{(2)} \mathbf{Y}_s^{(2)T}} \end{pmatrix} \mathbf{P}_s^{-1} - \hat{\boldsymbol{\mu}}\hat{\boldsymbol{\mu}}^T,$$

$$\hat{\boldsymbol{\theta}}_\sigma = (\hat{a}_{11}, \ldots, \hat{a}_{nn}, \hat{a}_{21}, \ldots, \hat{a}_{n,n-p})^T,$$

$$\hat{\boldsymbol{\Sigma}} = \boldsymbol{\Sigma}(\hat{\boldsymbol{\theta}}_\sigma),$$

where elements of $\hat{\boldsymbol{\Sigma}}$ on off-diagonals $p + 1, \ldots, n$ are obtained recursively through the use of equation (2.32) of Theorem 2.4.

The iterative procedure can be initialized by replacing the right-hand sides of expressions (5.43) through (5.45) with vectors and matrices of zeros in the first prediction step.

The EM algorithm just described can deal with any pattern of missingness, provided that all elements of the unstructured AD(p) complete-data covariance matrix are estimable. However, if the pattern is nearly monotone, a more computationally efficient implementation of the EM algorithm can be constructed, which utilizes the closed-form expressions for the maximum likelihood estimators under monotone sampling given by Theorem 5.7, in place of those given by Theorem 5.1.

If there is little or no duplication of measurement times across subjects, however, then one or more elements of the unstructured AD(p) covariance matrix are non-estimable. In this case, more structure must be imposed on the antedependence model to render its parameters estimable, unless similar measurement times for different subjects are grouped together and treated as equal within groups.

An alternative to the EM algorithm and its corresponding observed-data likelihood-based approach to inference from incomplete data is a multiple imputation approach. Zhang (2005) describes and implements such an approach for antedependence models.

Example: Fruit fly mortality data
Recall that roughly 22% of the fruit fly mortality data are missing, and that the missingness is not monotone (Tables 1.10 through 1.12). Recall also that a preliminary model identification exercise (Section 4.3.1) suggested that a first-order antedependence model is plausible for these data. Accordingly, we use the EM algorithm described in this section to obtain maximum likelihood estimates of an unstructured AD(1) model; for comparison purposes we obtain such estimates for the AD(10) (general multivariate dependence) model as well. Table 5.6 gives maximum likelihood estimates of the variances and correlations for both models. Estimates for the two models appear to match reasonably well except in cases of correlations with mortalities at times 9 and 11; the AD(1) model overestimates the former and underestimates the latter. More formal methods are needed to determine if these discrepancies would

Table 5.6 *Maximum likelihood estimates of marginal variances (along the main diagonal) and correlations (off the main diagonal) for the fruit fly mortality data, under normal saturated-mean unstructured AD(p) models: (a) p = 10; (b) p = 1. Entries in (b) are differences between the estimates under the two models (first-order minus tenth-order).*

(a)

0.69										
.59	1.10									
.54	.71	1.69								
.46	.61	.78	2.75							
.37	.55	.63	.82	2.63						
.30	.45	.48	.64	.86	1.94					
.22	.36	.32	.40	.65	.81	1.14				
.12	.29	.28	.25	.45	.59	.77	0.68			
.02	.16	.02	.03	.16	.19	.43	.64	0.48		
.11	.28	-.01	.13	.20	.21	.37	.43	.52	0.36	
.23	.30	.09	.37	.44	.40	.35	.35	.29	.43	0.49

(b)

.00										
.01	-.02									
-.11	.01	.02								
-.12	-.05	.01	.06							
-.09	-.09	.01	.00	-.04						
-.06	-.05	.08	.07	.00	.01					
-.02	-.03	.14	.18	.06	.01	.06				
.04	-.03	.08	.21	.11	.06	.02	.04			
.08	.01	.22	.27	.21	.24	.08	.02	.01		
-.05	-.19	.12	.02	-.01	.01	-.11	-.10	-.02	-.01	
-.22	-.27	-.04	-.30	-.37	-.32	-.26	-.23	-.11	-.06	-.07

lead us to favor a higher-order AD model. As an aside, we compare the estimates for the AD(10) model to the corresponding sample variances and correlations computed previously from the available data [see Table 4.4(a)] to examine the effects that adjusting for imbalances in the data across time have on the estimates. We see that the greatest effect on the variances is at time 5, where the relative difference is about 20%; some of the time-11 correlations are considerably affected also.

5.5 Structured antedependence models

Next we consider maximum likelihood estimation of the parameters of structured antedependence models. For these models, closed-form expressions for the estimators do not exist (with a few exceptions), regardless of whether the data are balanced, so the advantages of balance are not as compelling as they were for unstructured antedependence models. Similarly, it is less of an advantage for the mean to be of multivariate regression form, for in this case estimates of at least some of the covariance parameters have to be obtained numerically, even if those of the mean structure do not. Therefore, unless noted otherwise, in this section we allow the data to be unbalanced and the mean structure to be of arbitrary linear form. Nevertheless, we require, as previously to Section 5.4.4, that the observed data be AD(p) for each subject. Furthermore, we consider only maximum likelihood estimation; the results are easily extended to REML estimation by replacing the log-likelihood or profile log-likelihood with $\log L_R(\boldsymbol{\theta})$, as given by (5.14).

Generally, maximum likelihood estimates of the mean and covariance parameters of structured antedependence models may be obtained by iterating between the equation

$$\hat{\boldsymbol{\beta}} = \left(\sum_{s=1}^{N} \mathbf{X}_s^T [\boldsymbol{\Sigma}_s(\hat{\boldsymbol{\theta}})]^{-1} \mathbf{X}_s \right)^{-1} \sum_{s=1}^{N} \mathbf{X}_s^T [\boldsymbol{\Sigma}_s(\hat{\boldsymbol{\theta}})]^{-1} \mathbf{Y}_s \qquad (5.46)$$

(where $\hat{\boldsymbol{\theta}}$ is the current estimate of $\boldsymbol{\theta}$) and numerical maximization of the function

$$
\begin{aligned}
\log L(\hat{\boldsymbol{\beta}}, \boldsymbol{\theta}) &= -\frac{n_+}{2} \log 2\pi - \frac{1}{2} \sum_{s=1}^{N} \log |\boldsymbol{\Sigma}_s(\boldsymbol{\theta})| \\
&\quad -\frac{1}{2} \sum_{s=1}^{N} (\mathbf{Y}_s - \mathbf{X}_s \hat{\boldsymbol{\beta}})^T [\boldsymbol{\Sigma}_s(\boldsymbol{\theta})]^{-1} (\mathbf{Y}_s - \mathbf{X}_s \hat{\boldsymbol{\beta}}) \\
&= -\frac{n_+}{2} \log 2\pi - \frac{1}{2} \sum_{s=1}^{N} \log |\boldsymbol{\Sigma}_s(\boldsymbol{\theta})| \\
&\quad -\frac{1}{2} \sum_{s=1}^{N} \operatorname{tr}\{ (\mathbf{Y}_s - \mathbf{X}_s \hat{\boldsymbol{\beta}}) (\mathbf{Y}_s - \mathbf{X}_s \hat{\boldsymbol{\beta}})^T [\boldsymbol{\Sigma}_s(\boldsymbol{\theta})]^{-1} \}.
\end{aligned}
$$

$$(5.47)$$

Here $\boldsymbol{\theta}$ represents either $\boldsymbol{\theta}_\sigma$, $\boldsymbol{\theta}_\xi$, $\boldsymbol{\theta}_\gamma$, or $(\boldsymbol{\theta}_\delta^T, \boldsymbol{\theta}_\phi^T)^T$, depending on the formulation (marginal, intervenor-adjusted, precision matrix, or autoregressive) of the SAD model.

The numerical maximization of $L(\hat{\boldsymbol{\beta}}, \boldsymbol{\theta})$ can be a non-trivial computational problem when the number of measurement times is large for one or more subjects, for then the determinants and inverses of some large covariance matrices must be evaluated, or so it would appear. Computing the determinant and inverse only once for each distinct pattern of measurement times that occurs across subjects will result in some savings in computation; indeed, if the data are balanced, these operations need be performed only once (on each iteration), regardless of the number of subjects. Additional savings in computation can be realized through the use of efficient formulae for the required determinants and inverses. The specifics of these savings depend on which formulation the SAD model is based upon, so we consider each of them in turn.

5.5.1 Marginal formulation

For marginally formulated SAD models, in general there is unfortunately no way to avoid computing the inverses and determinants of $n_s \times n_s$ matrices, as required by (5.46) and (5.47). However, in some special cases these computations can be performed relatively efficiently. In the first-order case, for instance, the inverses and determinants may be evaluated via expressions (2.40) and (2.41), with substantial savings in computation (Zimmerman, Núñez-Antón, and El-Barmi, 1998). In higher-order cases, the determinants may be evaluated using expression (2.33) in Theorem 2.5(d); furthermore, if the data are balanced and the mean is of multivariate regression form, the inversion of $\Sigma(\boldsymbol{\theta}_\sigma)$ can be avoided altogether, for in this case the maximum likelihood and ordinary least squares estimators of $\boldsymbol{\beta}$ coincide [cf. (5.30)] and the last sum in (5.47) may be evaluated by the formula for $\mathrm{tr}\{\mathbf{A}[\Sigma(\boldsymbol{\theta}_\sigma)]^{-1}\}$ given by Theorem 2.6, with $\mathbf{A}$ defined as $\sum_{s=1}^{N}(\mathbf{Y}_s - \mathbf{X}_s\hat{\boldsymbol{\beta}})(\mathbf{Y}_s - \mathbf{X}_s\hat{\boldsymbol{\beta}})^T$. These formulae allow one to trade computation of the determinant of the $n_s \times n_s$ matrix $\Sigma_s(\boldsymbol{\theta}_\sigma)$ for computation of the determinants of $n_s - p$ matrices of dimensions $(p+1) \times (p+1)$ and $n_s - p - 1$ matrices of dimensions $p \times p$, and trade computation of $\mathrm{tr}\{\mathbf{A}[\Sigma_s(\boldsymbol{\theta}_\sigma)]^{-1}\}$ for computations of traces involving inverses of the same smaller matrices whose determinants are evaluated. When, as usual, p is considerably smaller than n_s for many s, the overall computational effort can be reduced significantly using these formulae.

Example 1: Treatment A cattle growth data
Recall, from the analysis of the Treatment A cattle growth data presented in Section 5.3.1, that a first-order unstructured AD model appears to fit the covariance structure of the cattle weights quite well. Recall further that the marginal sample variances and lag-one correlations of these data both tend to increase over time [see Table 5.1(a)], suggesting that one or more structured AD models may also fit well. Here, for illustration we consider a marginally formulated power law SAD(1) model with a saturated mean structure. This would seem to be a reasonable candidate model since its variances and lag-one correlations can exhibit monotone increasing behavior (see Figure 3.3).

The marginally formulated power law SAD(1) model fitted here is as follows:

$$\sigma_{ii} = \sigma^2(1 + \psi_1 t_i + \psi_2 t_i^2 + \psi_3 t_i^3), \quad i = 1, \dots, 11,$$

$$\rho_{i,i-1} = \rho^{t_i^\lambda - t_{i-1}^\lambda}, \quad i = 2, \ldots, 11,$$

where $\sigma^2 > 0$ and $0 \le \rho < 1$. For simplicity we scale time in two-week units, starting at time zero; thus $t_1 = 0, t_2 = 1, \ldots, t_{10} = 9, t_{11} = 9.5$. Again, correlations for lags beyond one, i.e., $\{\rho_{i,i-j} : i = 3, \ldots, n; j = 1, \ldots, i - 2\}$, are taken to equal the appropriate products of the lag-one correlations, in accordance with (2.37). Observe that this model parameterizes the power law dependence of lag-one correlations on time a bit differently than the original power law SAD(1) model (3.13) does; the present parameterization, though simpler than the original, should not be used when the lag-one sample correlations decrease over time, as the case $\lambda = 0$ corresponds to a singular covariance matrix. Since the lag-one sample correlations of these data clearly do not decrease over time, however, we may as well use this simpler parameterization. REML estimates of model parameters are

$$\begin{pmatrix} \tilde{\sigma}^2 \\ \tilde{\psi}_1 \\ \tilde{\psi}_2 \\ \tilde{\psi}_3 \\ \tilde{\rho} \\ \tilde{\lambda} \end{pmatrix} = \begin{pmatrix} 110.0889 \\ 0.25534 \\ 0.02416 \\ -0.00206 \\ 0.81208 \\ 0.55020 \end{pmatrix}.$$

Figure 5.2 comprises plots of the REML estimates of marginal variances and lag-one correlations against time, and includes the sample variances and lag-one correlations for comparison. Based on this informal assessment, the model appears to fit very well.

Another example of the use of the marginally formulated power law SAD(1) model is provided by Hou et al. (2005), who use it to model the covariance structure of growth trajectories in studies of genetic determinants that affect complex phenotypes undergoing developmental changes over time.

Example 2: Fruit fly mortality data
In Section 4.3.1 we determined, using informal model identification tools, that plausible models for the fruit fly mortality data would include structured first-order AD models in which the marginal variances are quadratic or cubic functions of time and the marginal correlations are quadratic functions of time. Here we estimate such a model, which we label as SAD-QM3(1): "Q" for quadratic correlations, "M" for marginal, and "3" for cubic variances. Thus, the fitted model is

$$\sigma_{ii} = \sigma^2(1 + \psi_1 t_i + \psi_2 t_i^2 + \psi_3 t_i^3), \quad i = 1, \ldots, 11,$$
$$\rho_{i,i-1} = \lambda_1 + \lambda_2 t_i + \lambda_3 t_i^2, \quad i = 2, \ldots, 11.$$

REML estimates of its parameters are as follows:

$$
\begin{pmatrix} \tilde{\sigma}^2 \\ \tilde{\psi}_1 \\ \tilde{\psi}_2 \\ \tilde{\psi}_3 \\ \tilde{\lambda}_1 \\ \tilde{\lambda}_2 \\ \tilde{\lambda}_3 \end{pmatrix} = \begin{pmatrix} 0.68973 \\ 1.61269 \\ -0.37628 \\ 0.02136 \\ 0.41577 \\ 0.13367 \\ -0.01130 \end{pmatrix}.
$$

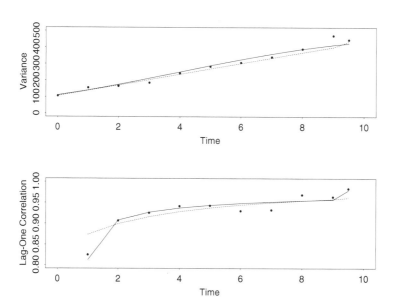

Figure 5.2 *REML estimates of marginal variances (top panel) and lag-one correlations (bottom panel) for the Treatment A cattle growth data plotted against time. Points are the sample variances and sample lag-one correlations; solid line connects estimates corresponding to the marginally formulated power law SAD(1) model; and dotted line connects estimates corresponding to the autoregressively formulated power law SAD(1) model.*

Figure 5.3 displays plots of REML estimates of this model's marginal variances and lag-one correlations against time. Sample variances and lag-one correlations, as given previously in Figure 4.3, are also plotted for comparison. Although the SAD-QM3(1) model is not wholly adequate, it does appear to reproduce the gross features of the variances and lag-one correlations reasonably well.

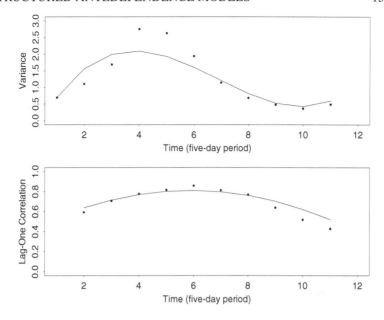

Figure 5.3 *REML estimates of marginal variances (top panel) and lag-one correlations (bottom panel) for the fruit fly mortality data plotted against time. Points are the sample variances and sample lag–one correlations; solid line connects REML estimates corresponding to the SAD-QM3(1) model described in Section 5.5.1.*

5.5.2 Intervenor-adjusted formulation

For an intervenor-adjusted formulation of an SAD model, the determinants in expression (5.47) may be evaluated very efficiently using Theorem 2.5(c), which yields

$$
\begin{aligned}
\log |\Sigma_s(\theta_\xi)| &= \sum_{i=1}^{n} \log \sigma_{sii} + \sum_{i=2}^{n} \log(1 - \rho_{si,i-1}^2) \\
&+ \sum_{i=3}^{n} \log(1 - \rho_{si,i-2\cdot i-1}^2) + \cdots \\
&+ \sum_{i=p+1}^{n} \log(1 - \rho_{si,i-p\cdot\{i-p+1:i-1\}}^2).
\end{aligned}
$$

The matrix inverses within this formulation, however, generally cannot be obtained efficiently; the best strategy for their computation appears to be to apply the inverse mapping, H^{-1}, to Ξ to obtain the marginal covariance matrix and then use the computational techniques for the marginal formulation described in the preceding section.

5.5.3 Precision matrix formulation

For an SAD model for which the elements of the precision matrix are parsimoniously modeled, it is clear from (5.46) and (5.47) that the computation of maximum likelihood estimators requires no inversions of $n_s \times n_s$ matrices. It does require the computation of determinants of such matrices, however. Noting that the first sum in (5.47) may be written as $-\sum_{s=1}^{N} \log |[\mathbf{\Sigma}_s(\boldsymbol{\theta}_\gamma)]^{-1}|$, we see that it suffices to compute the determinants of the $n_s \times n_s$ precision matrices. By Theorem 2.2, these matrices are banded, with zeros on off-diagonals $p+1, \ldots, n_s - 1$. In the first-order case, the precision matrices are tridiagonal so their determinants can be obtained with $O(n_s)$ computations (rather than the $O(n_s^3)$ computations required for an arbitrary $n_s \times n_s$ matrix) using the algorithm of El-Mikkawy (2004). In the second-order case, the precision matrices are pentadiagonal and their determinants may also be obtained with only $O(n_s)$ computations using the recursive algorithm of Sogabe (2008). For cases with order higher than two, no $O(n_s)$ algorithms for computing determinants are known to the authors, but several numerical routines exist for the computation of determinants, which can exploit, to some degree, the bandedness of the precision matrices.

5.5.4 Autoregressive formulation

For autoregressively formulated SAD models, versions of expressions (5.46) and (5.47) may be given in terms of $\boldsymbol{\theta}_\delta$ and $\boldsymbol{\theta}_\phi$, with the aid of (2.17) and Theorem 2.5(a), as follows:

$$
\hat{\boldsymbol{\beta}} = \left(\sum_{s=1}^{N} \mathbf{X}_s^T \mathbf{T}_s^T(\hat{\boldsymbol{\theta}}_\phi) [\mathbf{D}_s(\hat{\boldsymbol{\theta}}_\delta)]^{-1} \mathbf{T}_s(\hat{\boldsymbol{\theta}}_\phi) \mathbf{X}_s \right)^{-1}
$$

$$
\times \sum_{s=1}^{N} \mathbf{X}_s^T \mathbf{T}_s^T(\hat{\boldsymbol{\theta}}_\phi) [\mathbf{D}_s(\hat{\boldsymbol{\theta}}_\delta)]^{-1} \mathbf{T}(\hat{\boldsymbol{\theta}}_\phi) \mathbf{Y}_s, \qquad (5.48)
$$

$$
\log L(\hat{\boldsymbol{\beta}}, \boldsymbol{\theta}_\delta, \boldsymbol{\theta}_\phi) = -\frac{n_+}{2} \log 2\pi - \frac{1}{2} \sum_{s=1}^{N} \sum_{i=1}^{n_s} \log \delta_{si}(\boldsymbol{\theta}_\delta)
$$

$$
-\frac{1}{2} \sum_{s=1}^{N} \mathbf{r}_s^T(\hat{\boldsymbol{\beta}}) \mathbf{T}_s^T(\boldsymbol{\theta}_\phi) [\mathbf{D}_s(\boldsymbol{\theta}_\delta)]^{-1} \mathbf{T}_s(\boldsymbol{\theta}_\phi) \mathbf{r}_s(\hat{\boldsymbol{\beta}}),
$$

$$
(5.49)
$$

where $\mathbf{r}_s(\hat{\boldsymbol{\beta}}) = (r_{si}(\hat{\boldsymbol{\beta}})) = \mathbf{Y}_s - \mathbf{X}_s\hat{\boldsymbol{\beta}}$. Since $\mathbf{D}_s(\boldsymbol{\theta}_\delta)$ is diagonal, computing its inverse is trivial. Thus, maximum likelihood estimators of the parameters of autoregressively formulated SAD models may be computed without explicit computation of determinants or inverses of $n_s \times n_s$ matrices. This is a strong point in favor of this formulation.

As noted by Pourahmadi (1999), an additional simplification occurs when the nonzero autoregressive coefficients are modeled linearly, as in (3.10), i.e.,

$$\phi_{i,i-j} = \sum_{l=1}^{m_2} \theta_{\phi l} u_{i,i-j,l},$$
$$i = 2,\ldots,n_s,\ \ j = 1,\ldots,p_i.$$

Here the $u_{i,i-j,l}$'s are observed covariates, which will typically be functions of the measurement times. Observe that in this case the ith element of $\mathbf{T}_s(\boldsymbol{\theta}_\phi)\mathbf{r}_s(\hat{\boldsymbol{\beta}})$ is given by

$$r_{si} - \sum_{j=1}^{p_i} \phi_{i,i-j}(\boldsymbol{\theta}_\phi)r_{s,i-j}$$
$$= r_{si} - \sum_{j=1}^{p_i} (\mathbf{u}_{i,i-j}^T\boldsymbol{\theta}_\phi)r_{s,i-j}$$
$$= r_{si} - \mathbf{w}_{si}^T\boldsymbol{\theta}_\phi$$

where $\mathbf{u}_{i,i-j} = (u_{i,i-j,l})$, $\mathbf{w}_{si} = \sum_{j=1}^{p_i} r_{s,i-j}\mathbf{u}_{i,i-j}$, and we have temporarily suppressed the dependence of various quantities on $\hat{\boldsymbol{\beta}}$. Putting

$$\mathbf{W}_s = \begin{pmatrix} \mathbf{w}_{s1}^T \\ \vdots \\ \mathbf{w}_{sn_s}^T \end{pmatrix},$$

we have

$$\mathbf{T}_s(\boldsymbol{\theta}_\phi)\mathbf{r}_s = \mathbf{r}_s - \mathbf{W}_s\boldsymbol{\theta}_\phi,$$

whereupon (5.49) may be reexpressed as follows:

$$\log L(\hat{\boldsymbol{\beta}}, \boldsymbol{\theta}_\phi, \boldsymbol{\theta}_\delta) = -\frac{n_+}{2}\log 2\pi - \frac{1}{2}\sum_{s=1}^{N}\sum_{i=1}^{n_s}\log \delta_{si}(\boldsymbol{\theta}_\delta)$$
$$-\frac{1}{2}\sum_{s=1}^{N}[\mathbf{r}_s(\hat{\boldsymbol{\beta}}) - \mathbf{W}_s(\hat{\boldsymbol{\beta}})\boldsymbol{\theta}_\phi]^T[\mathbf{D}_s(\boldsymbol{\theta}_\delta)]^{-1}$$
$$\times [\mathbf{r}_s(\hat{\boldsymbol{\beta}}) - \mathbf{W}_s(\hat{\boldsymbol{\beta}})\boldsymbol{\theta}_\phi]. \tag{5.50}$$

Observe that for each fixed value $\boldsymbol{\theta}_{\delta 0}$ of $\boldsymbol{\theta}_\delta$, (5.50) is maximized with respect to $\boldsymbol{\theta}_\phi$ by

$$
\hat{\boldsymbol{\theta}}_\phi = \left(\sum_{s=1}^N \mathbf{W}_s^T(\hat{\boldsymbol{\beta}})[\mathbf{D}_s(\boldsymbol{\theta}_{\delta 0})]^{-1}\mathbf{W}_s(\hat{\boldsymbol{\beta}}) \right)^{-1} \sum_{s=1}^N \mathbf{W}_s^T(\hat{\boldsymbol{\beta}})[\mathbf{D}_s(\boldsymbol{\theta}_{\delta 0})]^{-1}\mathbf{r}_s(\hat{\boldsymbol{\beta}}).
$$
(5.51)

Thus, for autoregressively specified SAD models in which the autoregressive coefficients are modeled linearly, a possibly efficient algorithm for computing the maximum likelihood estimators is to iterate between equation (5.48), equation (5.51) with $\boldsymbol{\theta}_{\delta 0} = \hat{\boldsymbol{\theta}}_\delta$, and the numerical maximization of

$$
\begin{aligned}
\log L(\hat{\boldsymbol{\beta}}, \hat{\boldsymbol{\theta}}_\phi, \boldsymbol{\theta}_\delta) &= -\frac{n_+}{2}\log 2\pi - \frac{1}{2}\sum_{s=1}^N \sum_{i=1}^{n_s}\log \delta_{si}(\boldsymbol{\theta}_\delta) \\
&\quad -\frac{1}{2}\sum_{s=1}^N [\mathbf{r}_s(\hat{\boldsymbol{\beta}}) - \mathbf{W}_s(\hat{\boldsymbol{\beta}})\hat{\boldsymbol{\theta}}_\phi]^T [\mathbf{D}_s(\boldsymbol{\theta}_\delta)]^{-1} \\
&\quad \times [\mathbf{r}_s(\hat{\boldsymbol{\beta}}) - \mathbf{W}_s(\hat{\boldsymbol{\beta}})\hat{\boldsymbol{\theta}}_\phi].
\end{aligned}
$$
(5.52)

In the further special case in which $\delta_{si}(\boldsymbol{\theta}_\delta)$ is a saturated function of measurement time, even the numerical maximization of (5.52) can be circumvented, for in this case an explicit expression can be given for the maximizing value of $\boldsymbol{\theta}_\delta$ in (5.50) for each fixed value $\boldsymbol{\theta}_{\phi 0}$ of $\boldsymbol{\theta}_\phi$. For example, if the data are balanced and $\delta_{si}(\boldsymbol{\theta}_\delta) = \delta_i$ for all s, then for each fixed $\boldsymbol{\theta}_{\phi 0}$, (5.50) is maximized by

$$
\hat{\delta}_i = \frac{1}{N}\sum_{s=1}^N [\mathbf{r}_s(\hat{\boldsymbol{\beta}}) - \mathbf{W}_s(\hat{\boldsymbol{\beta}})\boldsymbol{\theta}_{\phi 0}]^T [\mathbf{r}_s(\hat{\boldsymbol{\beta}}) - \mathbf{W}_s(\hat{\boldsymbol{\beta}})\boldsymbol{\theta}_{\phi 0}].
$$

Example: Treatment A cattle growth data
To the analysis of the Treatment A cattle growth data based on the marginally formulated model presented in Section 5.5.1, we add here analyses based on two autoregressively formulated models: a power law SAD(1) model and an unconstrained linear model, each with saturated mean. The autoregressively formulated power law SAD(1) model fitted here is given by

$$
\begin{aligned}
\delta_i &= \delta, \quad i = 2, \ldots, 11, \\
\phi_{i,i-1} &= \phi^{t_i^\lambda - t_{i-1}^\lambda}, \quad i = 2, \ldots, 11,
\end{aligned}
$$

where $\delta > 0$ and $\phi \geq 0$; δ_1 is left unstructured; and time is scaled as it was in the marginally formulated model. This model takes the innovation variances, apart from the first, to be constant over time. Allowing the first innovation variance to be different than the rest is important for the sake of increased flexibility, as noted previously in Section 3.6. Also, we have used a simpler parameterization for the power law dependence of lag-one autoregressive coefficients on time than that originally prescribed in Section 3.6; this is permissible because the lag-one autoregressive coefficients corresponding to the sample covariance matrix [Table 4.1(c)] do not decrease over time. REML estimates of model parameters are

$$
\begin{pmatrix} \tilde{\delta}_1 \\ \tilde{\delta} \\ \tilde{\phi} \\ \tilde{\lambda} \end{pmatrix} = \begin{pmatrix} 102.0266 \\ 31.6731 \\ 0.99999 \\ 0.55136 \end{pmatrix}.
$$

These yield REML estimates of marginal variances and lag-one correlations that are plotted as dotted lines in Figure 5.2. When compared with estimates from the marginally formulated model fitted previously, it can be seen that the autoregressively formulated model does not fit the lag-one correlations quite as well; however, it fits the variances just about as well, despite having two fewer parameters.

The unconstrained linear model fitted here is motivated by the regressogram and innovariogram displayed in Figure 5.4 (which is an augmented version of Figure 4.8 given previously). These plots suggest using an unconstrained linear SAD(10) model for which the log innovation variances are a cubic function of time and the autoregressive coefficients are a cubic function of lag. Accordingly we fit the model

$$
\begin{aligned}
\log \delta_i &= \psi_1 + \psi_2 t_i + \psi_3 t_i^2 + \psi_4 t_i^3, \\
\phi_{ij} &= \theta_1 + \theta_2 (t_i - t_j) + \theta_3 (t_i - t_j)^2 + \theta_4 (t_i - t_j)^3,
\end{aligned}
$$

where for simplicity we again scale time in two-week units, except for the last observation which we pretend was also taken two weeks (rather than one) after the penultimate measurement on each cow. Also, for greater ease of interpretation we center the time axis by setting $t_1 = -5, t_2 = -4, \ldots, t_{10} = 4, t_{11} = 5$. REML estimates of this model's parameters are as follows:

$$
\begin{pmatrix} \tilde{\psi}_1 \\ \tilde{\psi}_2 \\ \tilde{\psi}_3 \\ \tilde{\psi}_4 \\ \tilde{\theta}_1 \\ \tilde{\theta}_2 \\ \tilde{\theta}_3 \\ \tilde{\theta}_4 \end{pmatrix} = \begin{pmatrix} 3.45036 \\ 0.08822 \\ 0.00848 \\ -0.01142 \\ 1.53842 \\ -0.88356 \\ 0.15117 \\ -0.00795 \end{pmatrix}.
$$

The fitted cubic functions, which yield REML estimates of the $\{\delta_i\}$ and $\{\phi_{ij}\}$, are plotted as curves in Figure 5.4. The corresponding REML estimates of innovation variances and autoregressive coefficients are displayed in Table 5.7, as are the resulting REML estimates of marginal variances and correlations. Comparison of the latter set of quantities with those for the unstructured AD(10) model in Table 5.1(a) suggests that this SAD(10) model may provide a reasonable fit to the data. We will carry out more formal comparisons of this model to other models in the next chapter.

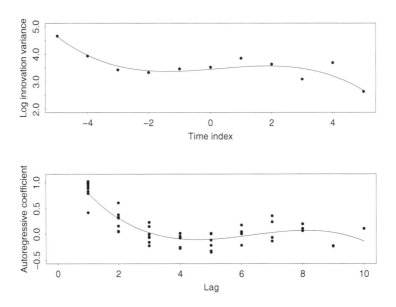

Figure 5.4 *Sample log innovariogram and regressogram for the Treatment A cattle growth data. Fitted curves correspond to REML estimates of cubic models for log innovation variances and autoregressive coefficients.*

For additional examples of fits of autoregressively formulated structured power law AD(1) models, both of which pertain to genetic studies, see Zhao et al. (2005a) and Jaffrézic et al. (2003, 2004).

5.6 Concluding remarks

This chapter has described, in considerable detail and for a variety of scenarios, likelihood-based estimation of the parameters of normal antedependence models. The scenarios are distinguished by whether or not the data are balanced and, if not, whether the missingness is monotone; by whether or not the mean structure is of multivariate regression form (including the saturated mean as a special case); and by whether the antedependence is unstructured or structured and, if the latter, how the antedependence is parameterized. In cases where the data are balanced or at worst the missingness is monotone, the mean structure is of multivariate regression form, and the antedependence is unstructured,

Table 5.7 *REML estimates of parameters of unconstrained linear SAD(10) model for the Treatment A cattle growth data: (a) log innovation variances (along the main diagonal) and autoregressive coefficients (below the main diagonal); (b) marginal variances (along the main diagonal) and correlations (below the main diagonal).*

(a)

104									
.80	53								
.31	.80	360							
.03	.31	.80	30						
−.09	.03	.31	.80	29					
−.09	−.09	.03	.31	.80	32				
−.04	−.09	−.09	.03	.31	.80	34			
.03	−.04	−.09	−.09	.03	.31	.80	35		
.07	.03	−.04	−.09	−.09	.03	.31	.80	33	
.04	.07	.03	−.04	−.09	−.09	.03	.31	.80	25
−.13	.04	.07	.03	−.04	−.09	−.09	.03	.31	.80	15

(b)

104									
.75	119								
.76	.87	163							
.73	.86	.92	213						
.68	.82	.90	.94	257					
.63	.77	.85	.91	.94	290				
.58	.72	.80	.86	.91	.94	318			
.54	.67	.75	.81	.86	.91	.94	348		
.53	.64	.71	.77	.82	.88	.92	.95	383	
.53	.63	.70	.75	.80	.85	.90	.94	.97	419
.48	.61	.67	.72	.77	.82	.87	.92	.96	.98	426

explicit expressions for the maximum likelihood (and REML) estimates exist and can be expressed in terms of regression coefficients and residual sums of squares from regressions of the response at a given time on a particular number of its predecessors plus covariates, that number being determined by the order of the model. Otherwise, the estimates must usually be obtained by numerical optimization. In this event the autoregressive parameterization of the model has several advantages, including the complete absence of constraints on the parameters (provided that we model the logs of the innovation variances rather than the innovation variances themselves), and trivial computation of the inverse and determinant of the covariance matrix.

In each scenario described in this chapter, we specified the minimum sample size needed for the likelihood-based estimates to exist. It should be noted that in every case, the required sample size is less than or equal to that required for existence of the maximum likelihood estimate of the covariance matrix under general multivariate dependence. For example, when the model is the normal multivariate-regression, unstructured AD(p) model, we require $N - m > p$, while for general multivariate dependence we require $N - m > n - 1$. Thus, relative to general multivariate dependence, AD(p) model parameters are estimable from fewer observations.

A final remark pertains to software for obtaining the estimates. PROC MIXED of SAS may be used for likelihood-based estimation of unstructured normal AD models of order 0, 1, and $n - 1$. The same procedure in SAS can fit three structured normal AD models: the first-order stationary, first-order heterogeneous, and first-order continuous-time autoregressive models. For estimating the parameters of unstructured normal AD models of arbitrary order, either constant or variable, R functions written by the first author are available for download from his Web page (see Section 4.4 for the address).

CHAPTER 6

Testing Hypotheses on the Covariance Structure

After a preliminary attempt to identify the data's mean and covariance structure using summary statistics and graphical diagnostics, followed perhaps by estimation of the parameters of a tentative model, the analyst will often want to formally test various hypotheses. Likelihood ratio tests (assuming normality) may be used for this purpose, provided that the hypotheses are nested. Likelihood ratio tests may be used to test hypotheses on either the mean structure or the covariance structure, or both. In this chapter we derive likelihood ratio tests for several hypotheses on antedependent covariance structures. Tests of hypotheses on the mean structure are deferred to the following chapter. This order of presentation corresponds to what we believe to be the most coherent model selection strategy, which is to determine the covariance structure prior to testing hypotheses on the mean structure. The assumed mean structure for the models whose covariance structures are tested should be the most general one possible; typically this will be saturated across time (within groups), but it may also include covariates identified by informal methods as possibly being important.

In practice, the likelihood ratio test statistic for comparing the fits of two nested antedependence models can be computed by maximizing the likelihood function under both models using procedures described at length in Chapter 5, and then forming the ratio of the two maxima; an explicit expression for it in terms of parameter estimates or other meaningful statistics is not strictly necessary. Nevertheless, such an expression can be helpful, as it often sheds light on the behavior of the test and provides a basis for improving the approximation of the test statistic's asymptotic distribution. It can also suggest efficient computing procedures. Consequently, in what follows we provide expressions for the test statistic for many such tests.

As a likelihood ratio test statistic is the ratio of maxima of the likelihood function under two models, so also a residual likelihood ratio test (RLRT) statistic is the ratio of maxima of the residual likelihood under two models. A RLRT may be used to compare two nested antedependence structures, analogously to a likelihood ratio test, provided that the mean structures of the two antedependence models are identical. Valid comparisons of models with different mean structures cannot be made within the REML framework because the error contrasts are different for two such models. For hypotheses about the covariance structure of unstructured antedependence models, the expression for the RLRT statistic generally may be obtained from that of the likelihood ratio test statistic by replacing the number of subjects with the number of subjects minus the number of unknown parameters in each subject's mean structure, and this difference is usually not sufficiently large to alter the conclusion reached by the likelihood ratio test. The relationship between the RLRT and likelihood ratio test statistics for hypotheses about the covariance structure of structured antedependence models is not as simple as it is in the unstructured case; nevertheless, in this case, also, the two test statistics do not often differ substantially. Hence, for the sake of brevity and consistency, in this chapter we feature likelihood ratio tests only.

We begin the chapter with likelihood ratio tests for properties satisfied by individual parameters of the covariance structure under antedependence. Following that, we give likelihood ratio tests for the order of an unstructured normal antedependence model, for a structured normal antedependence model versus an unstructured normal antedependence model of the same order, and for homogeneity of unstructured normal antedependence across treatment (or other) groups. Finally, for those situations where the hypotheses of interest are not nested, we describe an approach for covariance model selection based on penalized likelihood criteria.

6.1 Tests on individual parameters

6.1.1 Partial correlations

Recall from Section 2.3.1 that each partial correlation corresponding to responses lagged more than p units apart is equal to zero under pth-order antedependence. It may therefore be of interest to test the hypothesis that any one of these partial correlations is equal to zero (against a two-sided alternative hypothesis). In Section 4.2 we gave an informal rule of thumb for making this determination but now we consider a formal hypothesis test.

Assume that the data are balanced. Now, by Theorem 4.3.5 of Anderson (1984), the distribution of the sample partial correlation, $r_{ij \cdot rest}$, based on a random sample of size N from a normal population with corresponding partial correlation $\rho_{ij \cdot rest}$, is identical to the distribution of the sample marginal correlation, r_{ij}, based on a random sample of size $N - (n - 2)$ from a normal population with corresponding marginal correlation ρ_{ij}. Consequently, any inference procedure for a partial correlation based on its maximum likelihood estimate from a sample of size N is analogous to an inference procedure on a marginal correlation based on its maximum likelihood estimate from a sample of size $N - n + 2$. In particular, the well-known likelihood ratio test for zero marginal correlation and the well-known, Fisher transformation-based, approximate confidence interval for a correlation coefficient can be adapted to test for zero partial correlation and place confidence bounds on a partial correlation coefficient. Details are given in the following theorem.

Theorem 6.1. *Suppose that* $\mathbf{Y}_1, \ldots, \mathbf{Y}_N$ *are balanced and follow a normal multivariate regression, unstructured* $AD(n-1)$ *model, i.e.,*

$$\mathbf{Y}_s \sim \text{independent } \mathrm{N}_n\left((\mathbf{z}_s^T \otimes \mathbf{I}_n)\boldsymbol{\beta}, \boldsymbol{\Sigma}(\boldsymbol{\theta})\right), \quad s = 1, \ldots, N,$$

where $\boldsymbol{\Sigma}(\boldsymbol{\theta})$ *is unstructured* $AD(n-1)$ *and* $\mathbf{z}_s$ *is* $m \times 1$. *Assume that* $N - m > n - 1$, *and let* $r_{ij \cdot rest}$ *denote the sample partial correlation between ordinary least squares residuals (from the fitted mean structure) at the ith and jth measurement times. For any* $\alpha \in (0, 1)$, *the size-*α *likelihood ratio test for the null hypothesis that the corresponding population partial correlation,* $\rho_{ij \cdot rest}$, *is zero (versus the alternative that it is not) rejects the null hypothesis if and only if*

$$\left(\frac{N - m - n + 1}{1 - r_{ij \cdot rest}^2}\right)^{1/2} |r_{ij \cdot rest}| > t_{\alpha/2, N-m-n+1}, \tag{6.1}$$

where $t_{\alpha/2, N-m-n+1}$ *is the* $100(1 - \alpha/2)$*th percentile of Student's t distribution with* $N - m - n + 1$ *degrees of freedom. Moreover, the endpoints of an approximate* $100(1 - \alpha)\%$ *confidence interval for* $\rho_{ij \cdot rest}$ *are given by*

$$\tanh\left(U_{ij} \pm z_{\alpha/2}/(N - m - n)^{1/2}\right), \tag{6.2}$$

where

$$U_{ij} = \frac{1}{2} \log\left(\frac{1 + r_{ij \cdot rest}}{1 - r_{ij \cdot rest}}\right)$$

and $z_{\alpha/2}$ *is the* $100(1 - \alpha/2)$*th percentile of the standard normal distribution.*

Theorem 6.1 can be extended easily to handle dropouts. If N_i subjects are observed at time i, then we may simply replace N in expressions (6.1) and (6.2) with N_i.

Examples: Treatment A cattle growth data and 100-km race data
The test described in Theorem 6.1 may be applied to the partial correlations of the Treatment A cattle growth data and 100-km race split times, which are displayed in Table 4.1(b) and Table 4.3(b), respectively. Doing so, we find that the null hypothesis of zero partial correlation is rejected at the 0.05 level of significance for those partial correlations that were deemed to be significant

using the rule of thumb of Section 4.2, and only for those partial correlations. Thus the rule of thumb works very well for these data.

6.1.2 Intervenor-adjusted partial correlations

Like standard partial correlations corresponding to variables lagged p or more units apart, intervenor-adjusted partial correlations corresponding to those same variables are equal to zero under pth-order antedependence (Definition 2.3). Hence it is likewise of interest to test whether any of these quantities are equal to zero. In Section 4.2, we gave an informal rule of thumb for making this determination. The next theorem, which follows by the same line of reasoning as that which established Theorem 6.1, gives the likelihood ratio test for this hypothesis. It can be extended to handle dropouts in exactly the same way that Theorem 6.1 can.

Theorem 6.2. *Suppose that* $\mathbf{Y}_1, \ldots, \mathbf{Y}_N$ *are balanced and follow the normal multivariate regression, unstructured* $AD(n-1)$ *model specified in Theorem 6.1. Assume that* $N - m > n - 1$, *and let* $r_{ij \cdot \{j+1:i-1\}}$ *denote the sample intervenor-adjusted partial correlation between ordinary least squares residuals (from the fitted mean structure) at the* ith *and* jth *measurement times* $(i > j)$. *For any* $\alpha \in (0, 1)$, *the size-*α *likelihood ratio test for the null hypothesis that the corresponding population intervenor-adjusted partial correlation,* $\rho_{ij \cdot \{j+1:i-1\}}$, *is zero (versus the alternative that it is not) rejects the null hypothesis if and only if*

$$\left(\frac{N - m - i + j}{1 - r_{ij \cdot \{j+1:i-1\}}^2} \right)^{1/2} |r_{ij \cdot \{j+1:i-1\}}| > t_{\alpha/2, N-m-i+j},$$

where $t_{\alpha/2, N-m-i+j}$ *is the* $100(1 - \alpha/2)$th *percentile of Student's* t *distribution with* $N - m - i + j$ *degrees of freedom. Moreover, the endpoints of an approximate* $100(1 - \alpha)\%$ *confidence interval for* $\rho_{ij \cdot \{j+1:i-1\}}$ *are given by*

$$\tanh \left(V_{ij} \pm z_{\alpha/2} / (N - m - i + j - 1)^{1/2} \right),$$

where

$$V_{ij} = \frac{1}{2} \log \left(\frac{1 + r_{ij \cdot \{j+1:i-1\}}}{1 - r_{ij \cdot \{j+1:i-1\}}} \right).$$

Examples: Treatment A cattle growth data and 100-km race data
We apply the test described in Theorem 6.2 to the intervenor-adjusted partial correlations of the Treatment A cattle growth data and 100-km race split times, which are displayed in Table 4.1(a) and Table 4.3(a), respectively. We find, as was the case for the test for partial correlations, that this test rejects the null hypothesis of zero intervenor-adjusted partial correlation at significance level

0.05 for precisely those correlations that were deemed to be significant using the rule of thumb of Section 4.2.

6.1.3 Autoregressive coefficients

Since autoregressive coefficients corresponding to variables lagged p or more units apart are equal to zero under pth-order antedependence (Theorem 2.3), testing whether an autoregressive coefficient is equal to zero may also be of interest. The final theorem of this section gives the likelihood ratio test for this hypothesis. The theorem also gives the likelihood ratio test for the hypothesis that, for fixed i, *all* autoregressive coefficients in the regression of the ith response variable on its predecessors and the covariates are zero.

Theorem 6.3. *Suppose that* $\mathbf{Y}_1, \ldots, \mathbf{Y}_N$ *are balanced and follow the normal multivariate regression, unstructured* $AD(n-1)$ *model specified in Theorem 6.1. Assume that* $N - m > n - 1$, *let* ϕ_{ij} *(where* $i > j$*) denote the autoregressive coefficient on* Y_j *in the ordinary least squares regression of* Y_i *on its predecessors and the covariates, and let* $\hat{\phi}_{ij}$ *and* $\tilde{\delta}_i$ *denote the maximum likelihood estimator of* ϕ_{ij} *and the REML estimator of* δ_i, *respectively, as given by Theorem 5.6.*

(a) *For any* $\alpha \in (0,1)$, *the size-α likelihood ratio test for the null hypothesis that* ϕ_{ij} *is zero (versus the alternative that it is not) rejects the null hypothesis if and only if*

$$\frac{|\hat{\phi}_{ij}|}{(\tilde{\delta}_i c_{ii,jj})^{1/2}} > t_{\alpha/2, N-m-i+1}, \tag{6.3}$$

where $c_{ii,jj}$ *is the jth diagonal element of* $\mathbf{A}_{1:i-1,i}^{-1}$ *and* $t_{\alpha/2, N-m-n+1}$ *is defined as in Theorem 6.2;*

(b) *For any* $\alpha \in (0,1)$, $100(1-\alpha)\%$ *confidence intervals for* ϕ_{ij} *and* δ_i *are given by*

$$\hat{\phi}_{ij} \pm t_{\alpha/2, N-m-i+1} (\tilde{\delta}_i c_{ii,jj})^{1/2}$$

and

$$\left(\frac{(N-m-i+1)\tilde{\delta}_i}{\chi^2_{\alpha/2, N-m-i+1}}, \frac{(N-m-i+1)\tilde{\delta}_i}{\chi^2_{1-\alpha/2, N-m-i+1}} \right),$$

respectively, where $\chi^2_{\alpha/2, N-m-i_1}$ *is the* $100(1-\alpha/2)$*th percentile of the chi-square distribution with* $N-m-i+1$ *degrees of freedom;*

(c) *Let* $\boldsymbol{\phi}_i = (\phi_{i1}, \phi_{i2}, \ldots, \phi_{i,i-1})^T$ *and let* $\hat{\boldsymbol{\phi}}_i$ *be its maximum likelihood estimator. Then, for any* $\alpha \in (0,1)$, *the size-α likelihood ratio test for the null hypothesis that* $\boldsymbol{\phi}_i$ *is equal to* $\mathbf{0}$ *(versus the alternative that it is not) rejects* *the null hypothesis if and only if*

$$\left(\frac{N-m-i+1}{i-1}\right)\frac{\hat{R}^2_{i\cdot 1,\ldots,i-1}}{1-\hat{R}^2_{i\cdot 1,\ldots,i-1}} > F_{\alpha,i-1,N-m-i+1}, \qquad (6.4)$$

where $\hat{R}^2_{i\cdot 1,\ldots,i-1} = \hat{\boldsymbol{\phi}}_i^T \mathbf{A}_{1:i-1,i}\hat{\boldsymbol{\phi}}_i/a_{ii}$ *and* $F_{\alpha,i-1,N-m-i+1}$ *is the* $100(1-\alpha)th$ *percentile of the F distribution with* $i-1$ *and* $N-m-i+1$ *degrees of freedom.*

Proof. Let i be a fixed integer between 2 and n, inclusive. It follows from standard results in multiple linear regression theory that, conditional on the response and covariates up to time $i-1$ (inclusive), $\hat{\boldsymbol{\phi}}_i$ is distributed as N($\boldsymbol{\phi}_i$, $\delta_i\mathbf{A}^{-1}_{1:i-1,i}$). Thus, the conditional distribution of $\hat{\phi}_{ij}$ is N($\phi_{ij}, \delta_i c_{ii,jj}$). So under the null hypothesis, conditionally $\hat{\phi}_{ij}/(\delta_i c_{ii,jj})^{1/2}$ is distributed as N$(0,1)$. Also from standard regression theory, conditionally $(N-m-i+1)\tilde{\delta}_i/\delta_i$ is distributed as chi-square with $N-m-i+1$ degrees of freedom, and $\tilde{\delta}_i$ and $\hat{\phi}_{ij}$ are conditionally independent. Thus, under the null hypothesis and conditional on the response and covariates up to time $i-1$,

$$\frac{\hat{\phi}_{ij}}{(\tilde{\delta}_i c_{ii,jj})^{1/2}}$$

is distributed as Student's t with $N-m-i+1$ degrees of freedom. Since this distribution is free of the responses up to time $i-1$, it is also the unconditional distribution, and parts (a) and (b) of the theorem follow immediately. The proof of part (c) follows by a similar argument, upon noting that $\boldsymbol{\phi}_i = \mathbf{0}$ if and only if $R^2_{i\cdot 1,\ldots,i-1} = 0$ and that $\hat{R}^2_{i\cdot 1,\ldots,i-1}$ is the maximum likelihood estimator of $R^2_{i\cdot 1,\ldots,i-1}$. □

By the proof of Theorem 6.3, it is clear that the test given by (6.3) is the standard t-test for zero slope coefficient on Y_j in the ordinary least squares regression of Y_i on its predecessors and the covariates, and that this test may therefore be carried out using standard software for fitting regression models. If there are dropouts, we merely replace N with N_i in expressions (6.3) and (6.4). We remind the reader that we performed this test on the autoregressive coefficients of the Treatment A cattle growth data and 100-km race split times in Section 4.2, indicating in Table 4.1(c) and Table 4.3(c) which autoregressive coefficients were significantly different from zero at the 0.05 level of significance.

6.2 Testing for the order of antedependence

The analyst will often want to formally test for the order of antedependence of the data's covariance structure. Recall from (2.2) that antedependence models of increasing order are nested. Here we present the likelihood ratio test for the null hypothesis of pth-order unstructured antedependence against the alternative hypothesis of $(p + q)$th-order unstructured antedependence, where p and q are specified integers such that $0 \leq p \leq n - 2$ and $1 \leq q \leq n - p - 1$. Practical strategies for determining the order of antedependence using a series of such tests will be described subsequently. Observe that the alternative hypothesis of highest possible order, i.e., that for which $p + q = n - 1$, imposes no structure whatsoever (other than positive definiteness) on the covariance matrix, so in this case the test is one of pth-order antedependence versus general multivariate dependence.

For our main result, we assume that the observations are balanced and we take the mean structure to be of multivariate regression form. Recall that the multivariate regression form of the mean structure includes the saturated mean as a special case. Our main result is an extension of a result originally given by Gabriel (1962), who considered only the case of a saturated mean and $q = 1$. Eventually we will further extend this result to allow for dropouts and to test for order in nested variable-order antedependence models.

Theorem 6.4. *Suppose that* $\mathbf{Y}_1, \ldots, \mathbf{Y}_N$ *are balanced and follow a normal multivariate regression, unstructured* $AD(p + q)$ *model, i.e.,*

$$\mathbf{Y}_s \sim \text{independent } N_n \left((\mathbf{z}_s^T \otimes \mathbf{I}_n)\boldsymbol{\beta}, \boldsymbol{\Sigma}(\boldsymbol{\theta}) \right), \quad s = 1, \ldots, N,$$

where $\boldsymbol{\Sigma}(\boldsymbol{\theta})$ *is unstructured* $AD(p + q)$ *and* $\mathbf{z}_s$ *is* $m \times 1$. *Assume that* $N - m > p + q$, *and let* $r_{i,i-k\cdot\{i-k+1:i-1\}}$ *denote the intervenor-adjusted sample partial correlation between ordinary least squares residuals (from the fitted mean structure) at the* ith *and* $(i - k)$th *measurement times. The likelihood ratio test for the null hypothesis that* $\boldsymbol{\Sigma}(\boldsymbol{\theta})$ *is unstructured* $AD(p)$ *[versus the alternative that it is unstructured* $AD(p + q)$*] rejects the null hypothesis if and only if*

$$-N \sum_{j=1}^{q} \sum_{i=p+j+1}^{n} \log \left(1 - r^2_{i,i-p-j\cdot\{i-p-j+1:i-1\}} \right) > K, \qquad (6.5)$$

where K *is a constant. For any* $\alpha \in (0, 1)$, *an asymptotically (as* $N \to \infty$*) valid size-α test is obtained by taking* K *to be the* $100(1 - \alpha)$th *percentile of a chi-square distribution with* $(2n - 2p - q - 1)(q/2)$ *degrees of freedom.*

Proof. Let $\hat{\boldsymbol{\theta}}_0$ and $\hat{\boldsymbol{\theta}}_1$ be the maximum likelihood estimators of the marginal covariance parameter $\boldsymbol{\theta}_\sigma$ under the null and alternative hypotheses, respectively. It follows from (5.33) and Theorem 5.6(a) that the maximized profile log-likelihood under the alternative hypothesis is

$$\sup_{H_1} \log L^*(\boldsymbol{\theta}) = -\frac{nN}{2} \log 2\pi - \frac{N}{2} \log |\boldsymbol{\Sigma}(\hat{\boldsymbol{\theta}}_1)| - \frac{N}{2} \text{tr} \left(\mathbf{A}[\boldsymbol{\Sigma}(\hat{\boldsymbol{\theta}}_1)]^{-1} \right),$$

where $\mathbf{A} = (a_{ij})$ is given by (5.31). Now denote the elements of $\boldsymbol{\Sigma}(\hat{\boldsymbol{\theta}}_1)$ and

$[\boldsymbol{\Sigma}(\hat{\boldsymbol{\theta}}_1)]^{-1}$ by $(\hat{\sigma}_{ij1})$ and $(\hat{\sigma}_1^{ij})$, respectively, and note from Theorem 2.2 that $\hat{\sigma}_1^{ij} = 0$ for $|i-j| > p+q$. Therefore,

$$
\begin{aligned}
\operatorname{tr}\left(\mathbf{A}[\boldsymbol{\Sigma}(\hat{\boldsymbol{\theta}}_1)]^{-1}\right) &= \sum_{i=1}^n \sum_{j=1}^n a_{ij}\hat{\sigma}_1^{ij} \\
&= \sum\sum_{0\leq|i-j|\leq p+q} a_{ij}\hat{\sigma}_1^{ij} \\
&= \sum\sum_{0\leq|i-j|\leq p+q} \hat{\sigma}_{ij1}\hat{\sigma}_1^{ij} \\
&= \sum_{i=1}^n \sum_{j=1}^n \hat{\sigma}_{ij1}\hat{\sigma}_1^{ij} \\
&= \operatorname{tr}\left([\boldsymbol{\Sigma}(\hat{\boldsymbol{\theta}}_1)][\boldsymbol{\Sigma}(\hat{\boldsymbol{\theta}}_1)]^{-1}\right) \\
&= \operatorname{tr}(\mathbf{I}_n) \\
&= n.
\end{aligned}
$$

Thus

$$
\sup_{H_1} \log L^*(\boldsymbol{\theta}) = -\frac{nN}{2}\log 2\pi - \frac{N}{2}\log|\boldsymbol{\Sigma}(\hat{\boldsymbol{\theta}}_1)| - \frac{Nn}{2}. \tag{6.6}
$$

Similarly, it may be shown that

$$
\sup_{H_0} \log L^*(\boldsymbol{\theta}) = -\frac{nN}{2}\log 2\pi - \frac{N}{2}\log|\boldsymbol{\Sigma}(\hat{\boldsymbol{\theta}}_0)| - \frac{Nn}{2}.
$$

Now, upon letting Λ denote the usual likelihood ratio test statistic, i.e.,

$$
\Lambda = \frac{\sup_{H_0} L^*(\boldsymbol{\theta})}{\sup_{H_1} L^*(\boldsymbol{\theta})},
$$

we obtain

$$
-2\log\Lambda = N\log\left(\frac{|\boldsymbol{\Sigma}(\hat{\boldsymbol{\theta}}_0)|}{|\boldsymbol{\Sigma}(\hat{\boldsymbol{\theta}}_1)|}\right). \tag{6.7}
$$

The determinants in (6.7) may be expressed in terms of various quantities using Theorem 2.5. Here, we use part (c) of Theorem 2.5 to express them in terms of certain intervenor-adjusted partial correlations among residuals from the fitted mean structure. Denote the intervenor-adjusted partial correlations corresponding to $\hat{\boldsymbol{\theta}}_l$ ($l = 0$ or 1) as $\{\hat{\rho}_{i,i-k\cdot\{i-k+1:i-1\},l}\}$, and observe that

$$
\hat{\rho}_{i,i-k\cdot\{i-k+1:i-1\},0} = \hat{\rho}_{i,i-k\cdot\{i-k+1:i-1\},1} \quad \text{for } k = 1,\ldots,p
$$

and

$$
\hat{\rho}_{i,i-k\cdot\{i-k+1:i-1\},1} = r_{i,i-k\cdot\{i-k+1:i-1\}} \quad \text{for } p+1 \leq k \leq p+q;
$$

the sample size condition of the theorem ensures that these quantities are well defined. Thus

$$
\begin{aligned}
-2\log\Lambda &= N\log\left(\frac{\prod_{i=1}^{n}\hat{\sigma}_{ii0}\prod_{i=2}^{n}(1-\hat{\rho}_{i,i-1,0}^{2})\cdots}{\prod_{i=1}^{n}\hat{\sigma}_{ii1}\prod_{i=2}^{n}(1-\hat{\rho}_{i,i-1,1}^{2})\cdots}\right.\\
&\qquad\times\left.\frac{\prod_{i=p+1}^{n}(1-\hat{\rho}_{i,i-p\cdot\{i-p+1:i-1\},0}^{2})}{\prod_{i=p+q+1}^{n}(1-\hat{\rho}_{i,i-p-q\cdot\{i-p-q+1:i-1\},1}^{2})}\right)\\
&= -N\sum_{j=1}^{q}\sum_{i=p+j+1}^{n}\log\left(1-r_{i,i-p-j\cdot\{i-p-j+1:i-1\}}^{2}\right).
\end{aligned}
$$

This establishes that the likelihood ratio test has the form given by (6.5). Now, by standard asymptotic theory for likelihood ratio testing, $-2\log\Lambda$ converges in distribution (as $N\to\infty$) to a chi-square random variable with degrees of freedom equal to the difference in dimensionality of the parameter spaces under the two hypotheses, which is

$$
\frac{(2n-p-q)(p+q+1)}{2}-\frac{(2n-p)(p+1)}{2}=(2n-2p-q-1)(q/2).\quad\square
$$

The form of the likelihood ratio test given by Theorem 6.4 is eminently reasonable: the test statistic is essentially a scaled average of quantities, each the log of one minus the square of an intervenor-adjusted sample partial correlation, which are small with high probability under the null hypothesis but not necessarily so under the alternative hypothesis. In fact, since $-N\log(1-r^{2})$ exceeds a constant if and only if $|r|/(1-r^{2})^{1/2}$ exceeds another constant, the likelihood ratio test is the sum of test criteria which are equivalent to those test criteria of Theorem 6.2 that correspond to intervenor-adjusted partial correlations between variables lagged at least p units apart.

A modification to the test criterion which yields a better asymptotic approximation to the chi-square distribution specified in Theorem 6.4 results from the following considerations. A two-term Taylor series approximation of $\log(1-r_{i,i-p-j\cdot\{i-p-j+1:i-1\}}^{2})$ is given by

$$
\begin{aligned}
\log(1-r_{i,i-p-j\cdot\{i-p-j+1:i-1\}}^{2}) &\doteq -r_{i,i-p-j\cdot\{i-p-j+1:i-1\}}^{2}\\
&\quad -r_{i,i-p-j\cdot\{i-p-j+1:i-1\}}^{4}/2.
\end{aligned}
$$

Now using expressions given, for example, by Anderson (1984, p. 108) for the even moments of a sample correlation when the corresponding population correlation is equal to zero, in tandem with the simple relationship between inference for ordinary correlations and partial correlations (Anderson, 1984, Theorem 5.3.5), we obtain

$$
E\left(r_{i,i-p-j\cdot\{i-p-j+1:i-1\}}^{2m}\right)=\frac{\Gamma[\frac{1}{2}(N-p-j)]\Gamma(m+\frac{1}{2})}{\sqrt{\pi}\Gamma[\frac{1}{2}(N-p-j)+m]}
$$

for any positive integer m. (Here, Γ represents the standard gamma function, defined by $\Gamma(t) = \int_0^\infty e^{-u} t^{u-1}\, du$, for $t > 0$.) Thus we find that

$$E\left[\log(1 - r^2_{i,i-p-j\cdot\{i-p-j+1:i-1\}})\right] = -\frac{1}{N-p-j} + O[(N-p-j)^{-2}].$$

As a result, the size of the test prescribed by Theorem 6.4 is more closely approximated by α if we replace the test statistic in (6.5) by

$$-\sum_{j=1}^{q}\left((N-p-j)\sum_{i=p+j+1}^{n}\log\left(1 - r^2_{i,i-p-j\cdot\{i-p-j+1:i-1\}}\right)\right).$$

The likelihood ratio test criterion given by Theorem 6.4 is expressed in terms of certain intervenor-adjusted sample partial correlations. It is possible, however, to express the criterion in several alternative ways, as a consequence of the various ways in which the determinant of an antedependent covariance matrix can be represented (Theorem 2.5). For example, the criterion may be expressed in terms of maximum likelihood estimates of certain innovation variances [Theorem 2.5(a)], certain multiple correlation coefficients [Theorem 2.5(b)], or certain principal minors of the residual covariance matrix [Theorem 2.5(d)]. Of these, the first is especially useful and we now consider it further. This form of the likelihood ratio test was first given by Kenward (1987).

Theorem 6.5. *The likelihood ratio test criterion given by (6.5) may be expressed equivalently as*

$$N\sum_{i=p+2}^{n}[\log RSS_i(p) - \log RSS_i(p+q)] > K, \tag{6.8}$$

where $RSS_i(k)$ is the residual sum of squares from the ordinary least squares regression of the ith response variable on the $k_i \equiv \min(k, i-1)$ such variables immediately preceding it plus the m covariates, and K is defined as in Theorem 6.4.

Proof. From (6.7) and Theorem 2.5(a), we have

$$\begin{aligned}
-2\log\Lambda &= N\log\left(\frac{\prod_{i=1}^{n}\hat{\delta}_i^{(p)}}{\prod_{i=1}^{n}\hat{\delta}_i^{(p+q)}}\right) \\
&= N\sum_{i=1}^{n}\left(\log\hat{\delta}_i^{(p)} - \log\hat{\delta}_i^{(p+q)}\right)
\end{aligned}$$

where $\hat{\delta}_i^{(k)}$ is the maximum likelihood estimator of the ith diagonal element of matrix $\mathbf{D}$ of the modified Cholesky decomposition of $\mathbf{\Sigma}^{-1}$ under an AD(k) model. Now, it was shown in the proof of Theorem 5.1(d) that $\hat{\delta}_i^{(k)}$ under such a model is merely $(1/N)$ times the residual sum of squares from the ordinary least squares regression of the ith response variable on the $k_i \equiv \min(k, i - 1)$ such variables immediately preceding it plus the m covariates. The theorem follows immediately. $\square$

An important advantage of the test criterion (6.8) relative to (6.5) is that it is expressed in terms of residual sums of squares from several regression models, implying that it can be computed using standard regression software.

Like the first criterion, this one's actual size more closely approximates its nominal size if it is modified in such a way that its expectation more closely matches that of the limiting chi-square distribution. Using the fact that for $i \geq p + 2$, $RSS_i(p)/RSS_i(p + q)$ has a Beta distribution with parameters $[N - m - (p+q)_i]/2$ and $[(p+q)_i - p]/2$ under the null hypothesis, Kenward (1987) derived the following improved criterion:

$$\frac{\sum_{i=p+2}^{n} [(p + q)_i - p] \sum_{i=p+2}^{n} [\log RSS_i(p) - \log RSS_i(p + q)]}{\sum_{i=p+2}^{n} \psi [(p + q)_i - p, N - m - (p + q)_i]} > K,$$

(6.9)

where K is defined as in Theorem 6.4 and

$$\psi(x, y) = \begin{cases} 0 & \text{if } x = 0 \\ (2y - 1)/[2y(y - 1)] & \text{if } x = 1 \\ 2\sum_{l=1}^{x/2}(y + 2l - 2)^{-1} & \text{if } x > 0 \text{ and even} \\ \psi(1, y) + \psi(x - 1, y + 1) & \text{if } x > 1 \text{ and odd.} \end{cases}$$

Some important special cases of the likelihood ratio test given by either of Theorems 6.4 or 6.5 may be singled out. First consider the case $(p, q) = (0, n-1)$, which corresponds to testing for complete independence against arbitrary dependence. In this case the criterion for an asymptotically valid size-α likelihood ratio test reduces to

$$-N \log |\mathbf{R}| > K,$$

(6.10)

where $\mathbf{R}$ is the sample correlation matrix of residuals (from the fitted mean structure) and K is the $100(1 - \alpha)$th percentile of a chi-square distribution with $n(n - 1)/2$ degrees of freedom. This test coincides with the well-known likelihood ratio test for independence given, for example, by Anderson (1984, Theorem 9.2.1). Next consider the case in which p is arbitrary and $q = 1$, which corresponds to testing for an AD(p) model versus an AD($p + 1$) model. In this case the criterion for an asymptotically valid size-α likelihood ratio test reduces to

$$-N \sum_{i=p+2}^{n} \log \left(1 - r^2_{i,i-p-1 \cdot \{i-p:i-1\}}\right) > K_p$$

(6.11)

where K_p is the $100(1 - \alpha)$th percentile of a chi-square distribution with $n - p - 1$ degrees of freedom. It follows easily from Theorem 2.5(c) that the test criteria (6.11) for the sequence of tests of AD(p) versus AD($p + 1$)

$(p = 0, 1, \ldots, n - 2)$ sum to the criterion (6.10) for testing complete independence versus arbitrary dependence. Thus, this sequence can be viewed as a
decomposition of the test for complete independence into steps of degree one.
Furthermore, the sequence suggests two practical strategies for selecting the
order of antedependence. Analogously to the selection of order of a polynomial model for the mean structure of a regression model, the analyst may use
either a forward selection strategy [starting by testing AD(0) versus AD(1),
and if AD(0) is rejected then testing AD(1) versus AD(2), etc.] or a backward
elimination strategy [starting with a test of AD$(n - 2)$ versus AD$(n - 1)$, and
if AD$(n - 2)$ is not rejected then testing AD$(n - 3)$ versus AD$(n - 2)$, etc.].

The likelihood ratio test for order of antedependence can be extended easily
for use when there are dropouts. As a consequence of Theorem 5.8, one may
simply compute the intervenor-adjusted partial correlations (for the test criterion of Theorem 6.4) or residual sums of squares (for the test criterion of
Theorem 6.5) from the available data and replace N in the criteria with N_i
(and move it inside the summation) but in every other respect proceed as in the
complete-data case. The test can also be extended for use with variable-order
antedependence models. Suppose that we wish to test the null hypothesis that
the covariance matrix is AD$(p_1, \ldots, p_n)$ versus the alternative hypothesis that
it is AD$(p_1 + q_1, \ldots, p_n + q_n)$, where the p_i's and q_i's are specified integers
such that $0 \leq p_i \leq i - 1$, $0 \leq q_i \leq i - 1 - p_i$, and at least one q_i is strictly
greater than 0. For this test, asymptotically valid size-α test criteria analogous
to those of Theorems 6.4 and 6.5 are

$$-N \sum_{i=2}^{n} \sum_{j=1}^{q_i} \log \left(1 - r_{i,i-p_i-j\cdot\{i-p_i-j+1:i-1\}}^2 \right) > K$$

and

$$N \sum_{i=2}^{n} [\log RSS_i(p_i) - \log RSS_i(p_i + q_i)] > K,$$

respectively, where K is the $100(1-\alpha)$th percentile of a chi-square distribution
with $\sum_{i=2}^{n} q_i$ degrees of freedom.

Simulation results
To demonstrate the improvement in approximating the nominal size that results
from using the modified test criterion (6.9), and to give some indication of
the test's power, we present the results of a small simulation study. We take
$p = q = 1$, hence we test AD(1) versus AD(2). We first consider size. For
$N = 20$, 40 or 80 and $n = 5$ or 10, we simulated 100,000 data sets of size
N from n-variate normal distributions with zero mean and AD(1) covariance
matrix Σ, where the modified Cholesky decomposition of Σ^{-1} was given by

Table 6.1 *Estimates of actual sizes of the nominal size-0.05 likelihood ratio test (LRT) for AD(1) versus AD(2), and of its modification. Estimated standard errors of estimated sizes are approximately 0.0007.*

N	n	LRT	Modified LRT
20	5	0.0927	0.0502
20	10	0.1191	0.0501
40	5	0.0675	0.0498
40	10	0.0776	0.0495
80	5	0.0592	0.0508
80	10	0.0629	0.0503

$\mathbf{D} = \mathbf{I}$ and

$$\mathbf{T} = \begin{pmatrix} 1 & 0 & \cdots & 0 & 0 \\ -1 & 1 & \cdots & 0 & 0 \\ 0 & -1 & \cdots & 0 & 0 \\ \vdots & \vdots & \vdots & \vdots & \vdots \\ 0 & 0 & \cdots & -1 & 1 \end{pmatrix}.$$

(Thus, the lag-one autoregressive coefficients were all taken to equal 1.0.) Table 6.1 reports, for each case of N and n, the proportion of simulations for which the nominal size-0.05 likelihood ratio test criterion, and its modification (6.9), reject AD(1) in favor of AD(2). It is clear that the actual size of the likelihood ratio test is somewhat too high, and that this discrepancy gets worse as either n increases or N decreases. Also, we see that in each case the modification restores the actual size of the test to a level imperceptibly distinct from the nominal size.

Next we examine power. For this purpose we simulated 100,000 data sets according to exactly the same prescription as above, except that we took

$$\mathbf{T} = \begin{pmatrix} 1 & 0 & \cdots & 0 & 0 & 0 \\ -1 & 1 & \cdots & 0 & 0 & 0 \\ -\phi_2 & -1 & \cdots & 0 & 0 & 0 \\ 0 & -\phi_2 & \cdots & 0 & 0 & 0 \\ \vdots & \vdots & \vdots & \vdots & \vdots & \vdots \\ 0 & 0 & \cdots & -\phi_2 & -1 & 1 \end{pmatrix},$$

where ϕ_2, the common lag-two autoregressive coefficient, is equal to either 0.1, 0.2, 0.3, 0.4, or 0.5. Thus the simulated data are AD(2). Table 6.2 gives the proportion of simulations for which the size-0.05 modified likelihood ratio test rejects AD(1) in favor of AD(2) for each case of N and n. The results indicate, as expected, that the power of the test increases as either the number

Table 6.2 *Estimates of power of the size-0.05 modified likelihood ratio test for AD(1) versus AD(2). Estimated standard errors of estimated powers are less than 0.0016.*

		Lag-two autoregressive coefficient, ϕ_2				
N	n	0.1	0.2	0.3	0.4	0.5
20	5	0.0677	0.1227	0.2262	0.3740	0.5428
20	10	0.0825	0.2035	0.4335	0.6971	0.8813
40	5	0.0925	0.2347	0.4745	0.7254	0.8938
40	10	0.1326	0.4591	0.8324	0.9783	0.9987
80	5	0.1430	0.4622	0.8169	0.9705	0.9978
80	10	0.2593	0.8327	0.9952	0.9999	1.0000

of subjects, the number of measurement times, or the magnitude of the lag-two autoregressive coefficient increases.

Example 1: Treatment A cattle growth data

Results of likelihood ratio tests for determining the order of antedependence of the Treatment A cattle growth data, assuming a saturated mean structure, are summarized in Table 6.3. From an examination of the sample correlations (displayed previously in Table 1.2), it is clear that observations on the same subject are not independent [AD(0)]. Therefore, we begin a forward selection procedure with a test of AD(1) versus AD(2). The modified likelihood ratio test statistic, (6.9), is

$$[\psi(1,27)]^{-1} \sum_{i=3}^{11} [\log RSS_i(1) - \log RSS_i(2)] = 17.01$$

and the corresponding asymptotic chi-square distribution has 9 degrees of freedom. This yields a p-value of 0.049, which constitutes some, though barely statistically significant, evidence against the AD(1) model in favor of the AD(2) model. If the analyst decides to take the selection procedure a step further by testing AD(2) versus AD(3), a modified likelihood ratio test statistic of 8.58 is obtained, with a corresponding p-value of 0.38. So the procedure stops with the selection of either a first-order or second-order model, depending on the level chosen for the test between these two. A backward elimination procedure results in the same choice. As a final check, we test each of the AD(1) and AD(2) models versus the AD(10) model, the general multivariate covariance structure. Both tests are non-significant, leading us to conclude that among constant-order AD models, those of order higher than two do not fit significantly better than the first-order and second-order models. For further analyses, we would recommend the second-order model, as it is probably better to err on the side of overfitting the covariance structure than underfitting it.

Table 6.3 *Results of hypothesis tests for order of unstructured antedependence of the Treatment A cattle growth data.*

Null hypothesis	Alternative hypothesis	Modified likelihood ratio test statistic	P
AD(1)	AD(2)	17.01	0.049
AD(2)	AD(3)	8.58	0.38
AD(3)	AD(4)	4.98	0.66
AD(4)	AD(5)	3.29	0.77
AD(5)	AD(6)	1.40	0.92
AD(6)	AD(7)	2.64	0.62
AD(7)	AD(8)	1.58	0.66
AD(8)	AD(9)	2.23	0.33
AD(9)	AD(10)	0.67	0.41
AD(1)	AD(10)	41.00	0.64
AD(2)	AD(10)	25.15	0.91

The use of an antedependence model of order slightly higher than necessary reduces efficiency slightly, but retains validity, whereas too low an order can render inferences invalid.

Example 2: 100-km race data

Results of likelihood ratio tests for the order of antedependence of the complete 100-km race data set, assuming a saturated mean structure, are given in Table 6.4. We again begin a forward selection procedure with a test of AD(1) versus AD(2), in view of the obvious lack of independence among the split times. The procedure stops unambiguously with the selection of the third-order model, and a test of AD(3) versus AD(9) confirms this choice. On the other hand, a backward elimination procedure proceeds by unambiguously reducing the order to five, but the test of AD(4) versus AD(5) yields a p-value of 0.055. So at this stage it is a close call, but by the same considerations noted in the previous example, one might argue in favor of the fifth-order model.

Table 6.4 also includes results of tests comparing several variable-order un-structured antedependence models. Motivated by which intervenor-adjusted partial correlations in Table 4.3 were judged previously to be significant, we perform tests of AD(0,1,1,1,1,1,1,1,1,3) versus AD(0,1,1,1,1,1,1,2,1,3), and of AD(0,1,1,1,1,1,1,2,1,1) versus AD(0,1,1,1,1,1,1,2,1,3). In both cases the lower-order model was rejected in favor of the higher-order model. Later in this chapter we will see that an even higher-order model, AD(0,1,1,1,2,1,1,2,3,5), minimizes a certain penalized likelihood criterion. Consequently, we also perform a test of AD(0,1,1,1,1,1,1,2,1,3) versus AD(0,1,1,1,2,1,1,2,3,5) here, finding that the former model is rejected in favor of the latter.

Table 6.4 *Results of hypothesis tests for order of unstructured antedependence of the 80 competitors' split times. In the last three rows of the table, VAD_1 represents AD(0,1,1,1,1,1,1,1,3); VAD_2 represents AD(0,1,1,1,1,1,1,2,1,3); VAD_3 represents AD(0,1,1,1,1,1,2,1,1); and VAD_4 represents AD(0,1,1,1,2,1,1,2,3,5).*

Null hypothesis	Alternative hypothesis	Modified likelihood ratio test statistic	P
AD(1)	AD(2)	46.29	3×10^{-6}
AD(2)	AD(3)	20.64	0.0043
AD(3)	AD(4)	4.37	0.63
AD(4)	AD(5)	10.83	0.055
AD(5)	AD(6)	1.70	0.79
AD(6)	AD(7)	2.77	0.43
AD(7)	AD(8)	1.87	0.39
AD(8)	AD(9)	2.60	0.11
AD(3)	AD(9)	24.17	0.28
VAD_1	VAD_2	27.12	2×10^{-7}
VAD_3	VAD_2	12.86	0.0016
VAD_2	VAD_4	33.03	4×10^{-6}

6.3 Testing for structured antedependence

Once a constant order, p, or variable order, $(p_1, \ldots, p_n)$, of antedependence has been determined for a set of longitudinal data, the analyst may wish to fit various structured AD models of that order to the data, comparing these fits to each other and to that of the unstructured AD model of that order. Likelihood ratio tests may also be used for this purpose, provided that the models are nested. Of course, within any collection of normal antedependence models of given order having a common mean structure and containing the unstructured model, an SAD member of the collection is always nested within the unstructured AD member; however, it can happen that not all pairs of SAD models in the collection will be nested. In such a case, penalized likelihood criteria may be used for model comparisons; see Section 6.5. In the present section, we limit our consideration to testing the null hypothesis that the data follow a specified SAD(p) model versus the alternative hypothesis that they follow the unstructured AD(p) model, where p is specified. We again assume that the data are balanced and that they have a multivariate regression mean structure with m covariates, and that $N - m > p$. Furthermore, in this section we do not provide modifications that make the size of the test closer to its nominal level.

Consider the important case where the null hypothesis is that the data follow an AR(p) model. It is easy to show that the asymptotically valid size-α likelihood ratio test criterion in this case is given by

$$-N\left\{\sum_{i=1}^{n}(\log\hat{\sigma}_{ii}-\log\hat{\sigma}^2)+\sum_{j=1}^{p}\sum_{i=j+1}^{n}\left[\log\left(1-r^2_{i,i-j\cdot\{i-j+1:i-1\}}\right)\right.\right.$$
$$\left.\left.-\log\left(1-\hat{\rho}^2_{j\cdot\{intervenors\}}\right)\right]\right\} > K, \tag{6.12}$$

where $\hat{\sigma}^2$ and $\hat{\rho}_{j\cdot\{intervenors\}}$ are the maximum likelihood estimates, under the AR(p) model, of the marginal variance and lag-j intervenor-adjusted partial correlation among residuals from the fitted mean structure, and K is the $100(1-\alpha)$th percentile of a chi-square distribution with $(2n-p)(p+1)/2-(p+1)=(2n-p-2)(p+1)/2$ degrees of freedom. Thus, the AR(p) model will be rejected in favor of the unstructured AD(p) model if either the marginal variances or the same-lag intervenor-adjusted partial correlations (for one or more lags up to lag p) are sufficiently heterogeneous, or both.

A closely related case of interest is one for which the null hypothesis is that the data follow an ARH(p) model. For this case, the test criterion is identical to (6.12) except that the term involving the log variances is omitted and the degrees of freedom of the chi-square cut-off value change to $p(2n-p-3)/2$. Thus the ARH(p) model will be rejected in favor of the unstructured AD(p) model if there is a sufficiently great disparity among same-lag intervenor-adjusted partial correlations up to lag p, which makes perfect sense.

Explicit forms for the likelihood ratio test criterion may be given for testing the other structured AD(p) models introduced in Chapter 3 against their unstructured pth-order counterpart. Such forms are not particularly enlightening, however, so we do not give them here, though we do compute several of them in the following example.

Example: Treatment A cattle growth data
Recall from the previous section that the order of antedependence for the Treatment A cattle growth data was determined to be either one or two. Might structured antedependence models of these orders fit the data nearly as well as the unstructured ones? To answer this question, we perform likelihood ratio tests for several structured antedependence models of orders one and two against their unstructured counterparts. Each model's mean structure is saturated. Results of these tests are summarized in Table 6.5. The fitted first-order SAD models are the marginally formulated and autoregressively formulated power law models whose REML estimates were given in Section 5.5, plus the AR(1) and ARH(1) models. The fitted second-order SAD models are analogues of these four. The results indicate that the AR and ARH models of both orders are rejected in favor of unstructured antedependence models of the corresponding orders. On the other hand, the marginally formulated SAD power law models

Table 6.5 *Results of hypothesis tests for structured antedependence models versus unstructured antedependence models, for the Treatment A cattle growth data. Here, SAD-PM(p) and SAD-PA(p) refer to the pth-order marginally formulated and autoregressively formulated power law antedependence models described in Sections 5.5.1 and 5.5.4, respectively, and POU(10) refers to the tenth-order SAD model of Pourahmadi described in Section 5.5.4.*

Null hypothesis	Alternative hypothesis	Likelihood ratio test statistic	P
AR(1)	AD(1)	33.2	0.023
ARH(1)	AD(1)	21.0	0.013
SAD-PM(1)	AD(1)	11.0	0.75
SAD-PA(1)	AD(1)	20.6	0.24
AR(2)	AD(2)	46.2	0.012
ARH(2)	AD(2)	33.6	0.010
SAD-PM(2)	AD(2)	18.6	0.67
SAD-PA(2)	AD(2)	36.0	0.030
SAD-PM(1)	SAD-PM(2)	11.4	0.003
POU(10)	AD(10)	52.2	0.69

are not rejected and, in the first-order case, neither is the autoregressively formulated power law model. Thus, we would select either first-order power law model over the unstructured AD(1), and the marginally formulated second-order model over the unstructured AD(2). Since the marginally formulated power law SAD(1) model is a special case of the marginally formulated power law SAD(2) model, we can compare these models by another likelihood ratio test, the result of which is included in Table 6.5. The first-order model is rejected in favor of the second-order model. Also, we include a test of the particular unconstrained linear SAD(10) model described previously in Section 5.5.4, against the unstructured AD(10) model; this SAD(10) model is not rejected. Based on these results, our two preferred models for the data's covariance structure at this stage of the analysis are the marginally formulated SAD(2) power law and unconstrained linear SAD(10) models.

6.4 Testing for homogeneity across groups

The final likelihood ratio test we consider is a test for homogeneity of unstructured pth-order antedependence across $G \geq 2$ groups. Often the groups will correspond to treatments, but they could instead correspond to some other factor of classification. In this context we relabel the response vectors as $\{\mathbf{Y}_{gs} : s = 1, \ldots, N(g); g = 1, \ldots, G\}$, where $\mathbf{Y}_{gs}$ is the response vector for subject

s in group g and $N(g)$ is the sample size for group g. Similarly we relabel the covariates as $\{\mathbf{z}_{gs} : s = 1, \ldots, N(g); g = 1, \ldots, G\}$. We allow the mean parameters and (under the alternative hypothesis) the covariance parameters to differ across groups, and we label these as $\boldsymbol{\beta}_1, \ldots, \boldsymbol{\beta}_G$ and $\boldsymbol{\theta}_1, \ldots, \boldsymbol{\theta}_G$, respectively.

We present the test for homogeneity via the following theorem, which ties together tests given by Kenward (1987) and Johnson (1989). As usual, we assume initially that the data are balanced and that they have a common multivariate regression mean structure across groups. Extensions to accommodate dropouts and test for homogeneity of variable-order antedependence are straightforward and will be noted subsequently.

Theorem 6.6. *Suppose that $\{\mathbf{Y}_{gs} : g = 1, \ldots, G; s = 1, \ldots, N(g)\}$ follow the normal multivariate regression, unstructured AD(p) model*

$$\mathbf{Y}_{gs} \sim \text{independent } \mathrm{N}_n \left((\mathbf{z}_{gs}^T \otimes \mathbf{I}_n)\boldsymbol{\beta}_g, \boldsymbol{\Sigma}(\boldsymbol{\theta}_g)\right)$$

where $\boldsymbol{\Sigma}(\boldsymbol{\theta}_g)$ is unstructured AD(p) and $\mathbf{z}_{gs}$ is an $m \times 1$ vector of observed covariates. Assume that $N(g) - m > p$ for all g, and let

$$\mathbf{A}(g) = \frac{1}{N(g)} \sum_{s=1}^{N(g)} \left[\mathbf{Y}_{gs} - \left(\mathbf{z}_{gs}^T \otimes \mathbf{I}_n\right)\hat{\boldsymbol{\beta}}_g\right]\left[\mathbf{Y}_{gs} - \left(\mathbf{z}_{gs}^T \otimes \mathbf{I}_n\right)\hat{\boldsymbol{\beta}}_g\right]^T$$

and

$$\mathbf{A} = \frac{1}{N} \sum_{g=1}^{G} N(g)\mathbf{A}(g),$$

where $N = \sum_{g=1}^{G} N(g)$. The likelihood ratio test for testing the null hypothesis $H_0 : \boldsymbol{\Sigma}(\boldsymbol{\theta}_1) = \boldsymbol{\Sigma}(\boldsymbol{\theta}_2) = \cdots = \boldsymbol{\Sigma}(\boldsymbol{\theta}_G)$ versus $H_A : \text{not } H_0$, rejects H_0 if and only if any of the following three equivalent criteria are satisfied:

(a)

$$\sum_{i=1}^{n} \left[N \log\left(\frac{RSS_i(p)}{N}\right) - \sum_{g=1}^{G} N(g) \log\left(\frac{RSS_{ig}(p)}{N(g)}\right)\right] > K,$$

where $RSS_i(p)$ is the residual sum of squares from the ordinary least squares regression of the ith response variable on the $p_i \equiv \min(p, i-1)$ such variables immediately preceding it plus the covariates, and $RSS_{ig}(p)$ is the analogous quantity computed using only the data from the gth group;

(b)

$$\sum_{g=1}^{G} N(g) \left\{ \sum_{i=1}^{n} (\log \hat{\sigma}_{ii} - \log \hat{\sigma}_{iig}) \right.$$

$$+ \sum_{i=2}^{n} [\log(1 - \hat{\rho}_{i,i-1}^2) - \log(1 - \hat{\rho}_{i,i-1,g}^2)] + \cdots$$

$$+ \sum_{i=p+1}^{n} [\log(1 - \hat{\rho}_{i,i-p\cdot\{i-p+1:i-1\}}^2) - \log(1 - \hat{\rho}_{i,i-p\cdot\{i-p+1:i-1\},g}^2)] \Bigg\}$$

$$> K,$$

where parameter estimates without a subscript g are as defined previously and those with a subscript g are the analogous quantities computed using only the data from the gth group;

(c)

$$-2 \left(\sum_{i=1}^{n-p} \log \Lambda_{i,i+p} - \sum_{i=1}^{n-p-1} \log \Lambda_{i+1,i+p} \right) > K,$$

where

$$\Lambda_{i,i+p} = \frac{\prod_{g=1}^{G} |\mathbf{A}_{i:i+p}(g)|^{N(g)/2}}{|\mathbf{A}_{i:i+p}|^{N/2}}$$

is the classical likelihood ratio test statistic for testing for the homogeneity of $\Sigma_{i:i+p}(\boldsymbol{\theta}_1), \ldots, \Sigma_{i:i+p}(\boldsymbol{\theta}_G)$ under general multivariate dependence, and $\Lambda_{i+1,i+p}$ is defined analogously.

An asymptotically valid (as $N(g) \to \infty$ for all g) size-α test is obtained by taking K to be the $100(1 - \alpha)$th percentile of a chi-square distribution with $(G - 1)(2n - p)(p + 1)/2$ degrees of freedom.

Proof. We will prove part (a); parts (b) and (c) then follow easily using parts (c) and (d) of Theorem 2.5. The same development that leads to (6.7), followed by the use of Theorem 2.5(a), yields

$$-2 \log \Lambda = \sum_{g=1}^{G} N(g) \log \left(\frac{|\mathbf{A}|}{|\mathbf{A}(g)|} \right)$$

$$= \sum_{g=1}^{G} N(g) \log \left(\frac{\prod_{i=1}^{n} \hat{\delta}_i^{(p)}}{\prod_{i=1}^{n} \hat{\delta}_{ig}^{(p)}} \right),$$

where $\hat{\delta}_i^{(p)}$ is defined as in the proof of Theorem 6.5 and $\hat{\delta}_{ig}^{(p)}$ is the analogous quantity computed using only the data from the gth group. Upon noting that

$$\hat{\delta}_i^{(p)} = \frac{RSS_i(p)}{N} \quad \text{and} \quad \hat{\delta}_{ig}^{(p)} = \frac{RSS_{ig}(p)}{N(g)},$$

the theorem follows via routine manipulations. $\square$

Parts (b) and (c) of Theorem 6.6 are of mostly academic interest; the most useful part of the theorem from a computational standpoint is part (a). Furthermore, the adequacy of the chi-square cut-off point can be improved by modifying the test criterion of part (a) as follows (Kenward, 1987, 1991): reject H_0 if and only if

$$\frac{[(G-1)(2n-p)(p+1)/2]}{\sum_{i=1}^{n}\sum_{g=1}^{G}N(g)\{\psi[N-N(g)-G+1,N(g)-p_i-1]-\log[N/N(g)]\}}$$

$$\times \sum_{i=1}^{n}\left[N\log\left(\frac{RSS_i(p)}{N}\right)-\sum_{g=1}^{G}N(g)\log\left(\frac{RSS_{ig}(p)}{N(g)}\right)\right] > K, \quad (6.13)$$

where ψ was defined in Section 6.4.1. Finally, it is easy to see how to extend the test criterion to accommodate dropouts and to test for homogeneity of variable-order antedependence. To handle dropouts, merely replace N and $N(g)$ in all expressions above with N_i and $N_i(g)$, respectively, where N_i is (as in Section 5.4) the overall number of subjects with complete data at time i and $N_i(g)$ is the number of such subjects within the gth group. To test for the homogeneity of G AD$(p_1, \ldots, p_n)$ covariance matrices, we merely replace the test criterion given in part (a) of the theorem with the following:

$$\sum_{i=1}^{n}\left[N\log\left(\frac{RSS_i(p_i)}{N}\right)-\sum_{g=1}^{G}N(g)\log\left(\frac{RSS_{ig}(p_i)}{N(g)}\right)\right] > K$$

where $RSS_i(p_i)$ is the residual sum of squares from the ordinary least squares regression of the ith response variable on the p_i such variables immediately preceding it plus the covariates, and $RSS_{ig}(p_i)$ is the analogous quantity computed using only the data from the gth group; and we change the degrees of freedom for K to $(G-1)(n+\sum_{i=2}^{n}p_i)$.

Example: Speech recognition data
Recall that the speech recognition scores are measured on subjects belonging to one of two groups, which correspond to the type of cochlear implant the subject has. Allowing initially for the possibility that the within-group covariance matrices for these two groups are not homogeneous, we carry out separate forward selection and backward elimination procedures for determining the order of antedependence of these two matrices. We take the mean structure to be saturated within groups, as in (5.42). These procedures unequivocally select a first-order model for both groups (numerical details not shown). The natural next question is, are the two AD(1) covariance structures different, and if so, how? To address this question, we perform the homogeneity test presented in this section. The computed likelihood ratio test statistic (6.13), properly adjusted to account for the dropouts, is 3.56, and the corresponding p-value is 0.83. We conclude that the two groups' AD(1) covariance structures are not

Table 6.6 *Results of hypothesis tests for order of unstructured antedependence of the pooled covariance structure for the speech recognition sentence scores.*

Null hypothesis	Alternative hypothesis	Modified likelihood ratio test statistic	P
AD(1)	AD(2)	1.66	0.44
AD(2)	AD(3)	1.82	0.18
AD(1)	AD(3)	3.69	0.30

significantly different, and that it is therefore reasonable to assume homogeneity in further analyses (as indeed we did assume in Section 5.4.2).

Next, assuming homogeneity, we carry out stepwise selection procedures for the order of antedependence of the pooled covariance structure. Results are displayed in Table 6.6. Forward selection and backward elimination procedures unequivocally select a first-order model. This comes as no surprise, in light of the orders determined separately for each within-group covariance matrix.

6.5 Penalized likelihood criteria

Sometimes the antedependence models we wish to compare, or at least some of them, are not nested. For example, we may wish to compare the fits of two unstructured variable-order AD models, neither of which is nested by the other. Or we may wish to compare the fit of a marginally specified SAD(p) model to that of an autoregressively specified SAD(p) model. Moreover, we may wish to compare fits of antedependence models to fits of various other models, such as the vanishing correlation models or random coefficient models described in Section 3.9. For comparing models such as these, likelihood ratio testing is not applicable. However, we may use penalized likelihood criteria, as proposed by Macchiavelli and Arnold (1994, 1995). In fact, penalized likelihood criteria may be used for comparing nested models also, and most proponents of such criteria would probably favor using them consistently rather than switching between them and likelihood ratio tests on the basis of whether the models are nested; for relevant discussions of this issue by advocates of the criteria, see Akaike (1974) and Burnham and Anderson (2002). Here we take a neutral approach by presenting both methodologies.

A penalized likelihood criterion balances model fit, as measured by a model's maximized log-likelihood (or a multiple thereof), against model complexity, as measured by a penalty term equal to a multiple of the number of parameters in the model. There are many such criteria, which differ by using a different multiplier in the penalty term. We consider criteria of the form

$$IC(k) = -\frac{2}{N} \log L^*(\hat{\boldsymbol{\theta}}_k) + d_k \frac{c(N)}{N},$$

where k indexes the models under consideration, $\hat{\boldsymbol{\theta}}_k$ is the maximum likelihood estimator of $\boldsymbol{\theta}$ under the kth model, d_k is the number of unknown parameters in the covariance structure of the kth model, and $c(N)$ is the penalty term. Criteria of this form penalize a model only for the parameters in its covariance structure, as we assume in this section that all models under consideration have a common mean structure. Furthermore, for models for which that common mean structure is of multivariate regression form, we also consider these criteria's REML analogues,

$$IC_R(k) = -\frac{2}{N-m} \log L_R(\tilde{\boldsymbol{\theta}}_k) + d_k \frac{c(N-m)}{N-m},$$

where m is the number of covariates [cf. (5.28)] and $\tilde{\boldsymbol{\theta}}_k$ is the REML estimator of $\boldsymbol{\theta}$ under the kth model. Note that $IC(k)$ and $IC_R(k)$ are presented in "smaller is better" form here, so our objective is to minimize them.

Two of the most well-known and commonly used penalized likelihood criteria are Akaike's Information Criterion (AIC and AIC_R) and Schwarz's Information Criterion (BIC and BIC_R), which we feature here. For these criteria, $c(N) = 2$ and $c(N) = \log N$, respectively. Because $\log N > 2$ for $N > 8$, whenever $N > 8$ BIC will penalize a model for its complexity more severely than AIC. Since in practice N is usually larger than 8, BIC will tend to select more parsimonious models than AIC. The use of BIC results in strongly consistent model selection, whereas that of AIC does not (Rao and Wu, 1989); nevertheless AIC is very popular so we will provide both criteria in our examples.

Model comparison for unstructured variable-order AD models merits special consideration because there is potentially a very large number of such models to compare. Specifically, since p_i can be any nonnegative integer less than or equal to $i - 1$, the number of such models when there are n distinct measurement times is $n!$. Fortunately, the computation of penalized likelihood criteria for each model can be circumvented, so that the model that minimizes any penalized likelihood criterion can be determined very efficiently. We demonstrate this for the criteria based on the ordinary likelihood, but it is true of REML-based criteria as well. Suppose that the observations are balanced and that the mean structure is of multivariate regression form. From (6.6) we see that apart from an additive constant,

$$-\frac{2}{N} \log L^*(\hat{\boldsymbol{\theta}}_k) = \log |\boldsymbol{\Sigma}(\hat{\boldsymbol{\theta}}_k)|$$
$$= \log \prod_{i=1}^{n} \hat{\delta}_i^{(k)}$$

$$= \text{constant} + \sum_{i=1}^{n} \log RSS_i(k),$$

where k indexes the $n!$ variable-order AD models. Now recall that $N\hat{\delta}_i^{(k)}$ and $RSS_i(k)$ are two representations for the residual sum of squares from the ordinary least squares regression of the ith response variable on its $p_i(k)$ immediate predecessors and the m covariates, where $p_i(k)$ is the order of antedependence at time i for the kth model. Thus, for the unstructured AD$(p_1, \dots, p_n)$ model, we have

$$IC(p_1, \dots, p_n) = \text{constant} + \sum_{i=1}^{n} \log RSS_i(p_i) + \left(n + \sum_{i=2}^{n} p_i\right) \frac{c(N)}{N}$$

$$= \text{constant} + \sum_{i=1}^{n} IC_i(p_i), \qquad (6.14)$$

say, where

$$IC_i(p_i) = \log RSS_i(p_i) + (p_i + 1) \frac{c(N)}{N}.$$

Note that only the ith summand in (6.14) depends on p_i. Consequently, $IC(p_1, \dots, p_n)$ can be minimized by minimizing $IC_i(p_i)$ separately for each i, or equivalently by taking

$$\hat{p}_i = \operatorname*{argmin}_{p_i=0,1,\dots,i-1} \left(\log RSS_i(p_i) + p_i \frac{c(N)}{N} \right). \qquad (6.15)$$

If there are dropouts, we merely compute $RSS_i(p_i)$ from the N_i observations available at time i and replace N with N_i in (6.15).

Example 1: Treatment A cattle growth data
Table 6.7 gives the maximized profile log-likelihood and penalized likelihood criteria, in particular AIC and BIC, for various antedependence models fitted to the Treatment A cattle growth data. In all models the mean is saturated. The models are listed in order of increasing BIC, but the ordering based on AIC is not dramatically different. The orderings for the REML-based criteria (AIC_R and BIC_R, which are not included in the table) are very similar to those of their maximum likelihood-based counterparts. Using (6.15), the variable-order unstructured antedependence model that minimizes BIC is found to be AD(0,1,1,1,1,1,1,2,2,1,1), while the constant-order unstructured antedependence model that minimizes BIC is AD(2); these two models are the third and fifth best, respectively, based on BIC, among all models considered. The best two models by either criterion are the second-order marginally formulated power law SAD model and the tenth-order unconstrained autoregressively formulated SAD model introduced in Section 5.5.4, in that order.

Table 6.7 *Penalized likelihood criteria for various antedependence models fitted to the Treatment A cattle growth data. Here, VAD refers to the variable-order AD(0,1,1,1,1,1,2,2,1,1) model, SAD-PM(p) and SAD-PA(p) refer to the pth-order marginally formulated and autoregressively formulated power law antedependence models described in Sections 5.5.1 and 5.5.4, respectively, and POU(10) refers to the tenth-order SAD model of Pourahmadi described in Section 5.5.4.*

Model	$\max \log L^*$	AIC	BIC
SAD-PM(2)	−1045.4	70.23	70.09
POU(10)	−1046.0	70.27	70.13
VAD	−1037.6	70.71	70.31
SAD-PM(1)	−1051.1	70.47	70.37
AD(2)	−1036.1	71.07	70.55
SAD-PA(1)	−1055.9	70.66	70.59
SAD-PA(2)	−1054.1	70.81	70.67
AD(1)	−1045.6	71.11	70.74
AR(2)	−1059.2	70.81	70.76
ARH(2)	−1052.9	71.06	70.83
AR(1)	−1062.2	70.95	70.91
ARH(1)	−1056.1	71.21	71.00
AD(10)	−1019.9	72.39	71.24

Example 2: 100-km race data

Table 6.8 gives the maximized profile log-likelihood, AIC, and BIC for various antedependence models fitted to the complete 100-km race data. In all models the mean is saturated. The models are listed in order of increasing BIC; the ordering based on AIC is somewhat different, while those corresponding to AIC_R and BIC_R are identical to their ordinary likelihood-based counterparts. The best variable-order unstructured antedependence model is AD(0,1,1,1,2,1,1,2,3,5). This model's increase in order of antedependence over the last three sections of the race agrees with the increase in the number of conditionally dependent predecessors seen previously from informal analyses (Section 4.2). Among the constant-order unstructured antedependence models, the closest competitor to the variable-order model is AD(2).

Also included in Table 6.8 are results from fitting two structured variable-order antedependence models, both of order (0,1,1,1,2,1,1,2,3,5). The first of these, labeled as VAD-PM, is a marginally formulated power law model motivated by the behavior of the data's sample variances and correlations, as displayed in Table 4.3. This model has variances

$$\sigma_{ii} = \sigma^2(1 + \psi_1 i + \psi_2 i^2 + \psi_3 i^3), \quad i = 1, \ldots, 10$$

Table 6.8 *Penalized likelihood criteria for various antedependence models fitted to the 100-km race data. Here, VAD refers to the variable-order AD(0,1,1,1,2,1,1,2,3,5) model, and VAD-PM and VAD-AM refer to structured AD models of the same variable order, which are described in more detail in Section 6.5.*

Model	$\max \log L^*$	AIC	BIC
VAD-PM	−2350.6	59.12	59.53
VAD	−2328.8	58.90	59.70
VAD-AM	−2362.9	59.41	59.84
AD(2)	−2343.3	59.26	60.06
AD(3)	−2332.5	59.16	60.17
AD(1)	−2367.2	59.66	60.22
AD(4)	−2330.2	59.26	60.45
AD(5)	−2324.4	59.24	60.57
AD(9)	−2319.4	59.36	61.00

and lag-one correlations

$$\rho_{i,i-1} = \rho_i^{i^\lambda - (i-1)^\lambda}, \quad i = 2, \ldots, 10;$$

correlations corresponding to the nonzero entries of the lower half of the matrix $\mathbf{T}$ are left unstructured, and the remaining elements of the covariance matrix are evaluated using (2.44). The other structured variable-order AD model, labeled as VAD-AM, is an autoregressively formulated power law model motivated by the innovariogram and regressogram of the data (Figure 4.9). This model has log innovation variances

$$\log \delta_i = \begin{cases} \psi_0, & i = 1 \\ \psi_1 + \psi_2 i + \psi_3 i^2, & i = 2, \ldots, 10 \end{cases}$$

and lag-one autoregressive coefficients

$$\phi_{i,i-1} = \theta_1 + \theta_2 i, \quad i = 2, \ldots, 10;$$

the remaining autoregressive coefficients are left unstructured. The VAD-PM model is the better-fitting of the two; in fact, it fits better than all other models in Table 6.8. The unstructured variable-order model is second best.

Example 3: Speech recognition data
Finally, Table 6.9 gives the maximized profile log-likelihood, AIC, and BIC for various antedependence models fitted to the speech recognition data. In all models the mean is saturated within each implant group; i.e., the mean structure is that given by (5.42). The orderings with respect to both penalized likelihood criteria (as well as their REML counterparts) are identical. The variable-order unstructured antedependence model that minimizes these

Table 6.9 *Penalized likelihood criteria for various antedependence models fitted to the speech recognition data.*

Model	$\max \log L^*$	AIC	BIC
SAD-PMS(1)	−536.1	26.35	26.51
SAD-PMC(1)	−538.7	26.42	26.55
AD(1)	−534.8	26.43	26.72
AD(3)	−533.3	26.50	26.92

criteria is AD(0,1,1,1), which is equivalent to the AD(1) model. Thus, for these data, no variable-order unstructured AD model fits better than a constant first-order AD model. However, two first-order marginally formulated power law SAD models fit slightly better than either unstructured model. The formulation of these two SAD models was informed by the REML estimates of pooled marginal variances and correlations displayed in Table 5.5. Based on those estimates we took the variance to be either constant over time, or a particular step function of time defined as follows:

$$\sigma_{ii} = \begin{cases} \sigma^2 & \text{for } i = 1 \\ \sigma^2 \psi & \text{for } i = 2, 3, 4, \end{cases}$$

where σ^2 and ψ are positive parameters. We label these models as SAD-PMC(1) and SAD-PMS(1), respectively, where "C" stands for "constant" and "S" stands for "step." Of these, the model with the step function variance fits slightly better.

6.6 Concluding remarks

This chapter has presented likelihood ratio tests, and simple modifications to them which achieve better agreement with nominal size, for several important hypotheses on the covariance structure of normal antedependence models. Penalized likelihood criteria were also introduced. As was the case for the likelihood-based estimates derived in Chapter 5, many of the test statistics and criteria are expressible in terms of quantities determined by regressing observations of the response at a given time on a particular number of its predecessors, plus covariates. This makes them relatively easy to compute. R functions, written by the first author, for computing several of the modified likelihood ratio tests and penalized likelihood criteria presented herein are available for download from his Web page, at the address provided in Section 4.4.

CHAPTER 7

Testing Hypotheses on the Mean Structure

Suppose that an antedependent covariance structure has been selected for a set of longitudinal data, using either the informal methods of Chapter 4 or the hypothesis testing procedures or penalized likelihood criteria of Chapter 6. To the extent that the order of antedependence of the selected covariance model is low to moderate and is structured rather than unstructured, inferences on the mean structure can be more efficient if this structure is exploited than if the general multivariate dependence structure is adopted. This chapter describes hypothesis tests for the mean structure of normal linear antedependence models which exploit the antedependence.

We begin with several likelihood ratio tests for the mean structure under an unstructured antedependence model: a test that the mean of a single population is equal to a specified vector; a test for equality of the means of two populations (also known as comparison of profiles); and a test of the importance of a subset of covariates in a model with multivariate regression mean structure. These tests are analogues, to unstructured antedependence models, of the well-known one-sample and two-sample Hotelling's T^2 tests and Wilk's lambda test, respectively, for hypotheses on mean vectors under general multivariate dependence. In fact, it will be seen that likelihood ratio test statistics for the mean structure under unstructured antedependence of any order can be expressed as functions of likelihood ratio test statistics for testing hypotheses on certain subvectors of the mean vector(s) under general multivariate dependence. We also give tests for contrasts and other linear combinations of the elements of these mean vectors, which are not likelihood ratio tests but are asymptotically equivalent to them. Following that, we consider likelihood ratio testing for the mean structure of structured antedependence models. The chapter concludes with a description of penalized likelihood criteria for comparing antedependence models with different mean structures (and possibly different covariance structures as well).

7.1 One-sample case

Consider a situation in which the observations are balanced and follow the normal saturated-mean, unstructured $AD(p)$ model

$$\mathbf{Y}_s \sim \text{iid } \mathrm{N}_n\left(\boldsymbol{\mu}, \boldsymbol{\Sigma}(\boldsymbol{\theta})\right), \quad s = 1, \ldots, N, \tag{7.1}$$

and we wish to test the null hypothesis that $\boldsymbol{\mu} = \boldsymbol{\mu}_0$ against the alternative that $\boldsymbol{\mu} \neq \boldsymbol{\mu}_0$, where $\boldsymbol{\mu}_0 = (\mu_{0i})$ is a specified $n \times 1$ vector. Although this is not usually a practically important hypothesis for longitudinal data from a single population, it is important in the case of paired longitudinal sampling from two populations (e.g., matched case-control sampling), for which the $\mathbf{Y}_s$'s represent vectors of within-pair differences of responses at the n measurement times and interest lies in testing the null hypothesis that the mean vector of those differences is zero. From (5.22), the log-likelihood function for model (7.1) is given by

$$\log L(\boldsymbol{\mu}, \boldsymbol{\theta}_\delta, \boldsymbol{\theta}_\phi) = -\frac{nN}{2} \log 2\pi - \frac{1}{2} \sum_{i=1}^{n} \sum_{s=1}^{N} \left\{ \log \delta_i + \left[Y_{si} - \mu_i \right. \right.$$
$$\left. \left. - \sum_{k=1}^{p_i} \phi_{i,i-k}(Y_{s,i-k} - \mu_{i-k}) \right]^2 \middle/ \delta_i \right\}, \tag{7.2}$$

and this function may be rewritten, in terms of a transformed mean vector $\boldsymbol{\mu}^* = (\mu_i^*)$, as

$$\log L(\boldsymbol{\mu}^*, \boldsymbol{\theta}_\delta, \boldsymbol{\theta}_\phi) = -\frac{nN}{2} \log 2\pi - \frac{1}{2} \sum_{i=1}^{n} \sum_{s=1}^{N} \left\{ \log \delta_i + \left(Y_{si} - \mu_i^* \right. \right.$$
$$\left. \left. - \sum_{k=1}^{p_i} \phi_{i,i-k} Y_{s,i-k} \right)^2 \middle/ \delta_i \right\} \tag{7.3}$$

where

$$\mu_i^* = \mu_i - \sum_{k=1}^{p_i} \phi_{i,i-k} \mu_{i-k}.$$

As noted in the alternative proof of Theorem 5.1, standard regression theory tells us that (7.3) is maximized with respect to the $\{\mu_i^*\}$, $\{\phi_{i,i-k}\}$, and $\{\delta_i\}$ by the least squares estimates of intercepts and slopes and the residual sums of squares divided by N, respectively, from the n regressions of each response variable on its p_i predecessors. Furthermore, the log-likelihood function under the null hypothesis, obtained by evaluating (7.2) at $\boldsymbol{\mu}_0$, is

$$\log L(\boldsymbol{\mu}_0, \boldsymbol{\theta}_\delta, \boldsymbol{\theta}_\phi) = -\frac{nN}{2} \log 2\pi - \frac{1}{2} \sum_{i=1}^{n} \sum_{s=1}^{N} \left\{ \log \delta_i + \left[Y_{si} - \mu_{0i} \right. \right.$$

$$-\sum_{k=1}^{p_i} \phi_{i,i-k}(Y_{s,i-k} - \mu_{0,i-k})\Bigg]^2 \Bigg/ \delta_i\Bigg\}. \qquad (7.4)$$

It follows, again from standard regression theory, that (7.4) is maximized with respect to the $\{\phi_{i,i-k}\}$ and $\{\delta_i\}$ by the least squares estimates of slopes and the residual sums of squares divided by N, respectively, from the n regressions of each "null-mean-corrected" variable, $Y_i - \mu_{0i}$, on its p_i predecessors without an intercept. Furthermore, letting Λ represent the likelihood ratio test statistic, we have

$$-2\log\Lambda = N\sum_{i=1}^{n}\left(\log(\hat{\delta}_{0i}) - \log(\hat{\delta}_{1i})\right)$$

where $\hat{\delta}_{0i}$ and $\hat{\delta}_{1i}$ are the maximum likelihood estimators of the ith innovation variance under the null and alternative hypotheses, respectively. We therefore have established the following theorem.

Theorem 7.1. *Suppose that $\mathbf{Y}_1, \ldots, \mathbf{Y}_N$ are balanced and follow the normal saturated-mean, unstructured AD(p) model, and $N - 1 > p$. The likelihood ratio test for the null hypothesis that $\boldsymbol{\mu} = \boldsymbol{\mu}_0$ (against the alternative that $\boldsymbol{\mu} \neq \boldsymbol{\mu}_0$) rejects the null hypothesis if and only if*

$$N\sum_{i=1}^{n}\left[\log RSS_i(\boldsymbol{\mu}_0) - \log RSS_i(\boldsymbol{\mu})\right] > K, \qquad (7.5)$$

where $RSS_i(\boldsymbol{\mu}_0)$ is the residual sum of squares from the regression of $Y_i - \mu_{0i}$ on its p_i predecessors $\{Y_{i-k} - \mu_{0,i-k} : k = 1, \ldots, p_i\}$ without an intercept, $RSS_i(\boldsymbol{\mu})$ is the residual sum of squares from the regression of Y_i on its p_i predecessors $\{Y_{i-k} : k = 1, \ldots, p_i\}$ with an intercept, and K is a constant. For any $\alpha \in (0,1)$, an asymptotically (as $N \to \infty$) valid size-α test is obtained by taking K to be the $100(1-\alpha)$th percentile of a chi-square distribution with n degrees of freedom.

The likelihood ratio test statistic given by Theorem 7.1 has an interesting interpretation in terms of certain Hotelling's T^2 statistics, as we now describe. Let $\hat{\boldsymbol{\theta}}_0$ and $\hat{\boldsymbol{\theta}}_1$ be the maximum likelihood estimators of $\boldsymbol{\theta}$ under the null and alternative hypotheses, respectively. Then by parts (a) and (d) of Theorem 2.5, we have

$$N\sum_{i=1}^{n}\left[\log RSS_i(\boldsymbol{\mu}_0) - \log RSS_i(\boldsymbol{\mu})\right] = N\log\left(\frac{|\boldsymbol{\Sigma}(\hat{\boldsymbol{\theta}}_0)|}{|\boldsymbol{\Sigma}(\hat{\boldsymbol{\theta}}_1)|}\right)$$

$$= N\log\left(\frac{\dfrac{\prod_{i=1}^{n-p}|\boldsymbol{\Sigma}_{i:i+p}(\hat{\boldsymbol{\theta}}_0)|}{\prod_{i=1}^{n-p-1}|\boldsymbol{\Sigma}_{i+1:i+p}(\hat{\boldsymbol{\theta}}_0)|}}{\dfrac{\prod_{i=1}^{n-p}|\boldsymbol{\Sigma}_{i:i+p}(\hat{\boldsymbol{\theta}}_1)|}{\prod_{i=1}^{n-p-1}|\boldsymbol{\Sigma}_{i+1:i+p}(\hat{\boldsymbol{\theta}}_1)|}}\right)$$

$$= \quad N \log \left(\frac{\prod_{i=1}^{n-p} \frac{|\boldsymbol{\Sigma}_{i:i+p}(\hat{\boldsymbol{\theta}}_0)|}{|\boldsymbol{\Sigma}_{i:i+p}(\hat{\boldsymbol{\theta}}_1)|}}{\prod_{i=1}^{n-p-1} \frac{|\boldsymbol{\Sigma}_{i+1:i+p}(\hat{\boldsymbol{\theta}}_0)|}{|\boldsymbol{\Sigma}_{i+1:i+p}(\hat{\boldsymbol{\theta}}_1)|}} \right).$$

Now by Theorem 5.1(a), $\boldsymbol{\Sigma}_{i:i+p}(\hat{\boldsymbol{\theta}}_1) = \mathbf{A}_{i:i+p}$ and $\boldsymbol{\Sigma}_{i+1:i+p}(\hat{\boldsymbol{\theta}}_1) = \mathbf{A}_{i+1:i+p}$. Moreover, it can easily be shown [by the same method used to establish Theorem 5.1(a)] that

$$\boldsymbol{\Sigma}_{i:i+p}(\hat{\boldsymbol{\theta}}_0) = \mathbf{A}_{i:i+p} + (\overline{\mathbf{Y}}_{i:i+p} - \boldsymbol{\mu}_{0,i:i+p})(\overline{\mathbf{Y}}_{i:i+p} - \boldsymbol{\mu}_{0,i:i+p})^T.$$

By Theorem A.1.3, we have

$$|\boldsymbol{\Sigma}_{i:i+p}(\hat{\boldsymbol{\theta}}_0)| = |\mathbf{A}_{i:i+p}|[1 + (\overline{\mathbf{Y}}_{i:i+p} - \boldsymbol{\mu}_{0,i:i+p})^T \mathbf{A}_{i:i+p}^{-1} (\overline{\mathbf{Y}}_{i:i+p} - \boldsymbol{\mu}_{0,i:i+p})],$$

by which we obtain

$$
\begin{aligned}
\prod_{i=1}^{n-p} \frac{|\boldsymbol{\Sigma}_{i:i+p}(\hat{\boldsymbol{\theta}}_0)|}{|\boldsymbol{\Sigma}_{i:i+p}(\hat{\boldsymbol{\theta}}_1)|} &= \prod_{i=1}^{n-p} [1 + (\overline{\mathbf{Y}}_{i:i+p} - \boldsymbol{\mu}_{0,i:i+p})^T \mathbf{A}_{i:i+p}^{-1} \\
&\qquad \times (\overline{\mathbf{Y}}_{i:i+p} - \boldsymbol{\mu}_{0,i:i+p})] \\
&= \prod_{i=1}^{n-p} \left[1 + (\overline{\mathbf{Y}}_{i:i+p} - \boldsymbol{\mu}_{0,i:i+p})^T \left(\frac{N-1}{N} \mathbf{S}_{i:i+p} \right)^{-1} \right. \\
&\qquad \left. \times (\overline{\mathbf{Y}}_{i:i+p} - \boldsymbol{\mu}_{0,i:i+p}) \right] \\
&= \prod_{i=1}^{n-p} [1 + T_{i:i+p}^2 / (N-1)],
\end{aligned}
$$

where $\mathbf{S}$ is the sample covariance matrix and $T_{i:i+p}^2$ is Hotelling's T^2 statistic,

$$N(\overline{\mathbf{Y}}_{i:i+p} - \boldsymbol{\mu}_{0,i:i+p})^T \mathbf{S}_{i:i+p}^{-1} (\overline{\mathbf{Y}}_{i:i+p} - \boldsymbol{\mu}_{0,i:i+p}),$$

for testing the null hypothesis that $\boldsymbol{\mu}_{i:i+p} = \boldsymbol{\mu}_{0,i:i+p}$ (versus the alternative that these vectors are not equal) under general multivariate dependence. A similar development yields

$$\prod_{i=1}^{n-p-1} \frac{|\boldsymbol{\Sigma}_{i+1:i+p}(\hat{\boldsymbol{\theta}}_0)|}{|\boldsymbol{\Sigma}_{i+1:i+p}(\hat{\boldsymbol{\theta}}_1)|} = \prod_{i=1}^{n-p-1} [1 + T_{i+1:i+p}^2 / (N-1)],$$

where $T_{i+1:i+p}^2$ is Hotelling's T^2 statistic for testing the null hypothesis that $\boldsymbol{\mu}_{i+1:i+p} = \boldsymbol{\mu}_{0,i+1:i+p}$ (versus the alternative that these vectors are not equal) under general multivariate dependence. It follows that

$$-2 \log \Lambda = N \log \left(\frac{\prod_{i=1}^{n-p} [1 + T_{i:i+p}^2 / (N-1)]}{\prod_{i=1}^{n-p-1} [1 + T_{i+1:i+p}^2 / (N-1)]} \right). \qquad (7.6)$$

Form (7.6) of the likelihood ratio test statistic was derived by Johnson (1989), who extended it from an original result given by Byrne and Arnold (1983) for the unstructured AD(1) model. It can be computed easily using standard software for multivariate analysis; however, the form given by Theorem 7.1 can be computed just as easily, using standard software for regression analysis. A virtue of the form given by Theorem 7.1 is that it is more obvious how to extend it to a variable-order antedependence model or to handle dropouts: in the first case one merely redefines p_i as the order of the model at time i, and in the second case one replaces N with N_i [and moves it inside the summation in (7.5)]. Another virtue of this statistic, owing to its form as a constant multiple of a sum of logs of beta random variables, is that it can be modified, analogously to the likelihood ratio tests of Chapter 6, to attain better agreement between its nominal size and actual size. The modified criterion is given by

$$\frac{n \sum_{i=1}^{n} [\log RSS_i(\boldsymbol{\mu}_0) - \log RSS_i(\boldsymbol{\mu})]}{\sum_{i=1}^{n} \psi(1, N - 1 - p_i)} > K, \qquad (7.7)$$

where K is defined as in Theorem 7.1.

Still another advantage of the form given by Theorem 7.1 is that it suggests the possibility of testing sequentially for the first measurement time at which the mean response differs from the corresponding component of $\boldsymbol{\mu}_0$, in a manner that adjusts for the effects of appropriate predecessors. Consider the ith summand of the first sum in (7.4) but with the restriction lifted on μ_i, i.e.,

$$N \log \delta_i + \sum_{s=1}^{N} \left(Y_{si} - \mu_i - \sum_{k=1}^{p_i} \phi_{i,i-k}(Y_{s,i-k} - \mu_{0,i-k}) \right)^2 \Bigg/ \delta_i.$$

The maximizer of this summand with respect to μ_i is the least squares estimate of the intercept in the regression of Y_i on $\{Y_{i-1} - \mu_{0,i-1}, Y_{i-2} - \mu_{0,i-2}, \ldots, Y_{i-p_i} - \mu_{0,i-p_i}\}$ (with intercept). It follows that each null hypothesis in the sequence

$$\boldsymbol{\mu}_{1:i} = \boldsymbol{\mu}_{0,1:i}, \quad i = 1, \ldots, n,$$

may be tested against the respective alternative hypothesis in the sequence

$$\mu_i \neq \mu_{0i}, \ \boldsymbol{\mu}_{1:i-1} = \boldsymbol{\mu}_{0,1:i-1}, \quad i = 1, \ldots, n,$$

by likelihood ratio tests, the ith of which is the standard t-test that the intercept is equal to μ_{0i} in the regression of Y_i on $\{Y_{i-k} - \mu_{0,i-k} : k = 1, \ldots, p_i\}$. Using the fact that AD(p) variables are also AD(p) in reverse order, a similar sequential procedure may be devised to test for the last measurement time at which the mean differs from the corresponding component of $\boldsymbol{\mu}_0$. Whether applied in sequential or reverse sequential order, these tests are independent. In this respect they differ from the standard t-tests associated with the "time-by-time ANOVA" approach to the analysis of longitudinal data described in Chapter 1. The present tests also differ by conditioning on, or equivalently adjusting for, responses at previous times, so we refer to them as predecessor-adjusted tests. As a consequence of the adjustment for predecessors, these tests are much more sensitive than standard tests to changes in means, relative to their hypothesized values, at successive measurement times. A forthcoming example will demonstrate this.

An alternative to the likelihood ratio procedure for testing that $\mu = \mu_0$ is obtained by substituting the uniformly minimum variance unbiased estimator (or equivalently the REML estimator) given by Corollary 5.4.1, i.e., $\tilde{\Sigma}$, for Σ in the test that is uniformly most powerful invariant when Σ is known. This yields a multivariate *Wald test* criterion: reject the null hypothesis at level α if and only if

$$N(\overline{\mathbf{Y}} - \boldsymbol{\mu}_0)^T \tilde{\Sigma}^{-1}(\overline{\mathbf{Y}} - \boldsymbol{\mu}_0) > K,$$

where K is defined as in Theorem 7.1 and we assume until noted otherwise that no observations are missing. This test is equivalent to the likelihood ratio test when $p = n - 1$, and asymptotically equivalent otherwise. Letting $\mathbf{B} = (\overline{\mathbf{Y}} - \boldsymbol{\mu}_0)(\overline{\mathbf{Y}} - \boldsymbol{\mu}_0)^T$ and using Theorems 2.6 and 5.2, the test statistic can be rewritten as follows:

$$N(\overline{\mathbf{Y}} - \boldsymbol{\mu}_0)^T \tilde{\Sigma}^{-1}(\overline{\mathbf{Y}} - \boldsymbol{\mu}_0) = N\mathrm{tr}\left[(\overline{\mathbf{Y}} - \boldsymbol{\mu}_0)^T \tilde{\Sigma}^{-1}(\overline{\mathbf{Y}} - \boldsymbol{\mu}_0)\right]$$

$$= N\mathrm{tr}\left(\mathbf{B}\tilde{\Sigma}^{-1}\right)$$

$$= N\sum_{i=1}^{n-p}\mathrm{tr}\left(\mathbf{B}_{i:i+p}\tilde{\Sigma}^{-1}_{i:i+p}\right)$$

$$- \sum_{i=1}^{n-p-1}\mathrm{tr}\left(\mathbf{B}_{i+1:i+p}\tilde{\Sigma}^{-1}_{i+1:i+p}\right)$$

$$= N\sum_{i=1}^{n-p}(\overline{\mathbf{Y}}_{i:i+p} - \boldsymbol{\mu}_{0,i:i+p})^T \mathbf{S}^{-1}_{i:i+p}(\overline{\mathbf{Y}}_{i:i+p} - \boldsymbol{\mu}_{0,i:i+p})$$

$$- N\sum_{i=1}^{n-p-1}(\overline{\mathbf{Y}}_{i+1:i+p} - \boldsymbol{\mu}_{0,i+1:i+p})^T \mathbf{S}^{-1}_{i+1:i+p}(\overline{\mathbf{Y}}_{i+1:i+p} - \boldsymbol{\mu}_{0,i+1:i+p})$$

$$= N\left(\sum_{i=1}^{n-p}T^2_{i:i+p} - \sum_{i=1}^{n-p-1}T^2_{i+1:i+p}\right).$$

Thus, this test statistic, though generally different than the likelihood ratio test statistic, is a function of the same Hotelling's T^2 test statistics appearing in expression (7.6) for the likelihood ratio test statistic.

Often, it will be of interest to test whether certain contrasts or other linear combinations of the elements of $\boldsymbol{\mu}$, rather than the elements of $\boldsymbol{\mu}$ itself, are equal to specified constants. Unfortunately, and in contrast to what transpires in the situation described in Theorem 7.1, the likelihood ratio test statistic for this hypothesis is generally not a simple function of the residual sums of squares from various regressions on at most p variables. This is due to the fact that linear combinations of n AD(p) variables are generally not antedependent of any order less than $n - 1$. Although the likelihood ratio test for this hypothesis can be obtained by an argument similar to that which led to (7.6), for simplicity we will merely consider the asymptotically equivalent test obtained by substituting $\tilde{\Sigma}$ for Σ in the test for this hypothesis which is uniformly most powerful invariant when Σ is known. We give the test via the following theorem.

Theorem 7.2. *Let $\mathbf{C}$ and $\mathbf{c}$ be a specified $c \times n$ matrix and $c \times 1$ vector, respectively, where the rows of $\mathbf{C}$ are linearly independent. Also, let $\tilde{\Sigma}$ be the REML estimator of Σ under the model and conditions of Theorem 7.1. Under the same model and conditions, the test obtained by substituting $\tilde{\Sigma}$ for Σ in the uniformly most powerful invariant test of the null hypothesis $\mathbf{C\mu} = \mathbf{c}$ against the alternative hypothesis $\mathbf{C\mu} \neq \mathbf{c}$ rejects the null hypothesis if and only if*

$$N(\mathbf{C\overline{Y}} - \mathbf{c})^T (\mathbf{C\tilde{\Sigma}C}^T)^{-1}(\mathbf{C\overline{Y}} - \mathbf{c}) > K.$$

For any $\alpha \in (0, 1)$, an asymptotically (as $N \to \infty$) valid size-α test is obtained by taking K to be the $100(1 - \alpha)$th percentile of a chi-square distribution with c degrees of freedom.

Modifications to Theorem 7.2 to handle dropouts are straightforward: one simply replaces $\overline{\mathbf{Y}}$ with the maximum likelihood or REML estimate of μ, given by Theorem 5.7, and removes the multiplier of N and multiplies $\tilde{\Sigma}$ by diag($1/N_1$, $\ldots, 1/N_n$).

The tests given by Theorems 7.1 and 7.2, by virtue of incorporating pth-order antedependence, generally are more powerful (when $p < n - 1$) than Hotelling's T^2 test, which does not impose any structure on the covariance matrix.

Simulation results

In order to demonstrate the gains in power possible from using the tests of this section, relative to tests that impose less structure on the covariance matrix, we present the results of a small simulation study. The study is somewhat similar to those presented in Section 6.2. Without loss of generality we test the null hypothesis that $\mu = \mathbf{0}$. For several combinations of N, n, and α, we simulated 100,000 data sets of size N from n-variate normal distributions with mean vector $\alpha \mathbf{1}_n$ and AD(1) covariance matrix Σ, where the modified Cholesky decomposition of Σ^{-1} was identical to that used in the first simulation study of Section 6.2. We then test the null hypothesis using the modified likelihood ratio test given by (7.7), assuming first-order antedependence. We test the same hypothesis twice more, again using tests of form (7.7) but assuming antedependence of orders 2 and $n - 1$ instead of order 1. The third test is, of course, merely a modified version of Hotelling's T^2. Power curves corresponding to the three tests, taking the nominal size of each to be 0.05, are plotted in Figure 7.1 for $N = 40$, 20, or 12; for $n = 10$; and for values of α ranging between 0 (where the power equals the size) and 0.80. The curves show, as expected, that when n is held constant, the power increases as either N or α increases. More interestingly, they also show that the tests that do not fully exploit the AD(1) covariance structure are not as powerful as the test that does. In fact, the power of Hotelling's T^2 test, relative to the first test, deteriorates rapidly as N gets close to n from above. Note that Hotelling's T^2 cannot be computed when $N \leq n$, for in this case the sample covariance matrix is not positive definite. The power of the second test relative to the first also deteriorates as N decreases with n fixed, but not as drastically.

Example 1: 100-km race data
For the split times on the ten 10-km sections, there is no particular value, μ_0, for which the hypothesis that $\mu = \mu_0$ is of real interest. Nevertheless, for purposes of illustration we shall test such a hypothesis (against the alternative $H_A : \mu \neq \mu_0$) for the complete 100-km race data, setting

$$\mu_0 = (48, 51, 50, 54, 54, 60, 63, 68, 68, 68)^T$$

(the measurement units of each component being minutes). We take the covariance structure to be that of the best-fitting variable-order, saturated-mean unstructured antedependence model, namely AD(0,1,1,1,2,1,1,2,3,5), as determined in Section 6.5. The modified likelihood ratio test statistic given by (7.7) is equal to 19.64, and the corresponding p-value (based on a chi-square distribution with 10 degrees of freedom) is 0.033. Thus, there is some evidence against the null hypothesis. We follow up this "overall" test with a sequence of predecessor-adjusted t-tests for determining the first section on which the mean split time differs from its hypothesized value. Results of these t-tests are given in the fourth column of Table 7.1. They indicate that the earliest mean split time that is significantly different from its hypothesized value, after adjusting for the predecessors defined by the AD(0,1,1,1,2,1,1,2,3,5) model, is that of the fifth section. For comparison purposes, we also include, in the table's final column, results of "standard" t tests, i.e., t tests which make no adjustment for predecessors. None of the standard tests yield a statistically significant result. So here, adjustment for predecessors helps to explain the rejection of the overall equality hypothesis and reveals a noteworthy change, which standard testing does not detect, in mean split time from the fourth section to the fifth section when compared to the hypothesized value (zero) of this change.

Although the previous null hypothesis was contrived for purposes of illustration, there are several hypotheses of the form $C\mu = c$ that may be of real interest in this situation. Here we illustrate testing one such hypothesis, which is that the mean profile is flat over the last three sections of the race, against

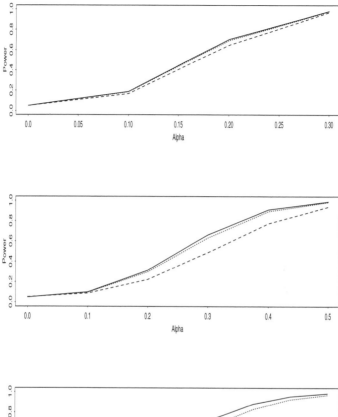

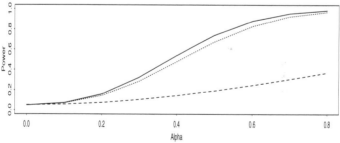

Figure 7.1 *Empirical power of three modified likelihood ratio tests for* $\mu = 0$, *as a function of* α, *for three combinations of* N *(number of subjects) and* n *(number of measurement times) used in the simulation study: top panel,* $N = 40$ *and* $n = 10$; *middle panel,* $N = 20$ *and* $n = 10$; *bottom panel,* $N = 12$ *and* $n = 10$. *Solid line, test which fully exploits AD(1); Dotted line, test which assumes AD(2); Dashed line, Hotelling's* T^2 *test.*

Table 7.1 *Predecessor-adjusted and standard t tests for illustrative hypothesis on mean split times in the 100-km race.*

Section (i)	μ_{0i}	$\overline{Y}_i$	Predecessor-adjusted t	Standard t
1	48	47.61	−0.67	−0.67
2	51	50.62	0.22	−0.57
3	50	49.40	−0.56	-0.76
4	54	53.03	−1.09	−1.13
5	54	54.48	3.52	0.45
6	60	59.91	−1.38	−0.07
7	63	62.41	−0.83	−0.51
8	68	69.07	1.78	0.78
9	68	68.42	0.13	0.31
10	68	66.99	−0.76	−0.69

the alternative of a saturated mean over these sections. Recall that the sample mean profile of these data (Figure 1.3) suggests that the means of the split times, which increase quite steadily through the first eight sections of the race, appear to level off and perhaps even decrease from the eighth to the tenth sections. For a test of flatness over the last three sections, an appropriate matrix $\mathbf{C}$ is

$$\mathbf{C} = \begin{pmatrix} 0 & 0 & 0 & 0 & 0 & 0 & 0 & 1 & -1 & 0 \\ 0 & 0 & 0 & 0 & 0 & 0 & 0 & 0 & 1 & -1 \end{pmatrix}.$$

Again we take the covariance structure to be that of the AD(0,1,1,1,2, 1,1,2,3,5) model. The Wald test statistic given by Theorem 7.2 is 3.46 which, when compared to percentiles of the chi-square distribution with 2 degrees of freedom, yields a p-value of 0.18. Thus there is insufficient evidence against the null hypothesis to reject it, and we conclude that the mean profile is not statistically different from a flat profile over the last three sections of the race.

Example 2: Treatment A cattle growth data

As a further example, we test for linearity of the mean profile of cattle weights over the 11 measurement times, against the alternative hypothesis of a saturated mean. Recall that the sample mean profile of these data (Figure 1.1) indicates that mean weight increases over the entire course of the study but growth appears to decelerate in its latter part, which suggests that mean growth is not linear. To test formally for linearity, we may use the Wald test of Theorem 7.2, taking the matrix C of the theorem to be, without loss of generality, as follows:

$$
C = \begin{pmatrix}
1 & -2 & 1 & 0 & 0 & \cdots & 0 & 0 \\
0 & 1 & -2 & 1 & 0 & \cdots & 0 & 0 \\
 & & & \vdots & & & & \\
0 & 0 & 0 & \cdots & 1 & -2 & 1 & 0 \\
0 & 0 & 0 & \cdots & & 1 & -3 & 2
\end{pmatrix}.
$$

Observe that the nonzero coefficients in the last row of this C are different than in the other rows, owing to the difference in elapsed time (one week versus two weeks) between the tenth and eleventh measurement times. We take the covariance structure for this test to be unstructured AD(2), which was the unstructured constant-order AD model we previously selected by both a forward selection and backward elimination procedure (Section 6.2). The Wald test statistic is 376.8 on 9 degrees of freedom. Thus, the hypothesis that mean growth of cattle receiving Treatment A over this time period is linear is rejected emphatically.

7.2 Two-sample case

Now consider a situation in which balanced observations are independently sampled from two groups; observations within the same group follow a normal saturated-mean, unstructured AD(p) model; and the covariance matrix is common to both groups. That is,

$$
\mathbf{Y}_{gs} \sim \text{independent } \mathrm{N}_n \left(\boldsymbol{\mu}_g, \boldsymbol{\Sigma}(\boldsymbol{\theta}) \right) \quad (g = 1, 2; \, s = 1, \ldots, N(g)) \quad (7.8)
$$

where $\boldsymbol{\Sigma}(\boldsymbol{\theta})$ is AD(p). Note that this is a special case of the multivariate regression-mean setting of Section 5.3.2 in which $\mathbf{z}_s^T$ is equal to either $(1,0)$ or $(0,1)$ depending on whether subject s belongs to group 1 or group 2. In this setting, we often wish to test the null hypothesis that the two mean vectors, $\boldsymbol{\mu}_1$ and $\boldsymbol{\mu}_2$, are equal against the alternative hypothesis that they are not equal. In the vernacular of longitudinal data analysis, tests of these and other related hypotheses are called *profile comparisons*.

The log-likelihood function for model (7.8) is given by

$$
\log L(\boldsymbol{\mu}_1, \boldsymbol{\mu}_2, \boldsymbol{\theta}_\delta, \boldsymbol{\theta}_\phi) = -\frac{nN}{2} \log 2\pi - \frac{1}{2} \sum_{i=1}^{n} \sum_{g=1}^{2} \sum_{s=1}^{N(g)} \left\{ \log \delta_i + \left[Y_{gsi} \right. \right.
$$

$$
\left. \left. - \mu_{gi} - \sum_{k=1}^{p_i} \phi_{i,i-k}(Y_{gs,i-k} - \mu_{g,i-k}) \right]^2 \middle/ \delta_i \right\},
$$

where $N = N(1) + N(2)$. This function may be rewritten in terms of transformed mean vectors, $\boldsymbol{\mu}_1^* = (\mu_{1i}^*)$ and $\boldsymbol{\mu}_2^* = (\mu_{2i}^*)$, as

$$
\log L(\boldsymbol{\mu}_1^*, \boldsymbol{\mu}_2^*, \boldsymbol{\theta}_\delta, \boldsymbol{\theta}_\phi) = -\frac{nN}{2} \log 2\pi - \frac{1}{2} \sum_{i=1}^{n} \sum_{g=1}^{2} \sum_{s=1}^{N(g)} \left\{ \log \delta_i + \left(Y_{gsi} \right. \right.
$$
$$
\left. \left. - \mu_{gi}^* - \sum_{k=1}^{p_i} \phi_{i,i-k} Y_{gs,i-k} \right)^2 \middle/ \delta_i \right\}
\tag{7.9}
$$

where

$$
\mu_{gi}^* = \mu_{gi} - \sum_{k=1}^{p_i} \phi_{i,i-k} \mu_{g,i-k} \quad (g = 1, 2).
\tag{7.10}
$$

By a development similar to that which led to Theorem 7.1 in the one-sample situation, we obtain the following theorem.

Theorem 7.3. *Suppose that observations are balanced and independently sampled from two groups, and that they follow model (7.8), where $N(g) - 1 > p$ for $g = 1, 2$. The likelihood ratio test for the null hypothesis that $\boldsymbol{\mu}_1 = \boldsymbol{\mu}_2$ (against the alternative that $\boldsymbol{\mu}_1 \neq \boldsymbol{\mu}_2$) rejects the null hypothesis if and only if*

$$
N \sum_{i=1}^{n} [\log RSS_i(\boldsymbol{\mu}) - \log RSS_i(\boldsymbol{\mu}_1, \boldsymbol{\mu}_2)] > K,
$$

where $RSS_i(\boldsymbol{\mu})$ is the residual sum of squares from the regression of Y_i on its p_i predecessors $\{Y_{i-k} : k = 1, \dots, p_i\}$ with a common intercept for the two groups; $RSS_i(\boldsymbol{\mu}_1, \boldsymbol{\mu}_2)$ is the pooled within-groups residual sum of squares from the regression of Y_i on its p_i predecessors $\{Y_{i-k} : k = 1, \dots, p_i\}$ with group-specific intercepts, and K is a constant. For any $\alpha \in (0, 1)$, an asymptotically (as $N \to \infty$) valid size-α test is obtained by taking K to be the $100(1 - \alpha)$th percentile of a chi-square distribution with n degrees of freedom.

The likelihood ratio test statistic given by Theorem 7.3, like that given by Theorem 7.1, may be extended to handle variable-order antedependence or dropouts in obvious ways. Likewise, this test statistic can be expressed as functions of certain Hotelling's T^2 statistics (Johnson, 1989). In this case we have

$$
-2 \log \Lambda = N \log \left(\frac{\prod_{i=1}^{n-p} [1 + T_{i:i+p}^2/(N-2)]}{\prod_{i=1}^{n-p-1} [1 + T_{i+1:i+p}^2/(N-2)]} \right)
$$

where $T_{i:i+p}^2$ is Hotelling's T^2 statistic for testing the null hypothesis that $\boldsymbol{\mu}_{1,i:i+p} = \boldsymbol{\mu}_{2,i:i+p}$ (versus the alternative that these vectors are not equal) under general multivariate dependence, given by

$$
T_{i:i+p}^2 = \left(\frac{N(1)N(2)}{N} \right) (\overline{\mathbf{Y}}_{1,i:i+p} - \overline{\mathbf{Y}}_{2,i:i+p})^T \mathbf{S}_{i:i+p}^{-1}
$$
$$
\times (\overline{\mathbf{Y}}_{1,i:i+p} - \overline{\mathbf{Y}}_{2,i:i+p}),
$$

and $T^2_{i+1:i+p}$ is Hotelling's T^2 statistic for testing the null hypothesis that $\mu_{1,i+1:i+p} = \mu_{2,i+1:i+p}$ (versus the alternative that these vectors are not equal) under general multivariate dependence, defined similarly. Here, $S_{i:i+p}$ is the indicated submatrix of the pooled within-groups sample covariance matrix.

A modification to the test criterion given by Theorem 7.3, which improves the correspondence between nominal size and actual size, is given by

$$\frac{n \sum_{i=1}^{n} \left[\log RSS_i(\mu) - \log RSS_i(\mu_1, \mu_2)\right]}{\sum_{i=1}^{n} \psi(1, N - 2 - p_i)} > K, \tag{7.11}$$

with K defined as in the theorem.

In some applications it may be of interest to determine the first or last times that the means of the two groups are different. From (7.9) and (7.10), and by analogy with the development in the previous section for the one-sample case, we see that each null hypothesis in the sequence

$$\mu_{1,1:i} = \mu_{2,1:i}, \quad i = 1, \ldots, n,$$

may be tested against the respective alternative hypothesis

$$\mu_{1i} \neq \mu_{2i}, \ \mu_{1,1:i-1} = \mu_{2,1:i-1}, \quad i = 1, \ldots, n,$$

by likelihood ratio tests, each of which is a t-test. The null hypothesis for this t-test is that the contrast $\mu_{1i} - \mu_{2i}$ is equal to 0 in the regression of Y_i on its p_i predecessors, assuming common slope coefficients on predecessors across groups but allowing intercepts to be group-specific. The first one of these tests that is statistically significant corresponds to the first measurement time at which the group means are judged to be different. The last measurement time at which the means of the two groups are different may be determined by re-ordering the observations in reverse time order, and then proceeding in exactly the same way.

It will often be of equal interest to test whether certain linear combinations of the elements of the two mean vectors are equal to zero or other specified constants. Again the likelihood ratio test statistic for such a hypothesis, though it can be derived, is generally not expressible as a simple function of residual sums of squares from certain pth-order regressions, like it is for the hypothesis of Theorem 7.3. Consequently, we give, via the following theorem, the asymptotically equivalent Wald test, which is obtained by substituting the REML estimator, $\tilde{\Sigma}$, for Σ in the test that is uniformly most powerful invariant when Σ is known.

Theorem 7.4. *Let* C *and* c *be a specified* $c \times n$ *matrix and* $c \times 1$ *vector, respectively, where the rows of* C *are linearly independent. Also, let* $\tilde{\Sigma}$ *be the REML estimator of* Σ *under the model and conditions of Theorem 7.3. Under the same model and conditions, the test obtained by substituting* $\tilde{\Sigma}$ *for* Σ *in the* *uniformly most powerful invariant test of the null hypothesis $\mathbf{C}(\boldsymbol{\mu}_1 - \boldsymbol{\mu}_2) = \mathbf{c}$ against the alternative hypothesis $\mathbf{C}(\boldsymbol{\mu}_1 - \boldsymbol{\mu}_2) \neq \mathbf{c}$ rejects the null hypothesis if and only if*

$$\left(\frac{N(1)N(2)}{N}\right)[\mathbf{C}(\overline{\mathbf{Y}}_1 - \overline{\mathbf{Y}}_2) - \mathbf{c}]^T(\mathbf{C}\tilde{\boldsymbol{\Sigma}}\mathbf{C}^T)^{-1}[\mathbf{C}(\overline{\mathbf{Y}}_1 - \overline{\mathbf{Y}}_2) - \mathbf{c}] > K.$$

$$(7.12)$$

For any $\alpha \in (0, 1)$, an asymptotically (as $N \to \infty$) valid size-α test is obtained by taking K to be the $100(1 - \alpha)$th percentile of a chi-square distribution with c degrees of freedom.

Example: Speech recognition data

In Section 5.4.2, we fitted a model to the speech recognition data in which the mean structure was saturated within groups and the within-groups covariance matrices were assumed to be homogeneous and unstructured AD(1). Our additional analysis of these data, given in Section 6.4, justified our assumptions on the covariance structure. It is of interest now to know whether the data provide sufficient evidence to conclude that the mean profiles of the two implant types are different, and if so, how they are different. Results of tests of five relevant hypotheses are given in Table 7.2. The modified likelihood ratio test for equality of the two mean profiles, as given by (7.11), indicates that this hypothesis cannot be rejected, and subsequent standard t tests and predecessor-adjusted t tests of implant differences at individual times agree with this (results not shown). Likewise, the Wald test for parallel profiles shows no evidence against parallelism. A subsequent Wald test for equality of mean profiles, assuming parallelism, yields some, though rather weak, evidence that the mean profile for implant A is shifted vertically from that of implant B. Note that this agrees with the relatively constant (across time) difference of about 10 to 15 units (implant A minus implant B) seen in the REML estimates of mean scores listed in Section 5.4.2. Additional Wald tests listed in Table 7.2 establish that both mean profiles increase significantly over time, indicating significant improvement in speech recognition for both implant types, and that the rate of increase is not linear, with greater improvement occurring from one month to nine months after connection than thereafter.

Since these data have dropouts, for all Wald tests the REML estimates of mean scores were used in place of sample means in (7.12), with a suitable modification to the constant multiplier as well.

7.3 Multivariate regression mean

Now consider a situation in which balanced normal AD(p) observations have a classical multivariate regression mean structure. That is,

$$\mathbf{Y}_s \sim \text{independent } \mathrm{N}_n\left((\mathbf{z}_s^T \otimes \mathbf{I}_n)\boldsymbol{\beta}, \boldsymbol{\Sigma}(\boldsymbol{\theta})\right), \quad s = 1, \ldots, N, \qquad (7.13)$$

Table 7.2 *Likelihood ratio and Wald tests for various null hypotheses about the mean profiles of the speech recognition data.*

Null hypothesis	Test statistic	P
Profiles equal	6.209	0.184
Profiles parallel	2.957	0.398
Profiles equal, assuming parallel	2.964	0.085
Profiles flat, assuming parallel	68.212	<0.001
Profiles linear, assuming parallel	35.087	<0.001

where $\Sigma(\theta)$ is an unstructured AD(p) covariance matrix, z_s is $m \times 1$, and

$$
Z = \begin{pmatrix} z_1^T \\ \vdots \\ z_N^T \end{pmatrix}
$$

is of full column rank m. Further, consider partitioning the covariate vector z_s into two parts as $z_s = (z_{s1}^T, z_{s2}^T)^T$, where z_{s1} is $m_0 \times 1$ and z_{s2} is $(m-m_0) \times 1$, with corresponding partitionings $\beta = (\beta_1^T, \beta_2^T)^T$ and $Z = (Z_1, Z_2)$. Suppose we wish to test whether all covariates in Z are important in explaining the variability of the response, or whether the covariates in Z_1 alone will suffice. That is, suppose we wish to test the null hypothesis that $\beta_2 = 0$ against the alternative hypothesis that $\beta_2 \neq 0$. Note that any linear hypothesis, $C\beta = 0$, can be put in this form via reparameterization. In particular, the hypotheses $\mu = \mu_0$ and $\mu_1 = \mu_2$ considered in the previous two sections are special cases. A development very similar to that of the previous section yields the following theorem and the additional results and comments subsequent to it. Some references providing more of this development are Kenward (1987), Patel (1991), Albert (1992), and Macchiavelli and Moser (1997).

Theorem 7.5. *Suppose that observations are balanced and follow the normal multivariate regression-mean, unstructured AD(p) model given by (7.13), with $N - m > p$. The likelihood ratio test for the null hypothesis that $\beta_2 = 0$ (against the alternative that $\beta_2 \neq 0$) rejects the null hypothesis if and only if*

$$
N \sum_{i=1}^{n} [\log RSS_i(\beta_1) - \log RSS_i(\beta_1, \beta_2)] > K,
$$

where $RSS_i(\beta_1)$ is the residual sum of squares from the regression of Y_i on its p_i predecessors and the m_0 covariates in z_1; $RSS_i(\beta_1, \beta_2)$ is the residual sum of squares from the regression of Y_i on its p_i predecessors and all m covariates in z, and K is a constant. For any $\alpha \in (0, 1)$, an asymptotically (as $N \to \infty$)

valid size-α test is obtained by taking K to be the $100(1 - \alpha)$th percentile of a chi-square distribution with $n(m - m_0)$ degrees of freedom.

Once again, the likelihood ratio test statistic given by Theorem 7.5 may be extended to handle variable-order antedependence or dropouts in obvious ways. It may be expressed as functions of certain Wilk's lambda statistics as follows:

$$-2 \log \Lambda = N \log \left(\frac{\prod_{i=1}^{n-p} \Lambda_{i:i+p}}{\prod_{i=1}^{n-p-1} \Lambda_{i+1:i+p}} \right)$$

where

$$\Lambda_{i:i+p} = \frac{|\hat{\Sigma}_{i:i+p}|}{|\hat{\Sigma}_{i:i+p} + N^{-1}\hat{\mathbf{B}}_2^T \mathbf{Z}_2^T [\mathbf{I} - \mathbf{Z}_1(\mathbf{Z}_1^T \mathbf{Z}_1)^{-1} \mathbf{Z}_1^T] \mathbf{Z}_2 \hat{\mathbf{B}}_2|}$$

and $\hat{\mathbf{B}}_2^T$ is the $n \times (m - m_0)$ matrix whose columns, when stacked one upon the other, yield $\hat{\beta}_2$. Also, better correspondence between nominal and actual size is obtained by using the modified likelihood ratio criterion,

$$\frac{n(m - m_0) \sum_{i=1}^{n} [\log RSS_i(\beta_1) - \log RSS_i(\beta_1, \beta_2)]}{\sum_{i=1}^{n} \psi(m - m_0, N - m - p_i)} > K,$$

where K is defined as in Theorem 7.5.

Furthermore, we can test the null hypothesis

$$\beta_{2j} = \mathbf{0}, \quad j = 1, \ldots, i,$$

against the alternative

$$\beta_{2i} \neq \mathbf{0}, \quad \beta_{2j} = \mathbf{0}, \quad j = 1, \ldots, i - 1,$$

by fitting the model that regresses Y_i on its p_i predecessors and $\mathbf{z}_i$ and using the F-test for $\beta_{2i} = \mathbf{0}$ in these fits. (If $m - m_0 = 1$, then the F-test is equivalent to a t-test.) These tests may be performed sequentially (starting with $i = 1$) to determine the first time at which the covariates in $\mathbf{Z}_2$ are important explanatory variables for the response, after adjusting for predecessors. An analogous procedure may be applied to the observations in reverse time order to determine the last time at which the covariates in $\mathbf{Z}_2$ are important explanatory variables for the response (after adjusting for successors).

Example: 100-km race data
We illustrate the methodology of this section with an analysis of the importance of age effects on the split times of the 76 competitors whose ages were recorded. Recall, from the exploratory analysis of the effects of age displayed in Figure 4.1 and the estimates of linear and quadratic age effects (with estimated standard errors) listed in Table 5.2, that there is some evidence that age has an effect on some of the split times, particularly late in the race, and that this effect may be quadratic. Thus we will take as our "full" mean structure the

Table 7.3 *Modified likelihood ratio tests for effects of age on 100-km race split times.*

Null hypothesis	Alternative hypothesis	$-2 \log \Lambda$	P
Saturated	Quadratic	32.58	0.037
Saturated	Linear	17.27	0.069
Linear	Quadratic	15.33	0.121

same model fitted in Section 5.3.2, which is

$$E(Y_{si}) = \beta_{0i} + \beta_{1i}[age(s)] + \beta_{2i}[age(s)]^2$$

where i indexes the 10-km sections and $[age(s)]$ represents the centered age of subject s. We take the covariance structure to be AD(0,1,1,1,2,1,1,2,3,5), the unstructured variable-order AD model found, in Section 6.5, to minimize *BIC* for the complete data set.

Results of three modified likelihood ratio tests for age effects are given in Table 7.3. The first test listed tests the saturated mean structure as a null hypothesis against the alternative of the full quadratic model. For this test, $m = 3$ and $m_0 = 1$. The test indicates that some evidence exists against the saturated mean model relative to of the full model. This suggests that age effects may be of some importance in explaining the variability of split times among competitors. To determine whether the important age effects are linear or quadratic (or both), we next test the saturated mean model as a null hypothesis in the context of a linear effects model, and the linear effects model as a null hypothesis in the context of the full quadratic model. The statistical evidence against each null hypothesis is slight, but suggestive. Our conclusion, based on these results and those given previously in Table 5.2, is that age has an important effect on some, but not all, of the split times during the course of the race, and that these effects are neither merely linear nor merely quadratic. A natural follow-up is to determine the first split time at which age effects manifest. We accomplish this with a sequence of predecessor-adjusted F tests on the quadratic model, the ith of which tests the null hypothesis

$$\beta_{1j} = \beta_{2j} = 0, \quad j = 1, \ldots, i, \tag{7.14}$$

against the alternative hypothesis

$$\beta_{1i} \neq 0 \text{ or } \beta_{2i} \neq 0, \quad \beta_{1j} = \beta_{2j} = 0, \quad j = 1, \ldots, i - 1.$$

The numerator degrees of freedom for each test is two, while the denominator degrees of freedom ranges from 68 to 73 as the number of predecessors included in the model ranges from 0 to 5. Results for these tests are given in Table 7.4. We find that the third section is the earliest for which the null hypothesis is rejected at the 0.05 level of significance ($P = 0.0464$), hence we

Table 7.4 *Predecessor-adjusted F tests for combined linear and quadratic effects of age on 100-km race split times.*

Section	Predecessor-adjusted F	P
1	0.96	0.39
2	1.14	0.32
3	3.20	0.05
4	1.03	0.36
5	1.95	0.15
6	0.24	0.79
7	0.11	0.90
8	5.00	0.01
9	0.61	0.55
10	2.62	0.08

conclude that when the immediately preceding split time is taken into account, age effects first become important on the third section. Furthermore, we find that while both the linear and quadratic effects contribute to this rejection, the linear effect is the more significant of the two; the t statistics for the two equality hypotheses in (7.14) are -2.01 and 1.48 for the linear and quadratic terms, respectively, and the associated p-values are 0.048 and 0.143. The negative linear effect implies that older runners run relatively faster than younger runners on the third section when split time on the second section is taken into account.

Note that the conclusions from the predecessor-adjusted F tests differ from, but do not contradict, the findings of the analysis presented in Section 5.3.2 (Table 5.2). In that analysis, which found that the effects of age were not statistically significant until the fifth section of the race and that the significant effects were quadratic rather than linear, the effects of age were determined without adjustment for previous split times. Such an adjustment may often lead to subtly different conclusions, as occurs in this case.

7.4 Other situations

The tests of hypotheses on mean structure considered so far in this chapter have allowed the covariance matrix of the antedependence model to be unstructured (of order p, where $0 \le p \le n - 1$). When the antedependence model is structured, it is generally not possible to obtain expressions for the likelihood ratio test statistics of hypotheses on mean vectors which are simple functions of residual sums of squares from various regressions. This is also true in situations in which there are non-monotonic missing observations or heterogeneous within-group covariance matrices, and/or the hypotheses being tested, though linear in the mean parameters, do not correspond to mean structures of multivariate regression form. Nevertheless, likelihood ratio testing procedures may still be carried out in some of these situations, by numerical maximizing the likelihood function under each of the null and alternative hypotheses and comparing minus twice the log of the ratio of the two maxima to a percentile from a chi-square distribution with degrees of freedom equal to the difference in the number of parameters in the two models.

Example 1: 100-km race data
In Section 5.3.2 we fitted an unstructured AD(3) model with a multivariate regression mean structure, i.e., model (5.34), to the split times of the 76 competitors whose ages were recorded. That model had an overall intercept, a linear age effect, and a quadratic age effect corresponding to each of the 10 sections (30 mean parameters in all). Subsequently, in Section 5.3.3 we fitted an unstructured AD(3) model with mean structure that was a cubic function of centered section number plus common (across sections) linear and quadratic functions of age, i.e., model (5.37) (with 6 mean parameters), to the same data. The latter model, though its mean structure is not of multivariate regression form, is nested within the former, hence the two models may be compared via a likelihood ratio test. Such a test yields a likelihood ratio test statistic of 159.2, with a p-value of essentially 0. We conclude that the latter, more parsimonious mean structure is inadequate and we retain the former model.

Example 2: Cattle growth data, Treatments A and B
Figure 7.2 displays the mean profiles for each treatment group for the cattle growth data. It can be seen that the mean weight of Treatment A cattle is slightly larger at the experiment's outset and that the difference in means gradually increases until Week 14, when the mean profiles abruptly cross. The Treatment B cattle maintain their newly acquired weight advantage until the last week of the experiment, at which time the profiles cross again. We wish to test (initially) whether the two mean profiles are equal. A standard likelihood ratio test for homogeneity of the two unstructured [AD(10)] within-group covariance matrices rejects this hypothesis ($P = 0.02$). Consequently, to test for the equality of mean profiles we do not use the likelihood ratio test given by Theorem 7.3, but instead use a Wald test similar to that given by Theorem 7.4, with $\mathbf{C}$ taken to be the 11×11 identity matrix and $\left(\frac{1}{N(1)} \tilde{\boldsymbol{\Sigma}}_1 + \frac{1}{N(2)} \tilde{\boldsymbol{\Sigma}}_2 \right)$ substituted for $\left(\frac{1}{N(1)} + \frac{1}{N(2)} \right) \tilde{\boldsymbol{\Sigma}}$, where $\tilde{\boldsymbol{\Sigma}}_1$ and $\tilde{\boldsymbol{\Sigma}}_2$ are REML estimates of AD(3) covariance matrices under separate fits to the data from each group. We compute the test statistic as 77.3, which, when compared to the chi-square distribution with 11 degrees of freedom, unambiguously establishes an overall difference in mean profiles. To investigate the time(s) at which the profiles differ significantly, we carry out standard "time-by-time" t tests and two types

of predecessor-adjusted t tests of mean differences at individual time points. The first type of predecessor-adjusted test is based on an assumption of a homogeneous AD(3) covariance structure. The rationale for taking the order to be three for this test is that the highest order in the best-fitting variable-order models for the covariance structure within either treatment group was three. On the other hand, this test does not account for heterogeneity of the within-group covariance matrices, so we consider a second type of predecessor-adjusted t test that replaces $\left(\frac{1}{N(1)} + \frac{1}{N(2)} \right) \tilde{\delta}_i$ in the first predecessor-adjusted test with $\left(\frac{1}{N(1)} \tilde{\delta}_{i1} + \frac{1}{N(2)} \tilde{\delta}_{i2} \right)$ and makes a standard Satterthwaite adjustment for degrees of freedom. Here, $\tilde{\delta}_{i1}$ and $\tilde{\delta}_{i2}$ are REML estimates of the group-specific innovation variances under assumed AD(3) structures, and $\tilde{\delta}_i$ is the REML estimate of an assumed common innovation variance. Only one type of standard t test is considered, for despite the evidence for heterogeneous within-group covariance matrices there is no evidence for heterogeneous group variances at any time point, as can be seen by comparing the main diagonal elements in Table 4.1(a) and Table 8.1(a).

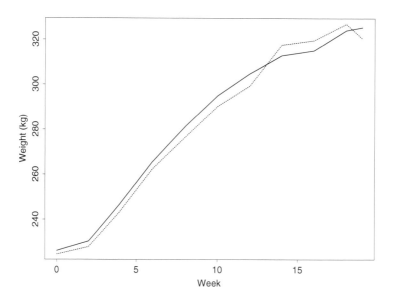

Figure 7.2 *Mean profiles for the cattle growth data. Solid line: Treatment A. Dotted line: Treatment B.*

Results of the t tests are given in Table 7.5. Note that the standard t tests are uninformative, as none of them find a significant difference in means. In contrast,

Table 7.5 *Standard t tests and two predecessor-adjusted t tests for treatment differences at individual time points for cattle growth data.*

Week	Standard t	Predecessor-adjusted t, first type	Predecessor-adjusted t, second type
0	0.60	0.60	0.60
2	0.82	0.57	0.59
4	1.03	0.72	0.68
6	0.88	−0.30	−0.33
8	1.21	1.07	1.08
10	1.12	0.15	0.16
12	1.28	0.60	0.64
14	−1.10	−7.26	−7.19
16	−0.95	−1.15	−1.14
18	−0.53	1.15	1.44
19	0.85	3.92	3.89

the predecessor-adjusted tests determine that when the three immediate predecessors are taken into account, the profiles are significantly different at week 14. As it happens, for these data there are not important differences in the conclusions drawn from the two predecessor-adjusted tests. However, in other situations with heterogeneous covariance matrices, the reduction in degrees of freedom produced by the Satterthwaite adjustment for tests of the second type could affect the assessment of significance.

Diggle et al. (2002) present an alternative time-by-time analysis of successive weight gains, $Y_{si} - Y_{s,i-1}$, instead of the weights themselves. The results of their analysis are very similar to those of the predecessor-adjusted analysis, which, in light of the strong positive correlation between successive weights, is not surprising. However, it is worth noting that time-by-time t tests of successive differences are correlated if the order of antedependence is higher than one at any measurement time (as is the case for these data), and they do not account for the possibility that successive differences have unequal group variances (as is also the case here at some of the measurement times). Therefore, for general use for data with heterogeneous covariance matrices, we recommend the second predecessor-adjusted approach.

7.5 Penalized likelihood criteria

In Section 6.5, we introduced the use of penalized likelihood criteria to compare models with different (and possibly non-nested) covariance structures.

Penalized likelihood criteria may also be used to compare models with different mean structures, or to compare models with different mean and covariance
structures. Of those criteria considered previously, only the likelihood-based
criteria AIC and BIC are applicable; REML-based criteria are not applicable for these purposes because the residual likelihood function is a function of
covariance parameters only. Thus we consider

$$IC(k) = -\frac{2}{N} \log L(\hat{\boldsymbol{\beta}}_k, \hat{\boldsymbol{\theta}}_k) + d_k \frac{c(N)}{N}$$

where k indexes the models under consideration, $\hat{\boldsymbol{\beta}}_k$ and $\hat{\boldsymbol{\theta}}_k$ are the maximum
likelihood estimators of $\boldsymbol{\beta}$ and $\boldsymbol{\theta}$ under the kth model, d_k is the number of
unknown parameters in the mean structure and covariance structure of the kth
model, and $c(N)$, the penalty term, is equal to 2 for AIC and equal to $\log N$
for BIC.

Example: Treatment A cattle growth data
Previously (Table 6.7) we compared various covariance structures for the Treatment A cattle growth data using penalized likelihood criteria, under the assumption of a saturated mean structure. Now we use AIC and BIC to compare
mean structures for some of those models. In particular, we compare polynomial mean structures of orders one through ten for the five best-fitting covariance structures listed in Table 6.7. Note that a tenth-order polynomial mean
coincides with a saturated mean for these data. Results for BIC are displayed
in Table 7.6. For the five antedependent covariance structures, the ordering of
mean structures is almost identical, with the saturated mean structure always
best. Similarly, the ordering of covariance structures is identical for all mean
structures, with the second-order marginally formulated power law SAD model
always best. Results for AIC (not shown) are identical with respect to ordering of mean structures and nearly identical with respect to ordering of covariance structures, the only differences being that the variable-order unstructured
antedependence model and first-order marginally formulated power law SAD
model are reversed from their order with respect to BIC.

Recall that for the POU(10) model defined previously (Section 5.5.4), the log-
innovation variances and autoregressive coefficients are modeled as cubic functions of time and lag, respectively. Of course, polynomials of other orders for
these quantities could be considered. When combined with comparisons of orders for polynomial mean structures, however, it would appear that there is a
large number, specifically $(n-1)^3$ in a general balanced longitudinal setting
with n measurement times, of such models that need to be fit to conduct an ex-
haustive search for the one with the smallest BIC (or AIC). Pan and MacKenzie (2003) conjectured that the optimal model, with polynomial orders d_{IV},
d_{AR}, and d_{mean} for the innovation variances, autoregressive coefficients, and
mean parameters, respectively, could alternatively be found using three BIC-
based searches involving the profile log-likelihoods obtained by saturating the

Table 7.6 *Bayesian information criteria (BIC) for comparing polynomial mean struc-*
tures of various orders for the Treatment A cattle growth data, for the five best-fitting
antedependence models from Table 6.7.

Polynomial order	SAD-PM(2)	POU(10)	VAD	SAD-PM(1)	AD(2)
1	74.89	75.19	75.93	76.11	76.28
2	73.53	73.76	73.80	73.99	74.14
3	73.35	73.46	73.73	73.90	74.08
4	71.77	71.90	72.07	72.30	72.41
5	70.80	70.90	71.07	71.29	71.41
6	70.80	70.86	71.13	71.27	71.34
7	70.46	70.53	70.73	70.86	71.14
8	70.21	70.26	70.40	70.54	70.81
9	70.16	70.19	70.33	70.47	70.67
10	70.09	70.13	70.31	70.37	70.55

three sets of parameters in pairs:

$$d_{IV} = \arg \min_{i=1,\ldots,n} \{BIC(i-1, n-1, n-1)\},$$

$$d_{AR} = \arg \min_{i=1,\ldots,n} \{BIC(n-1, i-1, n-1)\},$$

$$d_{mean} = \arg \min_{i=1,\ldots,n} \{BIC(n-1, n-1, i-1)\}.$$

This approach reduces the number of maximizations required to find the optimum model from $(n-1)^3$ to $3n$. Pan and MacKenzie (2003) reported that
their conjecture proved to be correct for the Treatment A cattle growth data, and
that the optimum model of this type for these data has $(d_{IV}, d_{AR}, d_{mean}) = (3, 4, 8)$. The same type of approach could be used with any autoregressively
formulated SAD model, but it is not known whether it will reliably yield the
optimum model.

7.6 Concluding remarks

We have given, in this chapter, likelihood ratio tests and Wald tests for linear
hypotheses on the mean structure of antedependence models. In comparison
to tests on the mean structure under general multivariate dependence, tests on
the mean structure under antedependence are more powerful, as our simulation
study demonstrated for the likelihood ratio test; similar results for the Wald
test are shown by Byrne and Arnold (1983) and Johnson (1989). Furthermore,
our tests on the mean structure under antedependence can be carried out in

situations in which the sample size is too small to test the hypothesis under general multivariate dependence. In every case, the likelihood ratio test statistic for testing a hypothesis about an n-dimensional mean vector(s) under pth-order antedependence turns out to be expressible as a function of likelihood ratio test statistics for hypotheses about subvectors of the mean vector(s) of dimension at most $p + 1$ under general dependence. It is also expressible in terms of residual sums of squares from regressions on predecessors (plus covariates), as were the likelihood ratio test statistics for hypotheses about the covariance structure described in Chapter 6.

Once again, R functions for performing the tests presented in this chapter are available from the first author's Web page.

The likelihood ratio and Wald test statistics are but two of several statistics that may be used to test hypotheses about the mean structure of antedependence models. Two others, which are asymptotically equivalent, are the Lawley-Hotelling trace and Pillai trace; details are given by Johnson (1989).

If one of the tests described in this chapter rejects the null hypothesis, then a natural follow-up is to perform predecessor-adjusted t tests (or F tests) for the mean(s) at each measurement time. Another follow-up analysis would be to obtain simultaneous confidence intervals for linear combinations of the mean parameters. A Wald test, with its elliptical acceptance region, leads directly to a set of Scheffé-based simultaneous confidence intervals for these combinations. Byrne and Arnold (1983) give such intervals, under first-order antedependence, for linear combinations of the elements of a saturated mean in the one-sample case. Johnson (1989) extends this to unstructured antedependence of arbitrary order and to the two-sample case as well. These authors show that, as expected, the simultaneous confidence intervals obtained under antedependence are narrower than their counterparts obtained under general multivariate dependence.

CHAPTER 8

Case Studies

In Chapters 4 through 7 of this book, we have illustrated various exploratory and inferential methods associated with antedependence models using four data sets introduced in Chapter 1: the cattle growth, 100-km race, speech recognition, and fruit fly mortality data. However, due to the specific focus on a particular method in each instance, the analysis of each data set was presented in a rather piecemeal fashion. Moreover, for obvious reasons only antedependence models were considered and we did not fit any alternatives often used for longitudinal data, such as random coefficient models or vanishing correlation models. In this chapter we present a concise summary of our previous analyses of each data set, adding to the analysis where it seems appropriate. Included among the supplemental analyses are fits and comparisons of alternative models. In each case we attempt to follow a coherent, data-driven approach to parametric modeling of the data's mean and covariance structure. We begin the chapter with a description of the components of this approach, and we close it out with a discussion of what the case studies tell us about the relative merits of antedependence models and other models for longitudinal data.

8.1 A coherent parametric modeling approach

The first stage of our modeling approach is to explore the data via summary statistics (e.g., means, variances, correlations, intervenor-adjusted correlations, innovation variances and autoregressive coefficients from the modified Cholesky decomposition of the precision matrix) and plots (e.g., profile plot, response-versus-covariate plots, ordinary scatterplot matrix and PRISM, innovariogram and regressogram). For these purposes, one ordinarily should take the mean to be saturated and, if the data come from discrete groups of subjects, one should compute summary statistics/plots separately for each group. In order to carry out this program, the data must be either balanced or sufficiently well-replicated across measurement times for the elements of the unstructured covariance matrix to be estimable; if this is not so, then the exploratory analyses may be based on residuals from a smooth fit to the mean structure across time, or measurement times may be grouped together until sufficient replication is achieved. The summary statistics and plots should help to determine whether parametric modeling of the mean structure or covariance structure or both is likely to be successful, and if so, which mean structures and covariance structures are plausible.

The second stage of our modeling approach is to fit saturated-mean models with plausible covariance structures to the data, estimating parameters by either maximum likelihood or its close relative, REML. The methods we present here assume that the observations are normally distributed, so this assumption should be checked and the data transformed if necessary to more closely satisfy it. The ordinary scatterplot matrix and PRISM can be helpful in this regard. For each fitted model, the maximized likelihood and various penalized likelihood criteria may be computed. In this book we feature only AIC and BIC and their REML analogues but there are several other possibilities; see, for example, Burnham and Anderson (2002). The maximized likelihoods may then be used to conduct formal likelihood ratio tests for comparing nested models, such as unstructured antedependence models of order 0 through $n-1$. The penalized likelihood criteria may be used to select the "best" model(s) among all those that were fitted, regardless of whether the models are nested. For data that come from discrete groups of subjects, another aspect of this stage of the analysis is a likelihood ratio test for homogeneity of the covariance structure across groups. This test may be based on an unstructured covariance matrix or, if both groups are determined to have an antedependent covariance matrix of the same order (less than $n-1$), this additional structure can be exploited to yield a more powerful test.

Once the best covariance structure(s) for a saturated-mean model has (have) been identified, then attention may shift to the third stage of the analysis: modeling the mean structure more parsimoniously. This will often involve fitting and comparing models with mean structures that are polynomial functions of time and/or other covariates. If the data come from discrete groups of subjects, it may also involve testing for equality, parallelism, etc. of the groups' mean profiles. For this, likelihood ratio testing is usually a viable option, since the mean structures of interest typically are nested; alternatively, models with different mean structures may, like models with different covariance structures, be compared using penalized likelihood criteria.

Finally, after completing the third stage of the analysis, revisiting previous stages may be worthwhile. For example, if, at the third stage, we determine that the mean structure is well-approximated by a low-order polynomial, say quadratic, function of time, then we may wish to compute the residuals from the fitted quadratic function and return to the first stage to compute correlations, etc. among these residuals rather than among those computed from a saturated mean.

8.2 Case study #1: Cattle growth data

The cattle growth data come from a designed experiment in which cattle receiving two treatments, generically labeled A and B, for intestinal parasites were weighed 11 times over a 133-day period. Thirty animals received Treatment A and thirty received Treatment B. The animals were weighed at two-week intervals except for the final measurement, which was made one week after the tenth measurement. Measurement times were common across animals and no observations were missing. The experimenter wishes to know if there is a difference in growth between treatment groups, and if so, the time of measurement at which it first occurs.

Profile plots of the data corresponding to Treatments A and B (Figures 1.1 and 1.2, respectively) have quite similar shapes, except near the very end of the experiment where Treatment B's profile dips down somewhat and Treatment A's does not. These show that mean growth is relatively slow in the first two weeks but then accelerates until the end of Week 8, where a long process of deceleration (more or less) in growth begins. They also indicate that the variability of responses across subjects increases over the course of the experiment. Figure 7.2 superimposes the mean profiles for the two treatments and indicates that the average weights of the animals receiving Treatment A were slightly higher at the experiment's outset and maintained or expanded this difference until Week 14, when the profiles suddenly crossed. The Treatment B cattle maintained their newfound weight advantage until the last week of the experiment, when the profiles crossed once more.

Given the experimenter's objectives and the shapes of the mean profiles, it is clear that attempting to model the mean structure of each treatment group as a polynomial function of time would be counterproductive. Thus, we retain a saturated mean structure within groups for the entire analysis.

An examination of the marginal variances and correlations for each treatment group separately [Table 1.2 and Table 8.1(a)] reveals that the variances within each group increase roughly fourfold over the course of the experiment and the correlations for both groups are positive and large, and though the correlations do broadly decrease as elapsed time between measurements increases, they remain relatively large (no smaller than 0.44). With such similar behavior across the two treatment groups it might be expected that we could pool the two within-group covariance matrices; however, the standard likelihood ratio test for the homogeneity of two unstructured covariance matrices rejects this hypothesis ($P = 0.02$). Accordingly, we further investigate each group's covariance structure separately.

For the Treatment A data, the sample intervenor-adjusted correlations and the autoregressive coefficients of the sample precision matrix's modified Cholesky decomposition (Table 4.1) suggest that a first-order unstructured antedependence model may fit the data well except possibly for observations taken on the eighth occasion (Week 14), for which second-order antedependence is indicated. The innovariogram and regressogram (Figure 4.8) reveal that the log-innovation variances and autoregressive coefficients may be adequately modeled as cubic functions of time and lag, respectively. Moreover, the same-lag marginal correlations appear to increase monotonically, more or less, as the experiment progresses (Table 1.2). The increase in the marginal variances does not appear to be linear or quadratic, but perhaps cubic (Table 1.2 also). The ordinary scatterplot matrix and PRISM corroborate these findings and do not point to any anomalies.

Informed by these results, we proceed to fit and compare many antedependence models to these data. Table 8.2 is the first of several tables in this chapter that list fitted models together with the number of covariance parameters, d, the maximized residual log-likelihood, $\max \log L_R$, and the penalized likelihood criteria, AIC_R and BIC_R, for each model. For some models we also list one or more "comparison models," which are tested as null hypotheses against the alternative hypothesis of that model via residual likelihood ratio tests. We also give the p-value, P, corresponding to the test. For Table 8.2, the fitted models include unstructured AD models of orders 1, 2, and 10; stationary and heterogeneous autoregressive models (AR and ARH) of orders 1 and 2; marginally and autoregressively specified power law SAD models (SAD-PM and SAD-AM) of orders 1 and 2; the unconstrained SAD(10) model of Pourahmadi in which the innovation variances and autoregressive coefficients are cubic functions of time and lag, respectively [POU(10)]; and the best-fitting variable-order AD model (VAD) as determined in Section 6.5. More specifics on these models may be found in Sections 5.5.1, 5.5.4, and 6.3.

From the results in Table 8.2, we see that the four best models are SAD models that are capable of accommodating nonstationarity in the variances and correlations. These models outperform both the unstructured AD models, which are overly flexible for these data, and the stationary and heterogeneous autoregressive models, which are too parsimonious. For the sake of comparison, we also include results for a few non-antedependence models: compound symmetry (CS), heterogeneous compound symmetry (CSH), linear random coefficient (RCL), quadratic random coefficient (RCQ), Toeplitz(10) [TOEP(10)], and heterogeneous Toeplitz(10) [TOEPH(10)] models. Of these, only the RCQ and TOEP(10) models are somewhat competitive, ranking seventh and ninth among all fitted models. We do not consider any vanishing correlation models

Table 8.1 *Summary statistics for the covariance structure of the Treatment B cattle growth data: (a) sample variances, along the main diagonal, and correlations, below the main diagonal; (b) sample innovation variances, along the main diagonal, and autoregressive coefficients, below the main diagonal. Autoregressive coefficients whose corresponding t-ratios are significant at the 0.05 level are set in bold type.*

(a)

105										
.86	108									
.83	.94	147								
.68	.89	.93	198							
.67	.84	.88	.95	218						
.66	.84	.87	.95	.98	250					
.61	.78	.82	.91	.93	.97	248				
.63	.81	.84	.92	.92	.95	.95	234			
.63	.79	.79	.89	.93	.95	.93	.96	287		
.48	.65	.67	.78	.78	.82	.76	.78	.83	405	
.44	.57	.62	.73	.68	.74	.71	.71	.75	.92	599

(b)

105										
.87	29									
.10	**1.02**	17								
-.56	.54	**1.04**	20							
.21	-.19	-.11	**1.09**	26						
.13	-.01	-.29	**.49**	**.76**	11					
.05	-.26	.00	.15	-.15	**1.09**	20				
.05	.06	-.11	.33	-.23	.37	**.51**	22			
.13	.27	**-.60**	-.05	.34	.45	-.20	**.76**	18		
-.11	-.53	.20	.64	-.92	1.61	-.73	-.54	1.12	142	
.41	-1.16	.51	.98	-1.17	-.04	.70	-.66	.28	**1.15**	86

for these data, as all the marginal correlations are too large for such models to be plausible.

Turning now to the covariance structure of the Treatment B data, the autoregressive coefficients [Table 8.1(b)] show that at each measurement time but the tenth, the response is significantly linearly associated with its immediate predecessor. In this respect the covariance structure for Treatment B is similar to that for Treatment A. However, in the present case some responses are partially associated with a few more of their predecessors. Thus, while unstructured antedependence models of orders one or two fit the Treatment A data best, we will not be surprised if the best unstructured antedependence model for the Treatment B data is of order three or higher. We see also that the innovation variances for Treatment B are about equal to or smaller than their Treatment A counterparts over the first nine measurement times, but the last two innovation variances are much larger.

The best constant order of antedependence, as determined by a backward elimination likelihood ratio testing procedure, is three, and the best (minimum BIC_R) variable-order antedependence model is VAD≡AD(0,1,1,3,1,2,1,2,3, 1,1). In addition to these models, we fit many of the same structured AD and non-antedependence models we fitted to the Treatment A data. Some results are displayed in Table 8.3. Recalling that several structured AD models fit better than unstructured AD models for Treatment A, it is interesting that here the situation is reversed: each structured AD model we fitted performs worse than the unstructured AD model of the same order (and worse than the best variable-order model also). Of the non-antedependent models, only TOEPH(10) is reasonably competitive, but it is not as good as the unstructured AD models of order three or less.

Finally, we test for equality of the two mean profiles, combining the data from both groups for this purpose but retaining distinct within-group covariance matrices: either SAD-PM(2) for Treatment A and VAD for Treatment B (the models determined to fit best), or AD(3) for both groups. In either case, the test indicates a highly significant difference between profiles ($P < 10^{-10}$). Predecessor-adjusted t tests for the Treatment A versus Treatment B mean comparison at each measurement time lead to the conclusion that the eighth measurement time (Week 14) is the first time at which the profiles are significantly different; see Section 7.4 for more details.

8.3 Case study #2: 100-km race data

The 100-km race data consist of split times for each of 80 competitors on each 10-km section of a 100-km race. Every competitor completed the race, so there are no dropouts or other missing responses. The data also include the ages of all but four of the competitors. The main analysis objective is to obtain a parsimonious model relating competitor performance to section and age.

The profile plot (Figure 1.3) indicates that the mean split time tends to increase at an ever-accelerating rate over the first eight sections of the race, but then level off over the remaining sections. The overall shape of the mean profile does not appear to be too amenable to parametric modeling by a low-order polynomial function. The plot also indicates that split time variances tend to

Table 8.2 *REML information criteria and residual likelihood ratio tests of covariance structures for the Treatment A cattle growth data. The horizontal line in the body of the table separates antedependence models (above the line) from non-antedependence models (below the line). Antedependence models are listed in order of increasing BIC_R.*

Model	d	max log L_R	AIC_R	BIC_R	Comparison model	P
SAD-PM(2)	8	−1034.6	109.75	110.15	SAD-PM(1)	0.00
					AR(2)	0.00
POU(10)	8	−1035.2	109.81	110.21		
SAD-PM(1)	6	−1040.2	110.13	110.42	AR(1)	0.00
SAD-PA(1)	4	−1044.9	110.41	110.61	AR(1)	0.00
AR(2)	3	−1049.0	110.74	110.89		
AR(1)	2	−1050.9	110.83	110.93		
SAD-PA(2)	8	−1043.1	110.64	111.04	SAD-PA(1)	0.49
					AR(2)	0.04
VAD	23	−1027.4	110.57	111.71		
ARH(2)	13	−1042.7	111.13	111.77		
ARH(1)	12	−1045.0	111.26	111.86	AR(1)	0.30
AD(1)	21	−1034.9	111.15	112.19	ARH(1)	0.02
					SAD-PM(1)	0.78
AD(2)	30	−1025.6	111.12	112.61	AD(1)	0.03
					SAD-PM(2)	0.71
AD(10)	66	−1009.7	113.23	116.51	AD(2)	0.67
CS	2	−1190.2	125.49	125.59		
CSH	12	−1160.3	123.40	124.00	CS	0.00
RCL	4	−1076.8	113.77	113.97	CS	0.00
RCQ	7	−1044.2	110.65	111.00	RCL	0.00
TOEP(10)	11	−1040.8	110.72	111.26		
TOEPH(10)	21	−1037.2	111.39	112.43	TOEP(10)	0.96

increase over the entire course of the race, and that the behavior of many runners on later sections of the race is more erratic, in the sense that consecutive same-runner split times fluctuate more. Section-specific scatterplots of split time versus age (Figure 4.1) suggest that age may have a quadratic effect on performance, especially later in the race. In particular, middle-aged runners appear to perform slightly better, on average, than either younger or older runners on later sections.

The sample correlations between split times (Table 4.2) are positive and tend to decline as the number of intervening sections increases, but remain relatively

Table 8.3 *REML information criteria and residual likelihood ratio tests of covariance structures for the Treatment B cattle growth data. The horizontal line in the body of the table separates antedependence models (above the line) from non-antedependence models (below the line). Antedependence models are listed in order of increasing BIC_R.*

Model	d	$\max \log L_R$	AIC_R	BIC_R	Comparison model	P
VAD	27	−1010.7	109.33	110.57		
AD(1)	21	−1022.9	109.88	110.93	ARH(1)	0.00
					SAD-PA(1)	0.00
ARH(1)	12	−1042.7	111.02	111.62	AR(1)	0.00
AD(2)	30	−1016.8	110.19	111.68	AD(1)	0.20
					SAD-PA(2)	0.00
AD(3)	38	−1007.0	110.00	111.89	AD(2)	0.01
					AD(1)	0.02
					SAD-PA(3)	0.00
AD(4)	45	−1002.5	110.26	112.50	AD(3)	0.26
SAD-PA(1)	4	−1070.0	113.05	113.25	AR(1)	0.00
SAD-PA(2)	8	−1069.3	113.40	113.80	SAD-PA(1)	0.84
SAD-PA(3)	13	−1063.4	113.31	113.95	SAD-PA(2)	0.04
AD(10)	66	−989.6	111.12	114.40	AD(3)	0.18
AR(1)	2	−1102.8	116.29	116.39		
CS	2	−1186.9	125.15	125.25		
CSH	12	−1128.9	120.09	120.69	CS	0.00
RCL	4	−1124.7	118.81	119.01	CS	0.00
RCQ	7	−1092.8	115.77	116.12	RCL	0.00
TOEP(10)	11	−1076.1	114.43	114.98		
TOEPH(10)	21	−1034.1	111.06	112.11	TOEP(10)	0.00

strong (greater than 0.4) even for the most widely separated sections. Furthermore, correlations between split times on consecutive sections tend to decrease late in the race. An examination of sample intervenor-adjusted partial correlations and autoregressive coefficients (Table 4.3) reveals that over the first seven sections, the split time immediately preceding any given split time is the only one with which it is significantly associated, when adjusted for the intervening or remaining split times. However, over the last three sections, some split times prior to the immediate predecessor also have significant partial associations with split time. In fact, the tenth split time is significantly partially associated with the fifth split time, which means that runners who perform well on the last section, relative to other runners and their own performance on previous sections, tend to be ones who performed relatively worse on the fifth section. These findings, which the PRISM (Figure 4.2, lower triangle) corroborates, are interesting in and of themselves, but they also suggest that antedependence models up to order five (at least) should be fitted and that antedependence models of constant order are unlikely to fit as well as those of variable order.

To carry out more formal inferences, we fit and compare many antedependence models, of which all have a saturated mean structure. Among the constant-order unstructured AD models, stepwise procedures for likelihood ratio testing select an AD(3), although an AD(5) is a worthy competitor (Section 6.2). On the basis of BIC_R, however, an AD(2) is the best constant-order model (Section 6.5). Furthermore, an AD(0,1,1,1,2,1,1,2,3,5) is the best variable-order AD model, and it is considerably better than its constant-order competitors (Section 6.5). That this VAD model has relatively higher orders late in the race agrees with the previously noted behavior of the sample intervenor-adjusted partial correlations and autoregressive coefficients. Among the structured AD models we fitted are two autoregressively specified power law SAD(2) models, SADL(2) and SADQ(2), for which the innovation variances are given by a linear or quadratic function, respectively, of section. Neither of these two models is competitive. Also included are two structured VAD models, VAD-PM and VAD-AM, of the same orders as the best unstructured VAD; detailed specifications of these two models may be found in Section 6.5. The VAD-PM model is superior to AD(0,1,1,1,2,1,1,2,3,5) on the basis of BIC_R, though it is worth noting that a residual likelihood ratio test emphatically rejects the VAD-PM in favor of the unstructured VAD model. Results for these models are given in Table 8.4.

In addition to the antedependence models, we fitted compound symmetric, heterogeneous compound symmetric, linear and quadratic random coefficient models, and Toeplitz(9) and heterogeneous Toeplitz(9) models. All of these, save TOEPH(9), prove to be vastly inferior to every antedependence model we fitted, save AR(1). Note that once again we do not consider vanishing correlation models, in light of the large magnitudes of all the marginal correlations.

The analysis to this point has taken the mean structure to be saturated. In light of the aforementioned possible quadratic effects of age, we next consider models with a multivariate regression mean structure consisting of section-specific intercepts, linear terms, and quadratic terms, as specified in Section 5.3.2. Due to the missing ages of four of the competitors, however, fitting these models is not completely straightforward. Assuming ignorability, various strategies of missing data analysis, e.g., the EM algorithm or multiple imputation, could be used, but here we opt instead for simply setting the data from those four competitors aside and fitting models to the data from the remaining 76 competitors. The results from fits of saturated mean models to the smaller data set, though slightly different numerically, yield the same ordering of models and hence we

Table 8.4 *REML information criteria and residual likelihood ratio tests of covariance structures for the 100-km race data, for models with saturated mean. The horizontal line in the body of the table separates antedependence models (above the line) from non-antedependence models (below the line). Antedependence models are listed in order of increasing* BIC_R.

Model	d	$\max \log L_R$	AIC_R	BIC_R	Comparison model	P
VAD-PM	14	−2341.4	67.29	67.75		
VAD	27	−2319.6	67.05	67.91	VAD-PM	0.00
VAD-AM	14	−2353.7	67.65	68.09		
AD(2)	27	−2334.1	67.46	68.33	AD(1)	0.00
AD(3)	34	−2323.3	67.35	68.44	AD(2)	0.00
AD(1)	19	−2358.0	67.91	68.52	ARH(1)	0.00
AD(4)	40	−2321.0	67.46	68.74	AD(3)	0.60
ARH(1)	11	−2384.7	68.45	68.80	AR(1)	0.00
AD(5)	45	−2316.1	67.43	68.88	AD(3)	0.21
AD(9)	55	−2310.2	67.58	69.34	AD(3)	0.20
SADQ(2)	10	−2412.0	69.20	69.52	SADL(2)	0.00
SADL(2)	9	−2460.5	70.56	70.85		
AR(1)	2	−2548.6	72.87	72.94		
CS	2	−2671.6	76.39	76.45		
CSH	11	−2547.6	73.10	73.46	CS	0.00
RCL	4	−2562.6	73.33	73.46	CS	0.00
RCQ	7	−2518.0	72.14	72.37	RCL	0.00
TOEP(9)	10	−2529.3	72.55	72.87		
TOEPH(9)	19	−2377.9	68.48	69.09	TOEP(9)	0.00

draw exactly the same conclusions that we drew from the fits of these models to the complete data set. Finally, we test for effects of age on split times, using VAD≡AD(0,1,1,1,2,1,1,2,3,5) as the covariance structure. A likelihood ratio test suggests that we opt for the full quadratic model rather than the smaller, saturated-mean or linear effects models (Table 7.3), and we therefore conclude that age has an important effect on at least some of the ten split times, and that this effect is not merely linear but quadratic. Standard t tests for significance of the linear and quadratic coefficients (Table 5.2) indicate that age has a significant quadratic effect on performance in five of the last six sections, but that no linear effects are significant. The quadratic effect takes the form of better performance among middle-aged runners than among their younger and older counterparts. On the other hand, predecessor-adjusted F tests reveal that when split times on previous sections are taken into account, the third section is the first at which age effects manifest significantly (Table 7.4). Moreover, the predecessor-adjusted linear effect on the third section is negative and more significant, both statistically and practically, than the quadratic effect.

Various antedependence models with mean structures consisting of linear and quadratic age effects and low-order polynomial functions of section were also fitted (e.g., Section 7.4), but they fit poorly and we retain the full quadratic model.

We should note that the PRISM (Figure 4.2, lower triangle) reveals the presence of an outlier and some other mildly interesting departures from the assumptions of the models fitted here. Additional analyses of these data could attempt to account for these features. For example, it is worth seeing what effect(s) the removal of the outlier would have on the analysis. We find that when the outlier is removed, the penalized likelihood approach for selecting the minimum-BIC_R variable-order AD model selects the model of order $(0,1,1,1,1,1,1,2,3,5)$, which differs from the result for the complete-data case only by one degree (order 1 rather than 2) on the fifth section. Furthermore, we find that removal of the outlier has very little effect on the ordering of models in Table 8.4.

Finally, we note that the sample size (number of competitors) is large enough here that tests for the mean structure under the assumption of the VAD-PM or any other well-fitting AD model probably do not have much greater power than their counterparts computed assuming general multivariate dependence. Nevertheless, the analysis from the antedependence-model perspective presented here is useful because it provides insights that analyses from other perspectives do not, such as the conditional dependence of split time on more immediately preceding split times in the last few sections of the race.

8.4 Case study #3: Speech recognition data

The speech recognition data consist of audiologic performance scores on a test administered to 41 profoundly deaf subjects who had received cochlear implants. The implants are of two types, A and B, with 20 subjects receiving type A and 21 receiving type B. Measurement times were scheduled at 1, 9, 18, and 30 months after implant connection, but some subjects dropped out of the study after the second or third measurement. The main analysis objective is to compare the mean profiles of the two implant types.

Figure 1.4 gives profile plots of the data for each type of implant. We observe that both mean profiles increase over time, with the largest increases occurring from $t = 1$ to $t = 9$ months after implantation. The variances appear to increase slightly from the first to second measurement, but remain relatively constant thereafter; furthermore, the variances appear to be similar across types.

Next we examine the two within-type covariance matrices, using a saturated mean for each type. Neither the ordinary scatterplot matrix nor the PRISM indicate any outliers or other gross departures from normality, so we proceed as though the data are multivariate normal. The standard likelihood ratio test for equality of the two within-type covariance matrices does not reject equality ($P = 0.35$). Furthermore, stepwise methods for determining the order of antedependence unequivocally select first-order models for each type (Section 6.4), and a subsequent test of equality across types, assuming first-order antedependence, does not reject it ($P = 0.83$) (Section 6.4). Taken together, these findings indicate that it is sensible to pool the two within-type covariance matrices and to use a first-order antedependence model.

To determine if any SAD(1) models are plausible, we examine in detail the REML estimates of correlations and variances corresponding to the pooled unstructured AD(1) covariance matrix [Table 5.5(b)]. The correlations are positive and large, and they decrease monotonically as the number of intervening measurements increases. Furthermore, correlations along the same subdiagonal tend to increase slightly over the course of the study, notwithstanding the longer interval between measurements as the study progresses. This phenomenon confirms a prior belief of the researchers who performed this study, which is that audiologic performance becomes more consistent over time. It also suggests that a marginally specified SAD(1) power law model may fit the data well. As for the variances, they do not differ greatly, though the first is roughly 70% as large as the average of the other three.

In Section 6.4 we fitted two SAD(1) models to these data, both of which were marginally specified power law models, but which differ with respect to their variance functions. The variance function for the SAD-PMS(1) model is a step function having one value for measurements taken one month after connection and another value for measurements taken at the three remaining times, whereas the variance function for the SAD-PMC(1) model is constant. REML model evaluation criteria for these models are given in Table 8.5. Also included in the table are values of the criteria for stationary and heterogeneous continuous-time AR(1) models, compound symmetric and heterogeneous compound symmetric models, and linear and quadratic random coefficient models. Vanishing correlation models were not included because they obviously do not comport with the large correlations in the correlation matrix. Based on AIC_R and BIC_R, the two SAD(1) models fit best, with a slight edge given to the one with nonconstant variance. The two closest competitors, depending on the criterion used for comparison, are the stationary AR(1) and unstructured AD(1) models. None of the non-antedependence models fit as well as the four models mentioned so far, but the random coefficient models, ARH(1), and AD(3) models fit about equally well.

It is of primary interest, of course, to test various hypotheses about the mean profiles for the two types of implants. In Section 7.2 we found that there is significant improvement in speech recognition for each implant type over time, but that this improvement is not linear: successive improvements get smaller over time. The number of measurement times (four) is too small for much to be gained by adopting a quadratic growth model, so we do not even bother testing for it, leaving the mean unstructured within each type. We also found that implant A has a nearly significantly larger mean profile than implant B, and that these profiles are parallel; that is, the relative improvement from using implant A rather than implant B does not depart significantly from a constant (over time) value of about 10 points on the sentence test. These conclusions are based on an assumed unstructured AD(1) model, but they have been found to be equally valid under the best-fitting model, SAD-PMS(1). The sample size is sufficiently small here that we may gain some power from the use of these first-order models, relative to the use of a general multivariate dependence model.

8.5 Case study #4: Fruit fly mortality data

These data are age-specific measurements of mortality for 112 cohorts of *Drosophila melanogaster*. The cohorts were derived from 56 recombinant inbred lines, each replicated twice. Cohorts consisted of approximately equivalent numbers (500 to 1000) of flies. Every day, dead flies were retrieved from the cage holding each cohort and counted, but these counts were pooled into 11 5-day intervals for analysis. Raw mortality rates were log-transformed to make the responses more normally distributed; the response variable is more precisely defined in Section 1.7.4. Approximately 22% of the data are missing, but all lags are well-replicated. The missingness is not monotone and we assume that it is ignorable. Our analysis objective is to find a parsimonious model that adequately describes how mortality, averaged over recombinant inbred lines, changes with age (time) and how mortality at any given age is related to mortality at previous ages.

From the profile plot (Figure 1.5), it can be seen that the overall mean profile is generally increasing; in fact it is rather sigmoidal. It is also clear that the marginal sample variances increase over the first four time periods and then decrease (Figure 4.3, top panel). The lag-one sample correlations (Figure 4.3, bottom panel) have an interesting quadratic behavior, increasing over the first half of the study and then decreasing. None of the intervenor-adjusted partial correlations or autoregressive coefficients of lag-two and higher are significantly different from zero [Table 4.4(b,c)].

Table 8.5 *REML information criteria and residual likelihood ratio tests of covariance structures for the speech recognition data. The horizontal line in the body of the table separates antedependence models (above the line) from non-antedependence models (below the line). Antedependence models are listed in order of increasing BIC_R.*

Model	q	$\max \log L_R$	AIC_R	BIC_R	Comparison model	P
SAD-PMS(1)	4	−520.1	31.76	31.95	SAD-PMC(1)	0.03
SAD-PMC(1)	3	−522.6	31.85	31.99	AR(1)	0.00
AR(1)	2	−527.1	32.07	32.16		
AD(1)	7	−519.0	31.88	32.20	SAD-PMS(1)	0.55
					ARH(1)	0.01
ARH(1)	5	−523.8	32.05	32.28	AR(1)	0.09
AD(3)	10	−517.7	31.98	32.44	AD(1)	0.44
CS	2	−532.6	32.40	32.49		
CSH	5	−531.1	32.49	32.72	CS	0.39
RCL	4	−527.1	32.19	32.37	CS	0.00
RCQ	7	−520.8	31.99	32.31	RCL	0.01

Stepwise likelihood ratio testing procedures for determining the order of unstructured antedependence by likelihood ratio testing unambiguously select an AD(1) model. Model selection of variable-order AD models by penalized likelihood criteria was not considered, due to the difficulty in implementing this methodology for data with non-monotone missingness. However, several structured first-order AD models were fitted to the data. The SAD-PM3(1), SAD-PA3(1), and SAD-PP3(1) models are all power law models, corresponding respectively to marginal, autoregressive, and precision matrix specifications; the relevant variances (marginal, innovation, or partial) in each model are given by a cubic function of time. The SAD-QM3(1) model is a marginally specified first-order AD model for which the marginal correlations are described by a quadratic function, and the marginal variances by a cubic function, of time. The SAD-QA3(1) model is defined similarly, but with autoregressive coefficients and innovation variances given by quadratic and cubic functions of time. Our consideration of these last two models was motivated by the behavior exhibited by the correlations and variances in Figure 4.3 and (to a lesser degree) by the innovation variances and autoregressive coefficients in Table 4.4(c).

Results of the fits of these and several other models, some antedependent and others not, are displayed in Table 8.6. The two best (minimum *BIC*) models are SAD-QM3(1) and SAD-PM3(1), the former having a slight edge. The best constant-order unstructured AD model is AD(1), which is not rejected at traditional levels of significance by a likelihood ratio test when pitted against AD(2) or AD(10). In light of the non-significance of the autoregressive coefficients at the last two measurement times [Table 4.4(c)], we also fit a variable-order AD(0,1,1,1,1,1,1,1,0,0) model, but it did not fit as well as the AD(1) model (its BIC is 22.71) and we do not include it in the table. The fits of the compound symmetric (homogeneous and heterogeneous) and random coefficient models are much inferior to all the antedependence models. Because of the presence of many near-zero elements in the higher-order subdiagonals of the marginal correlation matrix of these data [Table 4.4(a)], we include Toeplitz, heterogeneous Toeplitz, and unstructured banded (BUN) models of various orders among the non-antedependent alternatives. The best (smallest BIC_R) model of each type is included in the table. We see that TOEPH(6) is competitive, being sixth-best among all fitted models. The Toeplitz and unstructured banded models, however, do not fare as well as most of the antedependence models.

For the analyses described so far, the mean structure was taken to be saturated. Low-order polynomial mean structures could be considered, but the sigmoidal shape of the mean profile suggests that they are not likely to be suitable. This is substantiated by fits of models with several of the best covariance structures and polynomial mean structures up to order five, which are all strongly rejected in favor of an unstructured mean. Nonlinear models for the mean structure may be more suitable, but we do not consider them here; see, however, Section 9.2 and the references cited therein.

8.6 Other studies

A large number of additional studies have been published which have based inferences for continuous univariate longitudinal data on non-trivial antedependence models. Many, but not all, of these are genetic studies. Listing all such studies is not practicable, but we mention the following: Núñez-Antón and Zimmerman (2000) for a reanalysis of Jones' data (Jones, 1990); Jaffrézic and Pletcher (2000) for analyses of reproductive output in fruit flies and growth in beef cattle (different from the cattle growth data in this book); Jaffrézic et al. (2002) for an analysis of milk production in dairy cattle; Jaffrézic et al. (2004) and Albuquerque and Meyer (2005) for analyses of still more cattle growth data sets; Zhao et al. (2005a) for an analysis of growth in mice; Lin et al. (2007) for a study of drug response in humans; Kearsley et al. (2008) for an evaluation of competition data on showjumping and other eventing horses; White et al. (2008) for a study of the effects of fire on wetland water quality. Still more studies have developed and fitted antedependence models for longitudinal data on two or more response variables; see Section 9.4 for references.

Table 8.6 *REML information criteria and residual likelihood ratio tests of covariance structures for the fruit fly mortality data. The horizontal line in the body of the table separates antedependence models (above the line) from non-antedependence models (below the line). Antedependence models are listed in order of increasing BIC_R.*

Model	d	max $\log L_R$	AIC_R	BIC_R	Comparison model	P
SAD-QM3(1)	7	−1096.7	21.86	22.03		
SAD-PM3(1)	6	−1107.5	22.05	22.20		
SAD-QA3(1)	8	−1106.9	22.08	22.28		
AD(1)	21	−1083.2	21.87	22.41	ARH(1)	0.00
					SAD-QM3(1)	0.02
					SAD-PM3(1)	0.00
					SAD-QA3(1)	0.00
					SAD-PP3(1)	0.00
					SAD-PA3(1)	0.00
ARH(1)	12	−1104.5	22.11	22.42	AR(1)	0.00
AD(2)	30	−1080.2	21.98	22.76	AD(1)	0.74
SAD-PP3(1)	6	−1135.8	22.61	22.77		
AD(3)	38	−1073.8	22.02	23.00	AD(2)	0.12
AR(1)	2	−1161.1	23.03	23.08		
SAD-PA3(1)	7	−1158.3	23.08	23.26		
AD(10)	66	−1057.3	22.24	23.95	AD(1)	0.23
CS	2	−1297.5	25.73	25.78		
CSH	12	−1227.7	24.55	24.86	CS	0.00
RCL	4	−1287.0	25.56	25.67		
RCQ	7	−1272.2	25.33	25.51	RCL	0.00
TOEP(5)	6	−1161.9	23.13	23.28		
TOEPH(6)	17	−1105.0	22.22	22.66		
BUN(3)	38	−1084.9	22.24	23.22		

8.7 Discussion

Over the past several decades, many models that either explicitly or implicitly specify parametric covariance structures have been proposed for longitudinal data. Of these, stationary autoregressive (AR) models and random coefficient models seem to get the most attention and use; in contrast, antedependence models (other than AR models) have received very little press. The early focus on AR models is easy to understand: they were well-known from their longstanding importance as models for time series and, due to their extreme parsimony, they were relatively easy to fit. However, with experience proving that stationarity is only occasionally satisfied in practice, nowadays it is clear that AR models cannot serve as useful all-purpose models for longitudinal data, even when those data exhibit serial correlation. It is our view that antedependence models in general can and should play this role. Unstructured antedependence models, because they can be estimated without numerical maximization of the likelihood function, are actually easier to fit than AR models, and any computational obstacles that may once have existed to the fitting of structured antedependence models more general than AR models have long since been removed. In our four case studies, a structured antedependence model that was sufficiently general to accommodate nonstationarity in not merely the variances but also the correlations always performed better than AR models.

The more recent emphasis on random coefficient (RC) models, stimulated by the paper of Laird and Ware (1982) and culminating in the publication of entire books devoted to the subject (see, for example, Verbeke and Molenberghs, 2001), is, in our view, more justified. There is something quite appealing about the notion that the response is a function of covariates (possibly including time) with regression coefficients that vary from one subject to the next, and that this variability can be modeled by a probability distribution. This is particularly so if one wishes to make inferences about the effects of covariates on individual subjects rather than the population. However, RC models should not be used uncritically. We have encountered some analysts of longitudinal data who choose to fit RC models and nothing else; they never actually examine the sample covariance matrix to see how well it agrees with the covariance structure implied by their RC model. RC models in common use, such as the RCL and RCQ models included in our model comparisons in this chapter, impose a great deal of structure on the covariance matrix (see Section 3.9.2), and if the data do not exhibit this structure the RC model likely will not fit the data as well as other models. In the four case studies presented in this chapter, RC models usually did not fit nearly as well as an antedependence model of some kind. Likewise, RC models fared poorly relative to antedependence models in the published studies cited in the previous section. We admit to some selection bias in our choice of case studies, and we suppose that there are many instances where a RC model of some kind fits better than any AD model. Nevertheless, it seems to us a no-brainer that antedependence models should generally be accorded as much consideration as RC models for longitudinal data, especially when the data exhibit serial correlation.

It is natural to wonder whether it would be worthwhile to combine features of a RC model with those of an antedependence model. For example, we could extend the random coefficient model (3.21) to the model

$$\mathbf{Y}_s = \mathbf{X}_s \boldsymbol{\beta} + \mathbf{Z}_s \mathbf{u}_s + \mathbf{e}_s, \qquad s = 1, \ldots, N,$$

where all symbols are defined as they were for model (3.21) and all the same assumptions are made, except that the elements of each $\mathbf{e}_s$ are assumed to be antedependent rather than independent. In the vernacular of RC modeling, this would be a case of a "random coefficient plus residual dependence" model. For what it's worth, we have fitted several such models to the Treatment A cattle growth data and the 100-km race data, but none of them fit nearly as well as the best-fitting antedependence models.

There are many other non-antedependent covariance structures that can be used for longitudinal data. When the data are serially correlated, however, they generally are inferior to antedependence models. The compound symmetry model, with its constant correlation, is obviously a non-starter for such data. Toeplitz models, being stationary, are, like AR models, rarely as useful as AD models. Heterogeneous Toeplitz models, because they allow for arbitrary positive variances, have more potential, and banded models may be useful if the correlation decays very rapidly with elapsed time between measurements. However, if the serial correlation in the data is persistent, meaning that it is substantial even among observations made at the most widely separated times, then AD models will generally perform better than banded models.

CHAPTER 9

Further Topics and Extensions

This book has featured antedependence models for univariate, continuous longitudinal data and has considered likelihood-based estimation and hypothesis
testing for the parameters of these models. However, there are several related
topics that we have not mentioned, and several ways in which the models and
associated inferential methods may be extended. In this chapter, we briefly sur-
vey some of these additional topics and extensions.

9.1 Alternative estimation methods

9.1.1 Nonparametric methods

Nonparametric methods for the estimation of unstructured antedependence
models are possible, either as a guide to the formulation of structured antedependence models or as a basis for formal inference without imposing as much
structure. For AD(p) models, Wu and Pourahmadi (2003) and Huang, Liu, and
Liu (2007) propose nonparametric smoothing of the subdiagonals of the matrix $\mathbf{T}$ of (negative) autoregressive coefficients and a similar smoothing of the
logarithms of the innovation variances. That is, they model the jth subdiagonal
($j = 1, \ldots, p$) of $-\mathbf{T}$ as a smooth function f_j, i.e.,

$$\phi_{i,i-j} = f_j \left(\frac{i-j}{n-j+1} \right), \quad i = j+1, \ldots, n,$$

setting all other subdiagonals to zero, and they model the log innovation variances as another smooth function f_0, i.e.,

$$\log \delta_i = f_0 \left(\frac{i}{n+1} \right), \quad i = 1, \ldots, n.$$

Wu and Pourahmadi (2003) estimate $f_0, f_1, \ldots, f_p$ utilizing local polynomial
smoothing (Fan and Gijbels, 1996), whereas Huang, Liu, and Liu (2007) approximate these functions with B-spline basis functions (de Boor, 2001) and

then estimate them via maximum likelihood. A relative advantage of the second of these two approaches is that it is able to incorporate simultaneous non-parametric estimation of the mean more naturally than the first.

9.1.2 Penalized regression methods

Very recently, a great deal of research activity has focused upon extending penalized regression methods, developed originally for variable selection for the mean structure of regression models, to the estimation of covariance structures. These methods shrink, either partially or completely, the off-diagonal elements of either the sample covariance matrix or precision matrix or the subdiagonal elements of the matrix $\mathbf{T}$ of the modified Cholesky decomposition of the precision matrix. The L_1 penalty of the method known as Lasso is especially useful in this regard, because it shrinks some elements of these matrices all the way to zero, resulting in greater parsimony and interpretability. Yuan and Lin (2007), Friedman, Hastie, and Tibshirani (2008), and Rothman et al. (2008) consider a sparse precision matrix estimate (equivalently a covariance selection or graphical model) obtained by adding an L_1 penalty on the elements of the precision matrix to the normal likelihood function, while Huang et al. (2006) add a similar penalty on the elements of $\mathbf{T}$ instead. Since these two approaches can produce zeros in arbitrary off-diagonal locations in the precision and $\mathbf{T}$ matrices, for index-ordered variables they may or may not result in an antedependence model of any order less than $n - 1$. However, Bickel and Levina (2008) impose an AD(p) covariance structure (although they do not call it such) by forcing $\mathbf{T}$ to be banded, and they obtain theoretical results on consistency without requiring normality. Furthermore, Levina, Rothman, and Zhu (2008) demonstrate how variable-order antedependence (which they call "adaptive banding") may be imposed on the covariance structure through the use of a "nested Lasso" penalty, which requires that $\phi_{ij} = 0$ whenever $\phi_{i,j+1} = 0$ $(i = 2, \ldots, n; \ j = 1, \ldots, i - 1)$.

9.1.3 Bayesian methods

In the Bayesian approach to inference, in addition to specifying a likelihood function for the observed data given an unknown vector of parameters, it is supposed that the unknown parameter vector is a random quantity sampled from a prior distribution. Inference concerning the parameter vector is based entirely on its posterior distribution, obtained as the ratio of the likelihood times the prior, to the integral of the same with respect to the parameter vector. Furthermore, in a hierarchical Bayesian model, the prior distribution may itself depend on additional parameters, known as hyperparameters, which must be either known, integrated out, or, in the case of empirical Bayes inference, estimated from the observed data. Typically, the posterior distribution cannot be given in closed form, but nevertheless it may often be approximated numerically using Markov chain Monte Carlo methods. For a detailed description of these methods, including Gibbs sampling and the Metropolis-Hastings algorithm, we refer the reader to Chapter 5 of Carlin and Louis (2000).

Several implementations of the Bayesian paradigm to the estimation of antedependence model parameters for continuous longitudinal data have been proposed. Though they share a lot in common, these implementations differ with respect to which formulation of antedependence is used and how much structure is imposed on it, and with respect to the priors chosen. For estimating an arbitrary [unstructured AD($n - 1$)] positive definite covariance matrix, Daniels and Pourahmadi (2002) put a multivariate normal prior on the vector ϕ of nontrivial elements of $\mathbf{T}$ and independent inverse gamma priors on each innovation variance of the modified Cholesky decomposition of the precision matrix. These prior distributions yield explicit forms for the full conditional distributions, thereby facilitating Gibbs sampling from the posterior distribution. A similar approach is taken by Cepeda-Cuervo and Núñez-Antón (2007). Daniels and Pourahmadi (2002) also consider Bayesian estimation of the parameters of their structured autoregressively formulated AD($n - 1$) model, which we described in Section 3.5. In this model the autoregressive coefficients are linear functions of parameters θ and the natural logarithms of the innovation variances are linear functions of parameters ψ; and independent normal priors are placed on θ, ψ, and the mean parameters μ. Full conditional distributions of θ and μ are readily sampled via Gibbs steps, but the full conditional distribution of ψ is intractable so it is sampled using a Metropolis-Hastings algorithm. Cepeda and Gamerman (2004) take a similar approach.

The approach of Smith and Kohn (2002) is similar to the first approach of Daniels and Pourahmadi (2002), but an additional parameter γ is introduced, which is a vector of binary random variables of the same length as ϕ that dictates which elements of ϕ are zero and which are nonzero. Through its ability to flexibly identify zeroes in ϕ, this approach, which is motivated by the same considerations that lead to penalized likelihood estimation using the Lasso and similar penalties, allows for more parsimonious modeling of the covariance structure. Smith and Kohn (2002) put independent beta priors on each element of γ and independent inverse gamma priors on the innovation variances, and, conditional on these parameters, they take the prior for ϕ to be proportional to the nth root of the (normal) likelihood. In conjunction with a precision matrix formulation of an arbitrary positive definite matrix, Wong, Carter, and Kohn (2003) introduce a parameter analogous to γ, but which in this case allows for flexible identification of null partial correlations rather than null autoregressive coefficients. They place independent gamma priors on the partial precisions and priors on the (negatives of the) partial correlation coefficients too intricate to describe here.

None of the aforementioned Bayesian approaches explicitly impose antedependence (of order less than $n - 1$) on the model, but they can be modified to do so quite easily, by merely setting the appropriate autoregressive coefficients or partial correlations equal to zero, and placing on the remaining autoregressive coefficients or partial correlations the same types of priors that would otherwise be placed on the entire set of these quantities. Possibilities exist for Bayesian estimation of highly structured, low-order antedependence models as well. For example, for the first-order marginally specified power law model given by (3.12) and (3.13), one could consider putting normal priors on λ and the elements of ψ, an inverse gamma prior on σ^2, and a beta prior on ρ, all priors being independent.

9.2 Nonlinear mean structure

Throughout this book we assumed that the mean structure of the antedependence model was linear, i.e., that $E(Y_{si}) = \mathbf{x}_{si}^T \boldsymbol{\beta}$, where $\mathbf{x}_{si}$ is a vector of covariates observed on subject s at time i and $\boldsymbol{\beta}$ is an unknown parameter vector. This includes cases where the mean is saturated or is a polynomial function of the covariates, with or without classificatory variables indicating group membership or arising as a result of the treatment structure of an experimental design. However, the saturated mean may not be as parsimonious as possible, and polynomials and other parsimonious linear models may not be altogether satisfactory when the mean response asymptotes to an upper or lower bound or exhibits sudden changes in behavior during the study. For such situations a nonlinear mean structure may fit better. An example is given by the logistic mean function,

$$E(Y_{si}) = \frac{\beta_1}{1 + \beta_2 \exp(-\beta_3 t_{si})},$$

where here the only covariate is the time of measurement, t_{si}. Note that the mean response at time 0 under this model is $\beta_1/(1 + \beta_2)$, and that the mean response asymptotes to β_1 as time increases. The upper asymptote makes this model useful for longitudinal studies of animal and plant growth that include times of measurement beyond the organisms' age of maturation. Some additional nonlinear mean functions useful for animal growth studies are summarized by Zimmerman and Núñez-Antón (2001). Applications of a structured antedependence model with logistic mean structure to growth studies are presented by Zhao et al. (2005a,b) and Cepeda-Cuervo and Núñez-Antón (2009).

9.3 Discrimination under antedependence

In the statistical inference problem known as *discrimination*, subjects belong to a number, say G, of groups and the objective of the statistical analysis is to separate, or discriminate between, the groups on the basis of values of n variables observed on each subject. The closely related *classification* problem is concerned with classifying or allocating a subject whose group membership is unknown to one of the G groups, again on the basis of n variables observed on all subjects (including the one whose group membership is unknown). Classical approaches to both problems yield decision rules that involve the (pooled) within-groups sample covariance matrix,

$$\mathbf{S} = \frac{1}{N - g} \sum_{g=1}^{G} \sum_{s=1}^{N(g)} (\mathbf{Y}_{gs} - \overline{\mathbf{Y}}_g)(\mathbf{Y}_{gs} - \overline{\mathbf{Y}}_g)^T,$$

where $N(g)$ and $\overline{\mathbf{Y}}_g$ are the sample size and sample mean vector for group g and $N = \sum_{g=1}^{G} N(g)$; see, for example, Johnson and Wichern (2002, Chapter 11). For example, the classical approach to classification is to allocate a subject having observational vector $\mathbf{Y}$ to the group that yields the smallest value of the quantities

$$D_g^2 = (\mathbf{Y} - \overline{\mathbf{Y}}_g)^T \mathbf{S}^{-1} (\mathbf{Y} - \overline{\mathbf{Y}}_g), \quad g = 1, \ldots, G.$$

For D_g^2 to be well-defined it is necessary, of course, that $\mathbf{S}$ be nonsingular (with probability one), which requires that $N - g \geq n$. Thus, $\mathbf{S}$ will fail to be nonsingular for data of sufficiently high dimension relative to the sample size.

A common approach for circumventing this difficulty is to reduce the dimensionality of the data to its first few principal components, and then to discriminate/classify on the basis of these. An alternative approach, offered by Krzanowski (1993, 1999) and Krzanowksi et al. (1995) for use when the n variables can be ordered in time or along a one-dimensional transect, is to adopt a low-order antedependence model for the data and to replace $\mathbf{S}$ with the REML estimate of the covariance matrix under that model. Krzanowski (1999) found this approach to be very effective. For example, for an application involving $n = 100$ spectroscopic variables on $N = 62$ rice samples, the estimated classification error rate based on an AD(2) covariance matrix was even smaller than that of the standard approach based on principal components. In another example, Levina, Rothman, and Zhu (2008) found their nested Lasso estimate of a variable-order antedependence model to be similarly effective for classification.

9.4 Multivariate antedependence models

An extension of antedependence models to index-ordered random *vectors* (rather than random variables) is straightforward. We consider the normal case only. Let $\mathbf{Y}_1, \ldots, \mathbf{Y}_n$ be r-dimensional random vectors ordered in time or along a transect in space, whose joint distribution is rn-variate normal. Gabriel (1962) defined $\mathbf{Y}_1, \ldots, \mathbf{Y}_n$ to be pth-order antedependent if $\mathbf{Y}_i$ and $(\mathbf{Y}_{i-p-q-1}, \mathbf{Y}_{i-p-q-2}, \ldots, \mathbf{Y}_1)^T$ are independent given $(\mathbf{Y}_{i-1}, \ldots, \mathbf{Y}_{i-p-q})^T$ for all $q = 0, 1, \ldots, n-p-2$ and all $i = p+q+2, \ldots, n$, or in other words, if Definition 2.1 holds for the vectors $\mathbf{Y}_1, \ldots, \mathbf{Y}_n$ (in place of scalars). As in the univariate case, there are several equivalent definitions and representations. For example, if $\boldsymbol{\Sigma} = \text{var}[(\mathbf{Y}_1^T, \ldots, \mathbf{Y}_n^T)^T]$, and its inverse, $\boldsymbol{\Sigma}^{-1}$, is partitioned into $r \times r$ submatrices $\boldsymbol{\Sigma}^{ij}$, then under pth-order antedependence $\boldsymbol{\Sigma}^{ij} = \mathbf{0}$ whenever $|i - j| > p$. Furthermore, we may write such a model in autoregressive form as

$$\mathbf{Y}_i - \boldsymbol{\mu}_i = \sum_{k=1}^{p_i} \boldsymbol{\Phi}_{i,i-k}(\mathbf{Y}_{i-k} - \boldsymbol{\mu}_{i-k}) + \boldsymbol{\epsilon}_i, \quad i = 1, \ldots, n, \tag{9.1}$$

where $p_i = \min(p, i - 1)$ and the $\boldsymbol{\epsilon}_i$'s are independent and identically distributed r-variate normal vectors with mean zero and positive definite covariance matrix, analogous to equation (2.21).

A few specific structured multivariate antedependence models have previously been put forward. For example, Jaffrézic, Thompson, and Hill (2003) consider a bivariate, autoregressively formulated structured first-order antedependence model in which the autoregressive coefficients are constant over time and the innovations are contemporaneously (but not otherwise) cross-correlated. More specifically, they consider the special case of model (9.1) in which $p = 1$,

$$\boldsymbol{\Phi}_{i,i-1} = \begin{pmatrix} \phi_{11} & \phi_{12} \\ \phi_{21} & \phi_{22} \end{pmatrix} \quad \text{for } i = 2, \ldots, n,$$

and the $\boldsymbol{\epsilon}_i$'s are independent with

$$\text{var}(\boldsymbol{\epsilon}_i) = \begin{pmatrix} \sigma_{i1}^2 & \rho_i(\sigma_{i1}^2 \sigma_{i2}^2)^{1/2} \\ \rho_i(\sigma_{i1}^2 \sigma_{i2}^2)^{1/2} & \sigma_{i2}^2 \end{pmatrix},$$

where $\sigma_{ik}^2 = \exp(\theta_{1k} + \theta_{2k}i + \theta_{3k}i^2)$ for $i = 1, \ldots, n$ and $k = 1, 2$, and $\rho_i = \exp(-\lambda_1 i) - \exp(-\lambda_2 i)$. Zhao et al. (2005b) specialize this model further by assuming that: (a) the cross-autoregressive coefficients are zero, i.e., $\phi_{12} = \phi_{21} = 0$; (b) the innovation variances are constant over time, i.e., $\sigma_{i1}^2 \equiv \sigma_1^2$ and $\sigma_{i2}^2 \equiv \sigma_2^2$; and (c) the innovation cross-correlation is constant over time, i.e., $\rho_i \equiv \rho$. Under these additional assumptions, the marginal variances, correlations, and cross-correlation are given by the following expressions:

$$\text{var}(Y_{ik}) = \frac{1 - \phi_{kk}^{2i}}{1 - \phi_{kk}^2} \sigma_k^2 \quad \text{for } i = 1, \ldots, n \text{ and } k = 1, 2$$

$$\text{corr}(Y_{ik}, Y_{jk}) = \phi_{kk}^{i-j} \sqrt{\frac{1 - \phi_{kk}^{2j}}{1 - \phi_{kk}^{2i}}}, \quad i > j$$

$$\text{corr}(Y_{i1}, Y_{j2}) = \begin{cases} \frac{\rho(\phi_{11}^{i-j} - \phi_{11}^{i}\phi_{22}j)}{1 - \phi_{11}\phi_{22}} \sqrt{\frac{(1-\phi_{11}^{2})(1-\phi_{22}^{2})}{(1-\phi_{11}^{2i})(1-\phi_{22}^{2j})}} & \text{for } i > j, \\ \frac{\rho(\phi_{22}^{j-i} - \phi_{11}^{j}\phi_{22}i)}{1 - \phi_{11}\phi_{22}} \sqrt{\frac{(1-\phi_{11}^{2})(1-\phi_{22}^{2})}{(1-\phi_{11}^{2j})(1-\phi_{22}^{2i})}} & \text{for } j > i. \end{cases}$$

Observe that the cross-correlation function is not symmetric, i.e., $\text{corr}(Y_{i1}, Y_{j2}) \neq \text{corr}(Y_{i2}, Y_{j1})$ for $i > j$.

Methods of inference for multivariate antedependence models are straightforward extensions of methods for the univariate case. Byrne (1996), for example, describes maximum likelihood estimation under the unstructured multivariate AD(p) model and derives tests of a general linear mean structure under this model. Gabriel (1962) gives the likelihood ratio test for the order of a multivariate unstructured antedependence model. Jaffrézic, Thompson, and Hill (2003) obtain REML estimates of their bivariate, autoregressively formulated structured antedependence model for bivariate responses from two longitudinal studies; one of these is the fruit fly study, for which they analyzed reproductive output in addition to the mortality data that were among the four featured data sets in this book. Zhao et al. (2005b) fit their simplified version of this model to stem height and radial diameter of *Populus* trees by maximum likelihood; see also Wu and Hou (2006). Lin and Wu (2005) fit a slightly different bivariate SAD(1) model to measures of drug efficacy and toxicity over a sequence of doses or concentrations. Jaffrézic, Thompson, and Pletcher (2004) embed marginally formulated power law SAD(1) models for each response variable in their bivariate "character process" model for genetic evaluation, and apply it to the fruit fly reproductive output and mortality data.

9.5 Spatial antedependence models

As we have noted several times already, the antedependence models considered in this book are applicable to continuous observations taken over time or along a one-dimensional transect. It is natural to consider whether and how they may be extended to continuous observations taken over two-dimensional (or higher) space. In fact, such extensions already exist and are known as spatial autoregressive models. Though originally developed as stationary models on doubly infinite regular lattices (see, e.g., Whittle, 1954; Besag, 1974), they have been applied most often to finite irregular lattices, such as those formed by a division of a geographic region into political or administrative subregions, for which they yield nonstationary covariance structures (Haining, 1990; Besag and Kooperberg, 1995; Wall, 2004) — a fact that has not always been recognized. Thorough treatments of spatial autoregressive models can be found in many spatial statistics books (see, e.g., Cressie, 1993; Banerjee, Carlin, and Gelfand, 2004), so we do not attempt another one here. We confine ourselves in this section to a brief review of the Gaussian cases, with an emphasis on their relationships to normal antedependence models.

Gaussian spatial autoregressive models are of two types: the simultaneous autoregression (SAR) and conditional autoregression (CAR). Both types can be viewed as extensions of normal antedependence models to spatial data, but only the CAR has the distinctive Markov property of such models. In the following descriptions, let D be a geographic region and let subregions $A_1, \ldots, A_n$ be a finite partition of D, i.e., $A_1 \cup A_2 \cup \cdots \cup A_n = D$ and $A_i \cap A_j = \emptyset$ for all $i \neq j$. The response on subregion A_i, which is usually a count or areal average, is denoted by $Y(A_i)$.

The Gaussian SAR model postulates that

$$Y(A_i) = \mu_i + \sum_{j=1}^{n} b_{ij}[Y(A_j) - \mu_j] + \epsilon_i, \quad i = 1, \ldots, n, \qquad (9.2)$$

where $\epsilon = (\epsilon_1, \ldots, \epsilon_n)^T \sim N(\mathbf{0}, \mathbf{\Lambda})$ with $\mathbf{\Lambda}$ diagonal, $E[Y(A_i)] = \mu_i$, and the $\{b_{ij}\}$ are known or unknown constants satisfying $b_{ii} = 0$ for all i. Typically, b_{ij} is taken to equal 0 unless A_i and A_j are "neighbors," i.e., regions that are in sufficiently close proximity for the corresponding responses to influence each other. Putting $\mathbf{B} = (b_{ij})$, we may write (9.2) in matrix form as

$$\epsilon = (\mathbf{I} - \mathbf{B})(\mathbf{Y} - \boldsymbol{\mu}) \qquad (9.3)$$

where $\mathbf{Y} = (Y(A_1), \ldots, Y(A_n))^T$ and $\boldsymbol{\mu} = (\mu_1, \ldots, \mu_n)^T$. Upon comparison of (9.3) and (2.15), we see that $\mathbf{I} - \mathbf{B}$ plays the same role that the unit lower triangular matrix, $\mathbf{T}$, plays in (2.15), and that

$$\mathbf{Y} \sim \mathrm{N}\left(\boldsymbol{\mu}, (\mathbf{I} - \mathbf{B})^{-1}\mathbf{\Lambda}(\mathbf{I} - \mathbf{B}^T)^{-1}\right).$$

However, while $\mathbf{I} - \mathbf{B}$ has ones along its main diagonal, it is generally not lower triangular, and consequently $Y(A_i)$ and $\{Y(A_j) : b_{ij} = 0\}$ are generally not conditionally independent, given $\{Y(A_j) : b_{ij} \neq 0\}$. Thus, the distinctive Markov property of antedependence models does not hold for Gaussian SAR models.

On the other hand, the Gaussian CAR model specifies that

$$Y(A_i)|Y(A_{-i}) \sim \mathrm{N}\left(\mu_i + \sum_{j=1}^{n} c_{ij}[Y(A_j) - \mu_j], \kappa_i^2\right)$$

where $Y(A_{-i}) = \{Y(A_j) : j \neq i\}$, $E[Y(A_i)] = \mu_i$, κ_i^2 is the ith conditional variance, and the $\{c_{ij}\}$ are known or unknown constants satisfying $c_{ii} = 0$ for all i. Like b_{ij}, c_{ij} is typically taken to equal 0 unless A_i and A_j are neighbors.

It can be shown (see Besag, 1974) that these conditional distributions imply a joint distribution

$$\mathbf{Y} \sim \mathrm{N}\left(\boldsymbol{\mu}, (\mathbf{I} - \mathbf{C})^{-1}\mathbf{K}\right)$$

where $\mathbf{C} = (c_{ij})$ and $\mathbf{K} = \mathrm{diag}(\kappa_1^2, \ldots, \kappa_n^2)$, provided that $\mathbf{I} - \mathbf{C}$ is positive definite and that $(\mathbf{I} - \mathbf{C})^{-1}\mathbf{K}$ is symmetric, or equivalently that $c_{ij}\kappa_j^2 = c_{ji}\kappa_i^2$. Thus, the inverse of the covariance matrix of $\mathbf{Y}$ is $\boldsymbol{\Sigma}^{-1} = \mathbf{K}^{-1}(\mathbf{I} - \mathbf{C})$, the elements σ^{ij} of which satisfy $\sigma^{ij} = 0$ whenever $c_{ij} = 0$. This structure of the precision matrix is completely analogous to that associated with a variable-order normal antedependence model for index-ordered random variables in one dimension, save for the following differences, which are attributable to the lack of a linear ordering of the variables in two or more dimensions: the zero elements of a row of the precision matrix do not necessarily begin with the first element within that row, nor are they necessarily contiguous within rows, as they are for a (variable-order) AD model for longitudinal data (cf. the discussion immediately following Theorem 2.7).

Typically, only one realization of the process is observed on the lattice, so in practice only highly parsimonious (structured) CAR models have actually been fitted to data. For example, one commonly used CAR model takes $\mathbf{C} = \rho\mathbf{W}$, where ρ is a scalar-valued "spatial dependence" parameter and $\mathbf{W}$ is a user-defined spatial neighbor incidence matrix that indicates whether regions are neighbors or not; often, neighbors are defined by adjacencies and the (i, j)th element of $\mathbf{W}$ is therefore taken to equal 1 if A_i and A_j share a common border, and 0 otherwise. A slightly more general formulation allows for anisotropy by replacing ρ with either ρ_{EW} or ρ_{NS} according to whether A_i and A_j are in east–west or north–south alignment (approximately) with each other. In those uncommon situations in which multiple independent realizations of the spatial process are available, more complicated CAR models may be fitted. It may be possible, for example, to fit a CAR model for which the spatial dependence parameter varies as a smooth function of spatial location. With a sufficiently large number of realizations, it is even possible to fit a completely unstructured Gaussian CAR model of specified order (i.e., one with the null c_{ij}'s specified but the values of the non-null c_{ij}'s left unspecified) using, with only slight modifications, the estimation methodology for variable-order AD models presented in Chapter 5.

9.6 Antedependence models for discrete data

Antedependence models exist for discrete longitudinal data as well as for continuous data, though it seems they are never called by this name. Instead, some names used in the literature are discrete Markov or discrete transition models for the general case, and autologistic and auto-Poisson for some specific cases. Diggle et al. (2002, pp. 190–207) and Molenberghs and Verbeke (2005, pp. 236–238) describe a useful class of such models, which are obtained by specifying a generalized linear model for the conditional distribution of each response given its "history," i.e., its predecessors and the present and past values of observed covariates. We summarize this class here and then discuss a possible extension. Let

$$H_{si} \equiv \{Y_{s,i-1}, \dots, Y_{s1}, \mathbf{x}_{si}, \dots, \mathbf{x}_{s1}\}$$

denote the history of Y_{si}, and define

$$\mu_{si}^C = E(Y_{si}|H_{si}) \quad \text{and} \quad (\sigma_{si}^C)^2 = \text{var}(Y_{si}|H_{si});$$

here the superscript C refers to conditional. Assume that the data are balanced and that

$$h(\mu_{si}^C) = \mathbf{x}_{si}^T \boldsymbol{\beta} + \sum_{k=1}^{p} f_k(H_{si}; \boldsymbol{\phi}, \boldsymbol{\beta}), \quad i = p+1, \dots, n, \qquad (9.4)$$

where h is a "link" function, the f_k's are functions of previous responses, and $\boldsymbol{\phi}$ is a vector of p parameters. Assume further that

$$(\sigma_{si}^C)^2 = \delta g(\mu_{si}^C), \quad i = p+1, \dots, n, \qquad (9.5)$$

where g is a variance function and δ is an unknown dispersion parameter. The joint distribution of the first p responses is left unspecified.

Several special cases of model (9.4) and (9.5) are of interest. The case in which Y_{si} is Gaussian, h is the identity link, $f_k(H_{si}; \boldsymbol{\phi}, \boldsymbol{\beta}) = \phi_k(Y_{s,i-k} - \mathbf{x}_{s,i-k}^T\boldsymbol{\beta})$, and $g(\mu_{si}^C) \equiv 1$ is identical to the stationary normal autoregressive model of order p, given by (3.3) with $\mu_i = \mathbf{x}_{si}^T\boldsymbol{\beta}$. For binary data, letting $f_k(H_{si}; \boldsymbol{\phi}, \boldsymbol{\beta}) = \phi_k Y_{s,i-k}$ and using a logit link gives

$$\text{logit}(\mu_{si}^C) = \log\left(\frac{\mu_{si}^C}{1 - \mu_{si}^C}\right) = \mathbf{x}_{si}^T\boldsymbol{\beta} + \sum_{k=1}^{p} \phi_k Y_{s,i-k},$$

$g(\mu_{si}^C) = \mu_{si}^C(1 - \mu_{si}^C)$, and $\delta = 1$. This is the Markov model for a binary time series suggested by Cox and Snell (1989), extended for use with multiple subjects; by analogy with a similar CAR model on a two-dimensional spatial lattice, it might be called an *autologistic* model of order p (see Cressie, 1993, pp. 423–427). For count data, if one assumes that $Y_{si}|H_{si}$ is Poisson, uses a log link, and takes $f_k(H_{si}; \boldsymbol{\phi}, \boldsymbol{\beta}) = \phi_k Y_{s,i-k}$ as in the autologistic model, one obtains a longitudinal analogue of the classical *auto-Poisson model* for spatial lattice data (Cressie, pp. 427–431). Two alternative auto-Poisson models, obtained using different f_k's, are considered in detail by Zeger and Qaqish (1988). Analogously defined "auto-" models for multinomial and ordered categorical data are also considered by Diggle et al. (2002).

Note that the class of models proposed by Diggle et al. (2002) takes each function f_k and the innovation dispersion parameter δ to be time-invariant. These restrictions are the reason why, in the Gaussian case with identity link, an AR(p) model, rather than an unstructured normal antedependence model, is obtained. It seems natural to consider an extension of the class to one in which the Gaussian case with identity link yields the unstructured normal antedependence model of order p. Such a class is obtained if (9.4) is replaced by

$$h(\mu_{si}^C) = \mathbf{x}_{si}^T\boldsymbol{\beta} + \sum_{k=1}^{p_i} f_{ik}(H_{si}; \boldsymbol{\phi}, \boldsymbol{\beta}), \quad i = 1, \ldots, n, \qquad (9.6)$$

and (9.5) is replaced by

$$(\sigma_{si}^C)^2 = \delta_i g(\mu_{si}^C) \quad i = 1, \ldots, n. \qquad (9.7)$$

If we put $f_{ik}(H_{si}; \boldsymbol{\phi}, \boldsymbol{\beta}) = \phi_{i,i-k}(Y_{s,i-k} - \mathbf{x}_{s,i-k}^T\boldsymbol{\beta})$ in (9.6) and $g(\mu_{si}^C) \equiv 1$ in (9.7), we obtain, in the Gaussian case with identity link, the unstructured normal AD(p) model with linear mean structure given by (2.21) with $\mu_i = \mathbf{x}_{si}^T\boldsymbol{\beta}$, as desired. In addition to being more general than the class of models proposed by Diggle et al. (2002), (9.6) and (9.7) has the advantage of providing, in non-Gaussian cases, a model for each observation, not merely for those observed after time p.

For either class of models, likelihood-based inferences are possible. For the class proposed by Diggle et al. (2002), the full likelihood function is not available (except in the Gaussian case), but inferences may be based on the conditional (on the first p responses on each subject) likelihood function

$$\prod_{s=1}^{N} \prod_{i=p+1}^{n} l(Y_{si} | H_{si}; \boldsymbol{\phi}, \boldsymbol{\beta}, \boldsymbol{\delta}),$$

where l is the conditional (on history) probability mass function of Y_{si}. For further details see Diggle et al. (2002, pp. 192-194). The full likelihood function is available for the extended class.

One drawback of these classes of models, or of any class of Markov models in which the effect of covariates on responses is specified via a conditional mean (i.e., conditional on the previous history of responses), is that the population-averaged effect of covariates is specified indirectly and in such a way that the interpretation of regression coefficients ($\boldsymbol{\beta}$) changes as assumptions regarding the conditional dependence structure [e.g., the order p or the choice of functions f_{ik} in (9.6)] are modified. This is in contrast to so-called marginal models, in which the population-averaged effect of covariates on the response is directly specified and interpretation of regression coefficients does not depend on specification of the dependence in the model. However, it is possible to remove this practical shortcoming by separating the Markov model into two parts and reparameterizing. The first part is a marginal mean model that directly specifies the population-averaged effect of covariates on the response, while the second part is a conditional mean model that describes serial dependence and identifies the joint distribution of the responses but specifies the dependence on covariates only implicitly. Such a two-part, reparameterized version of the model is called a marginalized transition model (MTM) (Heagerty and Zeger, 2000). For example, for binary data, Heagerty (2002), building on earlier work of Azzalini (1994), proposes the following MTM, which he labels the MTM(p):

$$\text{logit}(\mu_{si}^M) = \mathbf{x}_{si}^T\boldsymbol{\beta}, \quad i = 1, \ldots, n$$

$$\text{logit}(\mu_{si}^C) = \Delta_{si} + \sum_{k=1}^{p} \phi_{sik}Y_{s,i-k}, \quad i = p+1, \ldots, n$$

$$\phi_{sik} = \mathbf{z}_{sik}^T\boldsymbol{\alpha}_k, \quad i = p+1, \ldots, n.$$

Here $\mu_{si}^M = E(Y_{si}|\mathbf{x}_{si})$ (the superscript M referring to marginal), Δ_{si} is an intercept parameter, ϕ_{sik} is a subject-specific autoregressive coefficient, $\mathbf{z}_{sik}$ is a vector of covariates on subject s which are a subset of the covariates in $\mathbf{x}_{si}$, and $\boldsymbol{\alpha}_k$ is a parameter vector. It is assumed that the regression model properly specifies the full covariate conditional mean (Pepe and Anderson, 1994), such that $E(Y_{si}|\mathbf{x}_{si}) = E(Y_{si}|\mathbf{x}_{s1}, \ldots, \mathbf{x}_{sn})$. Parameters $\boldsymbol{\beta}$ and $\boldsymbol{\alpha}_k$ in this model are unconstrained, but Δ_{si} is fully constrained and must yield the proper marginal expectation μ_{si}^M when μ_{si}^C is averaged over the distribution of H_{si}; for further details see Heagerty (2002). Estimation of the unconstrained parameters may proceed via maximizing the full likelihood, using lower order MTM model assumptions for the first p responses on each subject. For ordinal categorical data, Lee and Daniels (2007) propose two extensions of the MTM(p) and a similar maximum likelihood estimation approach.

In our view, marginalized transition models appear to be a very promising class of antedependence models for discrete longitudinal data. Because the marginal mean structure is specified directly, and separately from the dependence structure, interpretation of the regression coefficients is invariant to modifications of the assumptions regarding the autoregressive parameters. A further advantage of these models over ordinary transition models is that estimates of covariate effects are more robust to mis-specification of dependence (Heagerty and Zeger, 2000; Heagerty, 2002; Lee and Daniels, 2007). It would appear that the constant-order versions of these models could be extended without difficulty to variable-order models.

Appendix 1: Some Matrix Results

We give here some matrix results used in the book. For some of these results, proofs are provided; for others a reference is provided where a proof may be found.

Theorem A.1.1. *Let $\mathbf{A}$ be an $n \times n$ nonsingular matrix that is partitioned as*

$$\mathbf{A} = \left(\begin{array}{cc} \mathbf{A}_{11} & \mathbf{A}_{12} \\ \mathbf{A}_{21} & \mathbf{A}_{22} \end{array} \right),$$

where $\mathbf{A}_{11}$ is $n_1 \times n_1$, $\mathbf{A}_{12}$ is $n_1 \times n_2$, $\mathbf{A}_{21}$ is $n_2 \times n_1$, and $\mathbf{A}_{22}$ is $n_2 \times n_2$. Let $\mathbf{B} = \mathbf{A}^{-1}$ and partition $\mathbf{B}$ as

$$\mathbf{B} = \left(\begin{array}{cc} \mathbf{B}_{11} & \mathbf{B}_{12} \\ \mathbf{B}_{21} & \mathbf{B}_{22} \end{array} \right),$$

where the submatrices of $\mathbf{B}$ have the same dimensions as the corresponding submatrices of $\mathbf{A}$. Then, if $\mathbf{A}_{11}$, $\mathbf{A}_{22}$, $\mathbf{A}_{11\cdot2} \equiv \mathbf{A}_{11} - \mathbf{A}_{12}\mathbf{A}_{22}^{-1}\mathbf{A}_{21}$, and $\mathbf{A}_{22\cdot1} \equiv \mathbf{A}_{22} - \mathbf{A}_{21}\mathbf{A}_{11}^{-1}\mathbf{A}_{12}$ are nonsingular, we have

(a) $\mathbf{B}_{11} = \mathbf{A}_{11\cdot2}^{-1} = \mathbf{A}_{11}^{-1} + \mathbf{A}_{11}^{-1}\mathbf{A}_{12}\mathbf{B}_{22}\mathbf{A}_{21}\mathbf{A}_{11}^{-1}$,

(b) $\mathbf{B}_{22} = \mathbf{A}_{22}^{-1} + \mathbf{A}_{22}^{-1}\mathbf{A}_{21}\mathbf{B}_{11}\mathbf{A}_{12}\mathbf{A}_{22}^{-1} = \mathbf{A}_{22\cdot1}^{-1}$,

(c) $\mathbf{B}_{12} = -\mathbf{B}_{11}\mathbf{A}_{12}\mathbf{A}_{22}^{-1} = -\mathbf{A}_{11}^{-1}\mathbf{A}_{12}\mathbf{B}_{22}$,

(d) $\mathbf{B}_{21} = -\mathbf{A}_{22}^{-1}\mathbf{A}_{21}\mathbf{B}_{11} = -\mathbf{B}_{22}\mathbf{A}_{21}\mathbf{A}_{11}^{-1}$,

(e) $\mathbf{A}_{12}\mathbf{A}_{22}^{-1} = -\mathbf{B}_{11}^{-1}\mathbf{B}_{12}$ *and* $\mathbf{A}_{21}^{-1}\mathbf{A}_{11}^{-1} = -\mathbf{B}_{22}^{-1}\mathbf{B}_{21}$.

(f) $\mathbf{A}_{11}^{-1} = (\mathbf{B}_{11} - \mathbf{B}_{12}\mathbf{B}_{22}^{-1}\mathbf{B}_{21})$ *and* $\mathbf{A}_{22}^{-1} = (\mathbf{B}_{22} - \mathbf{B}_{21}\mathbf{B}_{11}^{-1}\mathbf{B}_{12})$.

Proof. A proof of parts (a) through (d) may be found in Schott (2005, pp. 256–257) or Harville (1997, pp. 99–100). Part (e) follows by multiplying the equation in part (c) by $-\mathbf{B}_{11}^{-1}$ and multiplying the equation in part (d) by $\mathbf{B}_{22}^{-1}$. For part (f), observe that $\mathbf{A} = \mathbf{B}^{-1}$ and thus interchanging the roles of $\mathbf{A}$ and $\mathbf{B}$ in part (a) yields $\mathbf{A}_{11} = (\mathbf{B}_{11} - \mathbf{B}_{12}\mathbf{B}_{22}^{-1}\mathbf{B}_{21})^{-1}$. Inverting the matrices on both sides of this equation yields the first equation in part (f). The second equation in part (f) can be verified in similar fashion using part (b). $\square$

Corollary A.1.1.1. *Suppose that $\mathbf{A} = (a_{ij})$ and $\mathbf{B} = (b_{ij})$ are defined as in Theorem A.1.1, with $n_1 = 1$, and that the nonsingularity conditions specified* *there hold. Then*

$$\mathbf{A}_{12}\mathbf{A}_{22}^{-1} = (-b_{12}/b_{11}, -b_{13}/b_{11}, \ldots, -b_{1n}/b_{11})^T.$$

Proof. The corollary follows immediately from Theorem A.1.1(e). □

Corollary A.1.1.2. *Suppose that* $\mathbf{A} = (a_{ij})$ *and* $\mathbf{B} = (b_{ij})$ *are defined as in Theorem A.1, with* $n_1 = 2$, *and that the nonsingularity conditions specified there hold. Suppose further that* $\mathbf{A}$ *is symmetric. Let* $a_{ij\cdot 2}$ *denote the* (i, j)*th element of* $\mathbf{A}_{11\cdot 2}$ $(i, j = 1, 2)$. *Then*

$$a_{12\cdot 2}/(a_{11\cdot 2}a_{22\cdot 2})^{1/2} = -b_{12}/(b_{11}b_{22})^{1/2}.$$

Proof. Let c_{ij} denote the (i, j)th element of $\mathbf{A}_{12}\mathbf{A}_{22}^{-1}\mathbf{A}_{21}$. Then by Theorem A.1.1(a) and the symmetry of $\mathbf{A}$,

$$\mathbf{A}_{11\cdot 2} = \begin{pmatrix} a_{11} - c_{11} & a_{12} - c_{12} \\ a_{12} - c_{12} & a_{22} - c_{22} \end{pmatrix} = \begin{pmatrix} b_{11} & b_{12} \\ b_{12} & b_{22} \end{pmatrix}^{-1}$$
$$\propto \begin{pmatrix} b_{22} & -b_{12} \\ -b_{12} & b_{11} \end{pmatrix},$$

from which the corollary immediately follows. □

Corollary A.1.1.3. *Let* $\mathbf{A}$ *be an* $n \times n$ *symmetric positive definite matrix that is partitioned as*

$$\mathbf{A} = \begin{pmatrix} \mathbf{A}_{11} & \mathbf{A}_{12} & \mathbf{A}_{13} \\ \mathbf{A}_{21} & \mathbf{A}_{22} & \mathbf{A}_{23} \\ \mathbf{A}_{31} & \mathbf{A}_{32} & \mathbf{A}_{33} \end{pmatrix},$$

where $\mathbf{A}_{11}$ *is* $n_1 \times n_1$, $\mathbf{A}_{12}$ *is* $n_1 \times n_2$, $\mathbf{A}_{13}$ *is* $n_1 \times n_3$, $\mathbf{A}_{21}$ *is* $n_2 \times n_1$, $\mathbf{A}_{22}$ *is* $n_2 \times n_2$, $\mathbf{A}_{23}$ *is* $n_2 \times n_3$, $\mathbf{A}_{31}$ *is* $n_3 \times n_1$, $\mathbf{A}_{32}$ *is* $n_3 \times n_2$, *and* $\mathbf{A}_{33}$ *is* $n_3 \times n_3$. *Let*

$$
\begin{aligned}
\mathbf{A}_{11\cdot 3} &= \mathbf{A}_{11} - \mathbf{A}_{13}\mathbf{A}_{33}^{-1}\mathbf{A}_{31}, \\
\mathbf{A}_{12\cdot 3} &= \mathbf{A}_{12} - \mathbf{A}_{13}\mathbf{A}_{33}^{-1}\mathbf{A}_{32}, \\
\mathbf{A}_{21\cdot 3} &= \mathbf{A}_{21} - \mathbf{A}_{23}\mathbf{A}_{33}^{-1}\mathbf{A}_{31}, \\
\mathbf{A}_{22\cdot 3} &= \mathbf{A}_{22} - \mathbf{A}_{23}\mathbf{A}_{33}^{-1}\mathbf{A}_{32}, \\
\mathbf{A}_{11\cdot 2,3} &= \mathbf{A}_{11} - (\mathbf{A}_{12}, \mathbf{A}_{13}) \begin{pmatrix} \mathbf{A}_{22} & \mathbf{A}_{23} \\ \mathbf{A}_{32} & \mathbf{A}_{33} \end{pmatrix}^{-1} \begin{pmatrix} \mathbf{A}_{31} \\ \mathbf{A}_{32} \end{pmatrix}.
\end{aligned}
$$

Then $\mathbf{A}_{11\cdot 2,3} = \mathbf{A}_{11\cdot 3} - \mathbf{A}_{12\cdot 3}\mathbf{A}_{22\cdot 3}^{-1}\mathbf{A}_{21\cdot 3}$.

Proof. Consider two partitions of $\mathbf{A}$: one in which the upper-left block is $\mathbf{A}_{11}$, and another in which the lower right block is $\mathbf{A}_{33}$. Applying Theorem A.1.1(a) to the first partition, we have that the upper left $n_1 \times n_1$ block of $\mathbf{A}^{-1}$ is equal to $\mathbf{A}_{11\cdot 2,3}^{-1}$. Applying the same theorem to the second partition, we have that the upper left $n_1 \times n_1$ block of $\mathbf{A}^{-1}$ is also equal to the upper left $n_1 \times n_1$ block of

$$\left(\begin{array}{cc} \mathbf{A}_{11 \cdot 3} & \mathbf{A}_{12 \cdot 3} \\ \mathbf{A}_{21 \cdot 3} & \mathbf{A}_{22 \cdot 3} \end{array} \right)^{-1}.$$

But by the same theorem once again, the latter block is equal to $(\mathbf{A}_{11 \cdot 3} - \mathbf{A}_{12 \cdot 3} \mathbf{A}_{22 \cdot 3}^{-1} \mathbf{A}_{21 \cdot 3})^{-1}$. Thus $\mathbf{A}_{11 \cdot 2,3}^{-1} = (\mathbf{A}_{11 \cdot 3} - \mathbf{A}_{12 \cdot 3} \mathbf{A}_{22 \cdot 3}^{-1} \mathbf{A}_{21 \cdot 3})^{-1}$, and the corollary then follows by matrix inversion. $\square$

Theorem A.1.2. (Modified Cholesky decomposition theorem). *Let* $\mathbf{A}$ *be a symmetric positive definite matrix. Then there exists a unique lower triangular matrix,* $\mathbf{T}$, *having a main diagonal of all ones, and a unique diagonal matrix,* $\mathbf{D}$, *having positive elements on the main diagonal, such that* $\mathbf{A}^{-1} = \mathbf{T}^T \mathbf{D}^{-1} \mathbf{T}$, *or equivalently* $\mathbf{T} \mathbf{A} \mathbf{T}^T = \mathbf{D}$.

Proof. A proof may be found in Harville (1997, pp. 228–229).

Theorem A.1.3. *Let* $\mathbf{A}$ *be an* $n \times n$ *nonsingular matrix that is partitioned as*

$$\mathbf{A} = \left(\begin{array}{cc} \mathbf{A}_{11} & \mathbf{A}_{12} \\ \mathbf{A}_{21} & \mathbf{A}_{22} \end{array} \right),$$

where $\mathbf{A}_{11}$ *is* $n_1 \times n_1$, $\mathbf{A}_{12}$ *is* $n_1 \times n_2$, $\mathbf{A}_{21}$ *is* $n_2 \times n_1$, *and* $\mathbf{A}_{22}$ *is* $n_2 \times n_2$. *If* $\mathbf{A}_{11}$ *is nonsingular, then*

$$|\mathbf{A}| = |\mathbf{A}_{11}||\mathbf{A}_{22} - \mathbf{A}_{21} \mathbf{A}_{11}^{-1} \mathbf{A}_{12}|.$$

Similarly, if $\mathbf{A}_{22}$ *is nonsingular, then*

$$|\mathbf{A}| = |\mathbf{A}_{22}||\mathbf{A}_{11} - \mathbf{A}_{12} \mathbf{A}_{22}^{-1} \mathbf{A}_{21}|.$$

Proof. A proof may be found in Harville (1997, pp. 188–189).

Next we give two theorems used to prove Theorem 5.1.

Theorem A.1.4. *Let* $\mathbf{A} = (a_{ij})$ *be an* $n \times n$ *symmetric matrix, and let* A_{ij} *be the cofactor of* a_{ij}, *i.e.,* $(-1)^{i+j}$ *times the determinant of the submatrix of* $\mathbf{A}$ *obtained by deleting the ith row and jth column. Then*

$$\frac{\partial |\mathbf{A}|}{\partial a_{ij}} = \left\{ \begin{array}{ll} A_{ii} & \text{if } i = j \\ 2A_{ij} & \text{if } i \neq j \end{array} \right.$$

Proof. A proof may be found in Anderson (1984, p. 599).

Theorem A.1.5. *Let* $\mathbf{A} = (a_{ij})$ *be an* $n \times n$ *nonsingular matrix, and let* A_{ij} *be the cofactor of* a_{ij}. *Let* a^{ij} *denote the element in the ith row and jth column of* $\mathbf{A}^{-1}$. *Then*

$$a^{ij} = \frac{A_{ij}}{|\mathbf{A}|}.$$

Proof. A proof may be found in Harville (1997, p. 192).

Theorem A.1.6. *Let* **A** *and* **B** *be* $m \times m$ *and* $n \times n$ *matrices, respectively. Then*

$$|\mathbf{A} \otimes \mathbf{B}| = |\mathbf{A}|^n |\mathbf{B}|^m.$$

Proof. A proof may be found in Harville (1997, p. 350).

Appendix 2: Proofs of Theorems 2.5 and 2.6

We provide here proofs of Theorems 2.5 and 2.6.

Proof of Theorem 2.5

Proof of (a). Since $\mathbf{T\Sigma T}^T = \mathbf{D}$ and the determinant of a lower (or upper) triangular matrix is the product of its diagonal elements, we have

$$\prod_{i=1}^{n} \delta_i = |\mathbf{D}| = |\mathbf{T\Sigma T}^T| = |\mathbf{T}||\mathbf{\Sigma}||\mathbf{T}^T| = |\mathbf{\Sigma}|.$$

Proof of (b). If $p = 0$ then $\mathbf{\Sigma}$ is diagonal and $|\mathbf{\Sigma}| = \prod_{i=1}^{n} \sigma_{ii}$ so the result holds trivially. Now suppose $p \geq 1$. Recall from (2.23) that

$$\delta_i = \begin{cases} \sigma_{11} & \text{for } i = 1 \\ \sigma_{ii} - \boldsymbol{\sigma}_{i-p_i:i-1,i}^T \mathbf{\Sigma}_{i-p_i:i-1}^{-1} \boldsymbol{\sigma}_{i-p_i:i-1,i} & \text{for } i = 2,\ldots,n. \end{cases}$$

Furthermore, using expression (2.4) for the multiple correlation coefficient, we have

$$R_{i\cdot\{i-p_i:i-1\}}^2 = \boldsymbol{\sigma}_{i-p_i:i-1,i}^T \mathbf{\Sigma}_{i-p_i:i-1}^{-1} \boldsymbol{\sigma}_{i-p_i:i-1,i}/\sigma_{ii} \quad \text{for } i = 2,\ldots,n.$$

Thus for $i = 2,\ldots,n$,

$$\begin{aligned} \delta_i &= \sigma_{ii} - \sigma_{ii} R_{i\cdot\{i-p_i:i-1\}}^2 \\ &= \sigma_{ii}(1 - R_{i\cdot\{i-p_i:i-1\}}^2). \end{aligned} \tag{A.2.1}$$

Part (b) of the theorem follows upon substituting σ_{11} for δ_1 and (A.2.1) for δ_i ($i = 2,\ldots,n$) in part (a).

Proof of (c). By (2.23) and (A.2.1),

$$\sigma_{ii\cdot\{i-p_i:i-1\}} = \sigma_{ii}(1 - R_{i\cdot\{i-p_i:i-1\}}^2).$$

But Lemma 2.1 yields

$$\sigma_{ii\cdot\{i-p_i:i-1\}} = \sigma_{ii\cdot\{i-p_i+1:i-1\}} - \frac{\sigma_{i,i-p_i\cdot\{i-p_i+1:i-1\}}^2}{\sigma_{i-p_i,i-p_i\cdot\{i-p_i+1:i-1\}}}$$

$$
\begin{aligned}
&= \; \sigma_{ii\cdot\{i-p_i+1:i-1\}}(1 - \rho^2_{i,i-p_i\cdot\{i-p_i+1:i-1\}}) \\
&= \; \sigma_{ii\cdot\{i-p_i+2:i-1\}}(1 - \rho^2_{i,i-p_i+1\cdot\{i-p_i+2:i-1\}}) \\
&\quad \times(1 - \rho^2_{i,i-p_i\cdot\{i-p_i+1:i-1\}}) \\
&\;\;\vdots \\
&= \; \sigma_{ii}\prod_{k=1}^{p_i}(1 - \rho^2_{i,i-k\cdot\{i-k+1:i-1\}}).
\end{aligned}
$$

Dividing both expressions for $\sigma_{ii\cdot\{i-p_i:i-1\}}$ by σ_{ii} yields

$$
1 - R^2_{i\cdot\{i-p_i:i-1\}} = \prod_{k=1}^{p_i}(1 - \rho^2_{i,i-k\cdot\{i-k+1:i-1\}}).
$$

Part (c) of the theorem may then be obtained by substituting the right-hand side of this last equation into part (b) for $1 - R^2_{i\cdot\{i-p_i:i-1\}}$, and rearranging terms.

Proof of (d). If $p = 0$ then $\boldsymbol{\Sigma}$ is a diagonal matrix, whence $|\boldsymbol{\Sigma}| = \prod_{i=1}^{p}\sigma_{ii}$, which coincides with (2.33). If $p = n - 1$ then (2.33) holds trivially. Therefore suppose that $1 \le p \le n-2$. Now for $i \in \{2 : n\}$, consider the positive definite matrix

$$
\boldsymbol{\Sigma}_{i-p_i:i} = \begin{pmatrix} \boldsymbol{\Sigma}_{i-p_i:i-1} & \boldsymbol{\sigma}_{i-p_i:i-1,i} \\ \boldsymbol{\sigma}^T_{i-p_i:i-1,i} & \sigma_{ii} \end{pmatrix}.
$$

Using a standard result for the determinant of a partitioned matrix (Theorem A.1.3) and expression (2.23), we obtain

$$
\begin{aligned}
|\boldsymbol{\Sigma}_{i-p_i:i}| &= \; |\boldsymbol{\Sigma}_{i-p_i:i-1}|(\sigma_{ii} - \boldsymbol{\sigma}^T_{i-p_i:i-1,i}\boldsymbol{\Sigma}^{-1}_{i-p_i:i-1}\boldsymbol{\sigma}_{i-p_i:i-1,i}) \\
&= \; |\boldsymbol{\Sigma}_{i-p_i:i-1}|\delta_i.
\end{aligned}
$$

By the positive definiteness of $\boldsymbol{\Sigma}$, all three terms in this equation are positive. Thus

$$
\delta_i = \frac{|\boldsymbol{\Sigma}_{i-p_i:i}|}{|\boldsymbol{\Sigma}_{i-p_i:i-1}|} \quad \text{for } i = 2, \ldots, n.
$$

Substituting this expression for δ_i into part (a) of this theorem yields

$$
\begin{aligned}
|\boldsymbol{\Sigma}| &= \; \prod_{i=1}^{n}\delta_i \\
&= \; \sigma_{11} \times \frac{|\boldsymbol{\Sigma}_{1:2}|}{\sigma_{11}} \times \cdots \times \frac{|\boldsymbol{\Sigma}_{1:p}|}{|\boldsymbol{\Sigma}_{1:p-1}|} \times \frac{|\boldsymbol{\Sigma}_{1:p+1}|}{|\boldsymbol{\Sigma}_{1:p}|} \\
&\quad \times \frac{|\boldsymbol{\Sigma}_{2:p+2}|}{|\boldsymbol{\Sigma}_{2:p+1}|} \times \frac{|\boldsymbol{\Sigma}_{3:p+3}|}{|\boldsymbol{\Sigma}_{3:p+2}|} \times \cdots \times \frac{|\boldsymbol{\Sigma}_{n-p:n}|}{|\boldsymbol{\Sigma}_{n-p:n-1}|} \\
&= \; \frac{\prod_{i=1}^{n-p}|\boldsymbol{\Sigma}_{i:i+p}|}{\prod_{i=1}^{n-p-1}|\boldsymbol{\Sigma}_{i+1:i+p}|}.
\end{aligned}
$$

$\square$

Proof of Theorem 2.6

Proof. Result (2.34) holds trivially if $p = 0$ or $p = n - 1$. Next suppose that $p = n - 2$. Partition the $n \times n$ matrices $\mathbf{A}$ and $\boldsymbol{\Sigma}^{-1}$ as

$$\mathbf{A} = \begin{pmatrix} a_{11} & \mathbf{a}_{1,2:n-1} & a_{1n} \\ \mathbf{a}_{2:n-1,1} & \mathbf{A}_{2:n-1} & \mathbf{a}_{2:n-1,n} \\ a_{n1} & \mathbf{a}_{n,2:n-1} & a_{nn} \end{pmatrix}$$

and

$$\boldsymbol{\Sigma}^{-1} = \begin{pmatrix} \sigma^{11} & \boldsymbol{\sigma}^{1,2:n-1} & \sigma^{1n} \\ \boldsymbol{\sigma}^{2:n-1,1} & \boldsymbol{\Sigma}^{2:n-1} & \boldsymbol{\sigma}^{2:n-1,n} \\ \sigma^{n1} & \boldsymbol{\sigma}^{n,2:n-1} & \sigma^{nn} \end{pmatrix}.$$

By Theorem 2.2, $\sigma^{1n} = \sigma^{n1} = 0$. Therefore,

$$\begin{aligned}
\mathrm{tr}(\mathbf{A}\boldsymbol{\Sigma}^{-1}) &= a_{11}\sigma^{11} + \mathbf{a}_{1,2:n-1}\boldsymbol{\sigma}^{2:n-1,1} \\
&\quad + \mathrm{tr}(\mathbf{a}_{2:n-1,1}\boldsymbol{\sigma}^{1,2:n-1} + \mathbf{A}_{2:n-1}\boldsymbol{\Sigma}^{2:n-1} + \mathbf{a}_{2:n-1,n}\boldsymbol{\sigma}^{n,2:n-1}) \\
&\quad + \mathbf{a}_{n,2:n-1}\boldsymbol{\sigma}^{2:n-1,n} + a_{nn}\sigma^{nn} \\
&= \mathrm{tr}(\mathbf{A}_{2:n-1}\boldsymbol{\Sigma}^{2:n-1}) + a_{11}\sigma^{11} + a_{nn}\sigma^{nn} \\
&\quad + 2(\mathbf{a}_{1,2:n-1}\boldsymbol{\sigma}^{2:n-1,1} + \mathbf{a}_{n,2:n-1}\boldsymbol{\sigma}^{2:n-1,n}). \tag{A.2.2}
\end{aligned}$$

Now by Theorem A.1.1(f),

$$\begin{aligned}
(\boldsymbol{\Sigma}_{1:n-1})^{-1} &= \begin{pmatrix} \sigma^{11} & \boldsymbol{\sigma}^{1,2:n-1} \\ \boldsymbol{\sigma}^{2:n-1,1} & \boldsymbol{\Sigma}^{2:n-1} \end{pmatrix} \\
&\quad - \begin{pmatrix} 0 \\ \boldsymbol{\sigma}^{2:n-1,n} \end{pmatrix}(\sigma^{nn})^{-1}\begin{pmatrix} 0 & \boldsymbol{\sigma}^{n,2:n-1} \end{pmatrix} \\
&= \begin{pmatrix} \sigma^{11} & \boldsymbol{\sigma}^{1,2:n-1} \\ \boldsymbol{\sigma}^{2:n-1,1} & \boldsymbol{\Sigma}^{2:n-1} - \boldsymbol{\sigma}^{2:n-1,n}(\sigma^{nn})^{-1}\boldsymbol{\sigma}^{n,2:n-1} \end{pmatrix},
\end{aligned}$$

and so

$$\begin{aligned}
\mathrm{tr}[\mathbf{A}_{1:n-1}(\boldsymbol{\Sigma}_{1:n-1})^{-1}] &= a_{11}\sigma^{11} + \mathbf{a}_{1,2:n-1}\boldsymbol{\sigma}^{2:n-1,1} + \mathrm{tr}[\mathbf{a}_{2:n-1,1}\boldsymbol{\sigma}^{1,2:n-1} \\
&\quad + \mathbf{A}_{2:n-1}(\boldsymbol{\Sigma}^{2:n-1} - \boldsymbol{\sigma}^{2:n-1,n}(\sigma^{nn})^{-1}\boldsymbol{\sigma}^{n,2:n-1})] \\
&= \mathrm{tr}(\mathbf{A}_{2:n-1}\boldsymbol{\Sigma}^{2:n-1}) + a_{11}\sigma^{11} + 2\mathbf{a}_{1,2:n-1}\boldsymbol{\sigma}^{2:n-1,1} \\
&\quad - \boldsymbol{\sigma}^{n,2:n-1}\mathbf{A}_{2:n-1}\boldsymbol{\sigma}^{2:n-1,n}/\sigma^{nn}. \tag{A.2.3}
\end{aligned}$$

Similarly,

$$\begin{aligned}
(\boldsymbol{\Sigma}_{2:n})^{-1} &= \begin{pmatrix} \boldsymbol{\Sigma}^{2:n-1} & \boldsymbol{\sigma}^{2:n-1,n} \\ \boldsymbol{\sigma}^{n,2:n-1} & \sigma^{nn} \end{pmatrix} \\
&\quad - \begin{pmatrix} \boldsymbol{\sigma}^{2:n-1,1} \\ 0 \end{pmatrix}(\sigma^{11})^{-1}\begin{pmatrix} \boldsymbol{\sigma}^{1,,2:n-1} & 0 \end{pmatrix} \\
&= \begin{pmatrix} \boldsymbol{\Sigma}^{2:n-1} - \boldsymbol{\sigma}^{2:n-1,1}(\sigma^{11})^{-1}\boldsymbol{\sigma}^{1,2:n-1} & \boldsymbol{\sigma}^{2:n-1,n} \\ \boldsymbol{\sigma}^{n,2:n-1} & \sigma^{nn} \end{pmatrix},
\end{aligned}$$

so that

$$
\begin{aligned}
\mathrm{tr}\big[\mathbf{A}_{2:n}(\mathbf{\Sigma}_{2:n})^{-1}\big] &= \mathrm{tr}\big[\mathbf{A}_{2:n-1}(\mathbf{\Sigma}^{2:n-1} - \boldsymbol{\sigma}^{2:n-1,1}(\sigma^{11})^{-1}\boldsymbol{\sigma}^{1,2:n-1}) \\
&\quad + \mathbf{a}_{2:n-1,n}\boldsymbol{\sigma}^{n,2:n-1}\big] + \mathbf{a}_{n,2:n-1}\boldsymbol{\sigma}^{2:n-1,n} \\
&\quad + a_{nn}\sigma^{nn} \\
&= \mathrm{tr}(\mathbf{A}_{2:n-1}\mathbf{\Sigma}^{2:n-1}) - \boldsymbol{\sigma}^{1,2:n-1}\mathbf{A}_{2:n-1}\boldsymbol{\sigma}^{2:n-1,1}/\sigma^{11} \\
&\quad + 2\mathbf{a}_{n,2:n-1}\boldsymbol{\sigma}^{2:n-1,n} + a_{nn}\sigma^{nn}. \qquad (\mathrm{A.2.4})
\end{aligned}
$$

Likewise, upon permuting the rows and columns of $\mathbf{\Sigma}^{-1}$ so that the $(n-2) \times (n-2)$ block $\mathbf{\Sigma}^{2:n-1}$ appears in the lower right corner of the partitioned matrix and then applying Theorem A.1.1(f) once more, we obtain

$$
\begin{aligned}
(\mathbf{\Sigma}_{2:n-1})^{-1} &= \mathbf{\Sigma}^{2:n-1} - \begin{pmatrix} \boldsymbol{\sigma}^{2:n-1,1} & \boldsymbol{\sigma}^{2:n-1,n} \end{pmatrix} \begin{pmatrix} \sigma^{11} & 0 \\ 0 & \sigma^{nn} \end{pmatrix}^{-1} \\
&\quad \times \begin{pmatrix} \boldsymbol{\sigma}^{1,2:n-1} \\ \boldsymbol{\sigma}^{n,2:n-1} \end{pmatrix} \\
&= \mathbf{\Sigma}^{2:n-1} - \boldsymbol{\sigma}^{2:n-1,1}(\sigma^{11})^{-1}\boldsymbol{\sigma}^{1,2:n-1} \\
&\quad - \boldsymbol{\sigma}^{2:n-1,n}(\sigma^{nn})^{-1}\boldsymbol{\sigma}^{n,2:n-1},
\end{aligned}
$$

yielding

$$
\begin{aligned}
\mathrm{tr}\big[\mathbf{A}_{2:n}(\mathbf{\Sigma}_{2:n})^{-1}\big] &= \mathrm{tr}(\mathbf{A}_{2:n-1}\mathbf{\Sigma}^{2:n-1}) - \boldsymbol{\sigma}^{1,2:n-1}\mathbf{A}_{2:n-1}\boldsymbol{\sigma}^{2:n-1,1}/\sigma^{11} \\
&\quad - \boldsymbol{\sigma}^{n,2:n-1}\mathbf{A}_{2:n-1}\boldsymbol{\sigma}^{2:n-1,n}/\sigma^{nn}. \qquad (\mathrm{A.2.5})
\end{aligned}
$$

Using (A.2.3), (A.2.4), and (A.2.5), we find that

$$
\begin{aligned}
\mathrm{tr}\big[\mathbf{A}_{1:n-1}(\mathbf{\Sigma}_{1:n-1})^{-1}\big] &+ \mathrm{tr}\big[\mathbf{A}_{2:n}(\mathbf{\Sigma}_{2:n})^{-1}\big] - \mathrm{tr}\big[\mathbf{A}_{2:n-1}(\mathbf{\Sigma}_{2:n-1})^{-1}\big] \\
&= \mathrm{tr}(\mathbf{A}_{2:n-1}\mathbf{\Sigma}^{2:n-1}) + a_{11}\sigma^{11} + a_{nn}\sigma^{nn} \\
&\quad + 2(\mathbf{a}_{1,2:n-1}\boldsymbol{\sigma}^{2:n-1,1} + \mathbf{a}_{n,2:n-1}\boldsymbol{\sigma}^{2:n-1,n}), \qquad (\mathrm{A.2.6})
\end{aligned}
$$

which is identical to (A.2.2). Thus, (2.34) holds for $p = n - 2$.

The same argument that yielded (A.2.6) may also be used to show that for any p and any $i = 1, \ldots, n - p - 1$,

$$
\begin{aligned}
\mathrm{tr}\big[\mathbf{A}_{i:i+p-1}(\mathbf{\Sigma}_{i:i+p-1})^{-1}\big] &= \mathrm{tr}\big[\mathbf{A}_{i:i+p}(\mathbf{\Sigma}_{i:i+p})^{-1}\big] + \mathrm{tr}\big[\mathbf{A}_{i+1:i+p+1}(\mathbf{\Sigma}_{i+1:i+p+1})^{-1}\big] \\
&\quad - \mathrm{tr}\big[\mathbf{A}_{i+1:i+p}(\mathbf{\Sigma}_{i+1:i+p})^{-1}\big].
\end{aligned}
$$

$$(\mathrm{A.2.7})$$

Finally, we show that if (2.34) holds for $p = n - j$ where $2 \leq j \leq n - 1$, then it also holds for $p = n - (j + 1)$, upon which the theorem follows by the method of induction. So suppose that (2.34) holds for $p = n - j$. Then

$$
\mathrm{tr}(\mathbf{A}\mathbf{\Sigma}^{-1}) = \sum_{i=1}^{j} \mathrm{tr}\big[\mathbf{A}_{i:i+n-j}(\mathbf{\Sigma}_{i:i+n-j})^{-1}\big]
$$

$$-\sum_{i=1}^{j-1} \text{tr}[\mathbf{A}_{i+1:i+n-j}(\mathbf{\Sigma}_{i+1:i+n-j})^{-1}]$$

$$= \sum_{i=1}^{j}\{\text{tr}[\mathbf{A}_{i:i+n-(j+1)}(\mathbf{\Sigma}_{i:i+n-(j+1)})^{-1}]$$

$$+ \text{tr}[\mathbf{A}_{i+1:i+n-j}(\mathbf{\Sigma}_{i+1:i+n-j})^{-1}]$$

$$- \text{tr}[\mathbf{A}_{i+1:i+n-(j+1)}(\mathbf{\Sigma}_{i+1:i+n-(j+1)})^{-1}]\}$$

$$- \sum_{i=1}^{j-1} \text{tr}[\mathbf{A}_{i+1:i+n-j}(\mathbf{\Sigma}_{i+1:i+n-j})^{-1}]$$

$$= \left\{\sum_{i=1}^{j}\text{tr}[\mathbf{A}_{i:i+n-(j+1)}(\mathbf{\Sigma}_{i:i+n-(j+1)})^{-1}]\right.$$

$$\left.- \text{tr}[\mathbf{A}_{i+1:i+n-(j+1)}(\mathbf{\Sigma}_{i+1:i+n-(j+1)})^{-1}]\right\}$$

$$+ \text{tr}[\mathbf{A}_{j+1:n}(\mathbf{\Sigma}_{j+1:n})^{-1}]$$

$$= \sum_{i=1}^{j+1}\text{tr}[\mathbf{A}_{i:i+n-(j+1)}(\mathbf{\Sigma}_{i:i+n-(j+1)})^{-1}]$$

$$- \sum_{i=1}^{j}\text{tr}[\mathbf{A}_{i+1:i+n-(j+1)}(\mathbf{\Sigma}_{i+1:i+n-(j+1)})^{-1}]$$

where we have used (A.2.7) for the second equality. But this last expression is merely the right-hand side of (2.34) when $p = n - (j + 1)$. Thus (2.34) holds for $p = n - (j + 1)$, and the theorem follows by induction. $\square$

References

Akaike, H. (1974). A new look at the statistical model identification. *IEEE Transactions on Automatic Control*, 19(6), 716–723.

Albert, J. M. (1992). A corrected likelihood ratio statistic for the multivariate regression model with antedependent errors. *Communications in Statistics – Theory and Methods*, 21(7), 1823–1843.

Albuquerque, L. G. and Meyer, K. (2005). Estimates of covariance functions for growth of Nelore cattle applying a parametric correlation structure to model within-animal correlations. *Livestock Production Science*, 93, 213–222.

Anderson, T. W. (1984). *An Introduction to Multivariate Statistical Analysis*, 2nd ed. New York: John Wiley & Sons, Inc.

Azzalini, A. (1994). Logistic regression for autocorrelated data with application to repeated measures. *Biometrika*, 81(4), 767–775.

Banerjee, S., Carlin, B. P., and Gelfand, A. E. (2004). *Hierarchical Modeling and Analysis for Spatial Data*. Boca Raton, Florida: Chapman & Hall/CRC.

Barrett, W. W. and Feinsilver, P. J. (1978). Gaussian families and a theorem on patterned matrices. *Journal of Applied Probability*, 15(3), 514–522.

Belcher, J., Hampton, J. S., and Tunnicliffe Wilson, G. (1994). Parameterization of continuous time autoregressive models for irregularly sampled time series data. *Journal of the Royal Statistical Society, Series B*, 56(1), 141–155.

Besag, J. (1974). Spatial interaction and the statistical analysis of lattice systems (with Discussion). *Journal of the Royal Statistical Society, Series B*, 36(2), 192–236.

Besag, J. and Kooperberg, C. (1995). On conditional and intrinsic autoregression. *Biometrika*, 82(4), 733–746.

Bickel, P. and Levina, E. (2008). Regularized estimation of large covariance matrices. *Annals of Statistics*, 36(1), 199–227.

Box, G. E. P. and Jenkins, G. M. (1976). *Time Series Analysis: Forecasting and Control*. San Francisco: Holden Day.

Brockwell, P. J. and Davis, R. A. (1991). *Time Series: Theory and Methods*, 2nd ed. New York: Springer-Verlag.

Burnham, K. P. and Anderson, D. R. (2002). *Model Selection and Multimodel*

Inference, 2nd ed. New York: Springer-Verlag.

Byrne, P. J. (1996). Multivariate analysis with an autoregressive covariance model. *Communications in Statistics – Theory and Methods*, 25(3), 555–569.

Byrne, P. J. and Arnold, S. F. (1983). Inference about multivariate means for a nonstationary autoregressive model. *Journal of the American Statistical Association*, 78(384), 850–855.

Carlin, B. P. and Louis, T. A. (2000). *Bayes and Empirical Bayes Methods for Data Analysis*, 2nd ed. Boca Raton, FL: Chapman and Hall/CRC Press.

Casella, G. and Berger, R. L. (2002). *Statistical Inference*, 2nd ed. Belmont, CA: Duxbury Press.

Cepeda, E. C. and Gamerman, D. (2004). Bayesian modeling of joint regression for the mean and covariance matrix. *Biometrical Journal*, 46(4), 430–440.

Cepeda-Cuervo, E. and Núñez-Antón, V. (2007). Bayesian joint modelling of the mean and covariance structures for normal longitudinal data. *SORT*, 31(2), 181–200.

Cepeda-Cuervo, E. and Núñez-Antón, V. (2009). Bayesian modelling of the mean and covariance matrix in normal nonlinear models. *Journal of Statistical Computation and Simulation*, 79(6), 837–853.

Chen, R. and Tsay, R. S. (1993). Functional-coefficient autoregressive models. *Journal of the American Statistical Association*, 88(421), 298–308.

Cox, D. R. and Snell, E. J. (1989). *Analysis of Binary Data*, 2nd ed. London: Chapman and Hall/CRC Press.

Cressie, N. (1993). *Statistics for Spatial Data*. New York: John Wiley & Sons, Inc.

Crowder, M. J. and Hand, D. J. (1990). *Analysis of Repeated Measures*. London: Chapman & Hall.

de Boor, C. (2001). *A Practical Guide to Splines* (Revised ed.), New York: Springer.

Daniels, M. J. and Pourahmadi, M. (2002). Bayesian analysis of covariance matrices and dynamic models for longitudinal data. *Biometrika*, 89(3), 553–566.

Davis, C. S. (2002). *Statistical Methods for the Analysis of Repeated Measurements*. New York: Springer-Verlag.

Davison, A. C. and Sardy, S. (2000). The partial scatterplot matrix. *Journal of Computational and Graphical Statistics*, 9(4), 750–758.

Dawson, K. S., Gennings, C., and Carter, W. H. (1997). Two graphical techniques useful in detecting correlation structure in repeated measures data. *The American Statistician*, 51(3): 275–283.

Dempster, A. P. (1972). Covariance selection. *Biometrics*, 28(1), 157–175.

Dempster, A. P., Laird, N. M., and Rubin, D. B. (1977). Maximum likelihood estimation from incomplete data via the EM algorithm (with Discussion).

Journal of the Royal Statistical Society, Series B, 39(1), 1–38.

Diggle, P. J. (1988). An approach to the analysis of repeated measurements. *Biometrics*, 44(4), 959–971.

Diggle, P. J., Heagerty, P. J., Liang, K. Y., and Zeger, S. L. (2002), *Analysis of Longitudinal Data*, 2nd ed. New York: Oxford University Press.

Dykstra, R. L. (1970). Establishing the positive definiteness of the sample covariance matrix. *Annals of Mathematical Statistics*, 41(6), 2153–2154.

El-Mikkawy, M. E. A. (2004). A fast algorithm for evaluating nth order tridiagonal determinants. *Journal of Computational and Applied Mathematics*, 166(2), 581–584.

Everitt, B. S. (1994a). Exploring multivariate data graphically: A brief review with examples. *Journal of Applied Statistics*, 21(3), 63–94.

Everitt, B. S. (1994b). *A Handbook of Statistical Analyses Using S-Plus*. London: Chapman and Hall.

Fan, J. and Gijbels, I. (1996). *Local Polynomial Modelling and its Applications*. London: Chapman and Hall.

Feller, W. (1968). *An Introduction to Probability Theory and its Applications, Volume I* (3rd edition). New York: John Wiley & Sons, Inc.

Friedman, J., Hastie, T., and Tibshirani, R. (2008). Sparse inverse covariance estimation with the graphical lasso. *Biostatistics*, 9(3), 432–441.

Fuller, W. A. (1976). Introduction to Statistical Time Series. New York: John Wiley & Sons, Inc.

Gabriel, K. R. (1962). Ante-dependence analysis of an ordered set of variables. *Annals of Mathematical Statistics*, 33(1), 201–212.

Greenberg, B. G. and Sarhan, A. E. (1959). Matrix inversion, its interest and application in analysis of data. *Journal of the American Statistical Association*, 54(288), 755–766.

Greenhouse, S. W. and Geisser, S. (1959). On methods in the analysis of profile data. *Psychometrika*, 24(2), 95–112.

Guttman, L. (1955). A generalized simplex for factor analysis. *Psychometrika*, 20(3), 173–192.

Haggan, V. and Ozaki, T. (1981). Modelling nonlinear random vibration using an amplitude-dependent autoregressive time series model. *Biometrika*, 68(1), 189–196.

Haining, R. (1990). *Spatial Data Analysis in the Social and Environmental Sciences*. Cambridge: Cambridge University Press.

Harville, D. A. (1975). Maximum likelihood approaches to variance component estimation and to related problems. Technical Report No. 75-0175, Aerospace Research Laboratories, Wright-Patterson AFB, Ohio.

Harville, D. A. (1977). Maximum likelihood approaches to variance component estimation and to related problems. *Journal of the American Statistical Association*, 72(358), 320–338.

Harville, D. A. (1997). *Matrix Algebra from a Statistician's Perspective*. New

York: Springer-Verlag.

Heagerty, P. J. (2002). Marginalized transition models and likelihood inference for longitudinal categorical data. *Biometrics*, 58(2), 342–351.

Heagerty, P. J. and Zeger, S. L. (2000). Marginalized multilevel models and likelihood inference. *Statistical Science*, 15(1), 1–19.

Hou, W., Garvan, C. W., Zhao, W., Behnke, M., Eyler, F. D., and Wu, R. (2005). A general model for detecting genetic determinants underlying longitudinal traits with unequally spaced measurements and nonstationary covariance structure. *Biostatistics*, 6(3), 420–433.

Huang, J. Z., Liu, L., and Liu, N. (2007). Estimation of large covariance matrices of longitudinal data with basis function approximations. *Journal of Computational and Graphical Statistics*, 16(1), 189–209.

Huang, J. Z., Liu, N., Pourahmadi, M., and Liu, L. (2006). Covariance matrix selection and estimation via penalised normal likelihood. *Biometrika*, 93(1), 85–98.

Huynh, H. and Feldt, L. S. (1976). Estimation of the Box correction for degrees of freedom from sample data in randomized block and split-plot designs. *Journal of Educational Statistics*, 1(1), 69–82.

Jaffrézic, F. and Pletcher, S. D. (2000). Statistical models for estimating the genetic basis of repeated measures and other function-valued traits. *Genetics*, 156, 913–922.

Jaffrézic, F., Thompson, R., and Hill, W. G. (2003). Structured antedependence models for genetic analysis of repeated measures on multiple quantitative traits. *Genetics Research*, 82, 55–65.

Jaffrézic, F., Thompson, R., and Pletcher, S. D. (2004). Multivariate character process models for the analysis of two or more correlated function-valued traits. *Genetics*, 168, 477–487.

Jaffrézic, F., Venot, E. Laloë, D., Vinet, A., and Renand, G. (2004). Use of structured antedependence models for the genetic analysis of growth curves. *Journal of Animal Science*, 82, 3465–3473.

Jaffrézic, F., White, I. M. S., Thompson, R., and Visscher, P. M. (2002). Contrasting models for lactation curve analysis. *Journal of Dairy Science*, 85, 968–975.

Johnson, K. L. (1989). Higher order antedependence models. Unpublished Ph.D. thesis, Department of Statistics, Pennsylvania State University.

Johnson, R. A. and Wichern, D. W. (2002). *Applied Multivariate Statistical Analysis*, 5th ed. Upper Saddle River, New Jersey: Prentice Hall.

Jones, R. H. (1981). Fitting continuous-time autoregressions to discrete data. In D. F. Findley (Ed.), *Applied Time Series Analysis II* (pp. 651–682). New York: Academic Press.

Jones, R. H. (1990). Serial correlation or random subject effects? *Communications in Statistics – Simulation and Computation*, 19(3), 1105–1123.

Jones, R. H. and Ackerson, L. M. (1990). Serial correlation in unequally spaced longitudinal data. *Biometrika*, 77(4), 721–731.

Kearsley, C. G. S., Woolliams, J. A., Coffey, M. P., and Brotherstone, S. (2008). Use of competition data for genetic evaluations of eventing horses in Britain: Analysis of the dressage, showjumping, and cross country phases of eventing competition. *Livestock Science*, 118, 72–81.

Kenward, M. C. (1987). A method for comparing profiles of repeated measurements. *Applied Statistics*, 36(3), 296–308.

Kenward, M. C. (1991). Corrigendum: A method for comparing profiles of repeated measurements. *Applied Statistics*, 40(2), 379.

Krzanowski, W. J. (1993). Antedependence modelling in discriminant analysis of high-dimensional spectroscopic data. In C. M. Cuadras and C. R. Rao (Eds.), *Multivariate Analysis: Future Directions 2* (pp. 87–95). Amsterdam: Elsevier.

Krzanowski, W. J. (1999). Antedependence models in the analysis of multigroup high-dimensional data. *Journal of Applied Statistics*, 26(1), 59–67.

Krzanowski, W. J., Jonathan, P., McCarthy, W. V., and Thomas, M. R. (1995). Discriminant analysis with singular covariance matrices: methods and applications to spectroscopic data. *Applied Statistics*, 44(1), 101–115.

Laird, N. M. and Ware, J. H. (1982). Random-effects models for longitudinal data. *Biometrics*, 38(4), 963–974.

Lauritzen, S. L. (1996). *Graphical Models*. Oxford: Clarendon Press.

Lee, K. and Daniels, M. J. (2007). A class of Markov models for longitudinal ordinal data. *Biometrics*, 63(4), 1060–1067.

Levina, E., Rothman, A., and Zhu, J. (2008). Sparse estimation of large covariance matrices via a nested lasso penalty. *Annals of Applied Statistics*, 2(1), 245–263.

Lin, M. and Wu, R. (2005). Theoretical basis for the identification of allelic variants that encode drug efficacy and toxicity. *Genetics*, 170, 919–928.

Lin, M. Li, H., Hou, W., Johnson, J. A., and Wu, R. (2007). Modeling sequence-sequence interactions for drug response. *Bioinformatics*, 23(10), 1251–1257.

Little, R. J. A. and Rubin, D. B. (2002). *Statistical Analysis with Missing Data*. Hoboken, New Jersey: John Wiley & Sons, Inc.

Macchiavelli, R. E. (1992). Likelihood-based procedures and order selection in higher order antedependence models. Unpublished Ph.D. thesis, Department of Statistics, Pennsylvania State University.

Macchiavelli, R. E. and Arnold, S. F. (1994). Variable-order antedependence models. *Communications in Statistics – Theory and Methods*, 23(9), 2683–2699.

Macchiavelli, R. E. and Arnold, S. F. (1995). Difficulties with the use of penalized likelihood criteria in antedependence and polynomial models. *Communications in Statistics – Theory and Methods*, 24(2), 501–522.

Macchiavelli, R. E. and Moser, E. B. (1997). Analysis of repeated measurements with ante-dependence covariance models. *Biometrical Journal*, 39(3), 339–350.

Mallik, R. K. (2001). The inverse of a tridiagonal matrix. *Linear Algebra and its Applications*, 325(1), 109–139.

Mauchly, J. W. (1940). Significance tests for sphericity of a normal n-variate distribution. *Annals of Mathematical Statistics*, 11(4), 204–209.

Molenberghs, G. and Verbeke, G. (2005). *Models for Discrete Longitudinal Data*. New York: Springer.

Nelder, J. A. and Mead, R. (1965). A simplex method for function minimization. *The Computer Journal*, 7(4), 308–313.

Núñez-Antón, V. and Zimmerman, D. L. (2000). Modeling nonstationary longitudinal data, *Biometrics*, 56(3), 699–705.

Pan, J. and MacKenzie, G. (2003). On modelling mean-covariance structures in longitudinal studies. *Biometrika*, 90(1), 239–244.

Patel, H. I. (1991). Analysis of incomplete data from a clinical trial with repeated measurements. *Biometrika*, 78(3), 609–619.

Pepe, M. and Anderson, G. (1994). A cautionary note on inference for marginal regression models with longitudinal data and general correlated response data. *Communications in Statistics, Part B – Simulation and Computation*, 23(4), 939–951.

Pourahmadi, M. (1999). Joint mean-covariance models with applications to longitudinal data: Unconstrained parameterisation. *Biometrika*, 86(3), 677–690.

Pourahmadi, M. (2000). Maximum likelihood estimation of generalised linear models for multivariate normal covariance matrix. *Biometrika*, 87(2), 425–435.

Pourahmadi, M. (2002). Graphical diagnostics for modeling unstructured covariance matrices. *International Statistical Review*, 70(3), 395–417.

Rao, C. R. and Wu, Y. (1989). A strongly consistent procedure for model selection in a regression problem. *Biometrika*, 76(2), 369–374.

Rothman, A. J., Bickel, P. J., Levina, E., and Zhu, J. (2008). Sparse permutation invariant covariance estimation. *Electronic Journal of Statistics*, 494–515.

Roy, S. N. and Sarhan, A. E. (1956). On inverting a class of patterned matrices. *Biometrika*, 43(1/2), 227–231.

Rue, H. and Held, L. (2005). *Gaussian Markov Random Fields*. Boca Raton, Florida: Chapman & Hall/CRC.

Schlegel, P. (1970). The explicit inverse of a tridiagonal matrix. *Mathematics of Computation*, 24(111), 665.

Schott, J. R. (2005). *Matrix Analysis for Statistics*. Hoboken, New Jersey: John Wiley & Sons, Inc.

Smith, M. and Kohn, R. (2002). Parsimonious covariance matrix estimation for longitudinal data. *Journal of the American Statistical Association*, 97(460),

1141–1153.

Sogabe, T. (2008). A fast numerical algorithm for the determinant of a penta-diagonal matrix. *Applied Mathematics and Computation*, 196(2), 835–841.

Tanabe, K. and Sagae, M. (1992). An exact Cholesky decomposition and the generalized inverse of the variance-covariance matrix of the multinomial distribution, with applications. *Journal of the Royal Statistical Society, Series B*, 54(1), 211–219.

Tong, H. (1990). *Non-linear Time Series*. Oxford: Clarendon Press.

Tyler, R. S., Abbas, P., Tye-Murray, N., Gantz, B. J., Knutson, J. F., McCabe, B. F., Lansing, C., Brown, C., Woodworth, G., Hinrichs, J., and Kuk, F. (1988). Evaluation of five different cochlear implant designs: Audiologic assessment and predictors of performance. *The Laryngoscope*, 98(10), 1100–1106.

Verbeke, G., Lesaffre, E., and Brant, L. (1998). The detection of residual serial correlation in linear mixed models. *Statistics in Medicine*, 17, 1391–1402.

Verbeke, G. and Molenberghs, G. (2001). *Linear Mixed Models for Longitudinal Data*. New York: Springer.

Verbyla, A. P. (1993). Modelling variance heterogeneity: Residual maximum likelihood and diagnostics. *Journal of the Royal Statistical Society, Series B*, 55(2), 493–508.

Wall, M. M. (2004). A close look at the spatial structure implied by the CAR and SAR models. *Journal of Statistical Planning and Inference*, 121, 311–324.

Weiss, R. E. (2005). *Modeling Longitudinal Data*. New York: Springer.

White, J. R. Gardner, L. M., Sees, M., and Corstanje, R. (2008). The short-term effects of prescribed burning on biomass removal and the release of nitrogen and phosphorus in a treatment wetland. *Journal of Environmental Quality*, 37, 2386–2391.

Whittaker, J. (1990). *Graphical Models in Applied Multivariate Statistics*. Chichester: John Wiley & Sons, Inc.

Whittle, P. (1954). On stationary processes in the plane. *Biometrika*, 41(3/4), 434–449.

Wong, F., Carter, C. K., and Kohn, R. (2003). Efficient estimation of covariance selection models. *Biometrika*, 90(4), 809–830.

Wu, R. and Hou, W. (2006). A hyperspace model to decipher the genetic architecture of developmental processes: allometry meets ontogeny. *Genetics*, 172, 627–637.

Wu, W. B. and Pourahmadi, M. (2003). Nonparametric estimation of large covariance matrices of longitudinal data. *Biometrika*, 90(4), 831–844.

Yuan, M. and Lin, Y. (2007). Model selection and estimation in the Gaussian graphical model. *Biometrika*, 94(1), 19–35.

Zeger, S. L. and Qaqish, B. (1988). Markov regression models for time series: a quasi-likelihood approach. *Biometrics*, 44(4), 1019–1031.

Zhang, P. (2005). Multiple imputation of missing data with ante-dependence covariance structure. *Journal of Applied Statistics*, 32(2), 141–155.

Zhao, W., Chen, Y. Q., Casella, G., Cheverud, J. M., and Wu, R. (2005a). A non-stationary model for functional mapping of complex traits. *Bioinformatics*, 21(10), 2469–2477.

Zhao, W., Hou, W., Littell, R. C., and Wu, R. (2005b). Structured antedependence models for functional mapping of multiple longitudinal traits. *Statistical Applications in Genetics and Molecular Biology*, 4(1), Article 33.

Zimmerman, D. L. (2000). Viewing the correlation structure of longitudinal data through a PRISM. *The American Statistician*, 54(4), 310–318.

Zimmerman, D. L. and Núñez-Antón, V. (1997). Structured antedependence models for longitudinal data. In T.G. Gregoire, D.R. Brillinger, P.J. Diggle, E. Russek-Cohen, W.G. Warren, and R. Wolfinger (Eds.), *Modelling Longitudinal and Spatially Correlated Data. Methods, Applications, and Future Directions* (pp. 63–76). New York: Springer-Verlag.

Zimmerman, D. L. and Núñez-Antón, V. (2001). Parametric modelling of growth curve data: An overview (with Discussion). *Test*, 10(1), 1–73.

Zimmerman, D. L., Núñez-Antón, V., and El-Barmi, H. (1998). Computational aspects of likelihood-based estimation of first-order antedependence models. *Journal of Statistical Computation and Simulation*, 60, 67–84.

Index

Printed and bound by CPI Group (UK) Ltd, Croydon, CR0 4YY

28/10/2024

01780160-0001